Climate Change Management

Series editor
W. Leal Filho, Hamburg, Germany

More information about this series at http://www.springer.com/series/8740

Walter Leal Filho • Kathryn Adamson •
Rachel M. Dunk • Ulisses M. Azeiteiro •
Sam Illingworth • Fátima Alves
Editors

Implementing Climate Change Adaptation in Cities and Communities

Integrating Strategies and Educational Approaches

Editors
Walter Leal Filho
Hamburg University of Applied Sciences
Research and Transfer Center
Faculty of Life Sciences
Hamburg, Germany

Rachel M. Dunk
Manchester Metropolitan University
School of Science and The Environment
Manchester, United Kingdom

Sam Illingworth
Manchester Metropolitan University
School of Research, Enterprise
Manchester, United Kingdom

Kathryn Adamson
Manchester Metropolitan University
School of Science and The Environment
Manchester, United Kingdom

Ulisses M. Azeiteiro
University of Coimbra
Centre for Functional Ecology
Coimbra, Portugal

University of Aberta
Porto, Portugal

Fátima Alves
University of Coimbra
Centre for Functional Ecology
Coimbra, Portugal

University of Aberta
Porto, Portugal

ISSN 1610-2010   ISSN 1610-2002   (electronic)
Climate Change Management
ISBN 978-3-319-80382-1   ISBN 978-3-319-28591-7   (eBook)
DOI 10.1007/978-3-319-28591-7

© Springer International Publishing Switzerland 2016
Softcover reprint of the hardcover 1st edition 2016
This work is subject to copyright. All rights are reserved by the Publisher, whether the whole or part of the material is concerned, specifically the rights of translation, reprinting, reuse of illustrations, recitation, broadcasting, reproduction on microfilms or in any other physical way, and transmission or information storage and retrieval, electronic adaptation, computer software, or by similar or dissimilar methodology now known or hereafter developed.
The use of general descriptive names, registered names, trademarks, service marks, etc. in this publication does not imply, even in the absence of a specific statement, that such names are exempt from the relevant protective laws and regulations and therefore free for general use.
The publisher, the authors and the editors are safe to assume that the advice and information in this book are believed to be true and accurate at the date of publication. Neither the publisher nor the authors or the editors give a warranty, express or implied, with respect to the material contained herein or for any errors or omissions that may have been made.

Printed on acid-free paper

This Springer imprint is published by Springer Nature
The registered company is Springer International Publishing AG Switzerland

Preface

Parallel to mitigation measures, which are essential to slow down the release of CO_2 emissions and hence reduce the speed with which global warming is taking place, there is a pressing need for climate change adaptation measures. Indeed, it is widely acknowledged that if we are to ameliorate the suffering of millions of people who are currently experiencing the impacts of climate change in sectors as varied as agriculture, land use, or infrastructure, adaptation is an important component part of the process.

This book has been prepared based on the need to better understand and explore ways to implement adaptation measures in cities and regions. It contains a set of papers presented at the World Symposium on Climate Change Adaptation, held in Manchester, United Kingdom, on 2–4 September 2015. The event, attended by over 200 experts from 40 countries, explored some theoretical components of climate change adaptation and introduced various practical examples of projects and initiatives, which are documented here.

The focus of this publication is a demonstration of how to pursue and implement adaptation initiatives in cities, communities, and regions.

Part I of the book deals with matters related to adaptation in cities and communities and offers an overview of issues pertaining to adaptation not only on an urban but also on a rural level. It also discusses the different degrees to which communities are exposed to climate variability and to climate change, with examples and case studies from across the world.

Part II focuses on both the integration of adaptation strategies and educational approaches. In this section, the papers discuss matters related to adaptation and livelihoods, describe a set of initiatives at the local and regional level, and introduce the role played by capacity-building as one of the various tools which can support climate change adaptation efforts. To this purpose, various projects and case studies aimed at raising awareness of local inhabitants of villages and farmers are presented.

We thank the authors for their willingness to share their knowledge, know-how, and the information they gathered from their work and from their field projects.

We hope that the experiences gathered here can be useful to others working in the field of climate change adaptation at city, community, and regional levels and that they may support current and future efforts towards helping people and communities to adapt to rapidly changing conditions, which affect both their well-being and their livelihood.

Enjoy the reading.

Winter 2015/2016
Hamburg, Germany
Manchester, UK
Porto, Portugal

Walter Leal Filho
Kathryn Adamson
Rachel Dunk
Ulisses Miranda Azeiteiro
Sam Illingworth
Maria de Fátima Pereira Alves

Contents

Part I Climate Change Adaptation in Cities and Communities

1. **Changes in Land Cover Categories within Oti-Kéran-Mandouri (OKM) Complex in Togo (West Africa) between 1987 and 2013** . . .   3
   Aniko Polo-Akpisso, Kpérkouma Wala, Soulemane Ouattara, Fousseni Foléga, and Yao Tano

2. **Exposure of Rural Communities to Climate Variability and Change: Case Studies from Argentina, Colombia and Canada**   23
   David Sauchyn, Jorge Julian Velez Upegui, Mariano Masiokas, Olga Ocampo, Leandro Cara, and Ricardo Villalba

3. **Small Scale Rain- and Floodwater Harvesting for Horticulture in Central-Northern Namibia for Livelihood Improvement and as an Adaptation Strategy to Climate Change**   39
   Alexander Jokisch, Wilhelm Urban, and Thomas Kluge

4. **Can Adaptation to Climate Change at All Be Mainstreamed in Complex Multi-level Governance Systems? A Case Study of Forest-Relevant Policies at the EU and Swedish Levels**   53
   E. Carina H. Keskitalo and Maria Pettersson

5. **A Novel Impact Assessment Methodology for Evaluating Distributional Impacts in Scottish Climate Change Adaptation Policy**   75
   Rachel M. Dunk, Poshendra Satyal, and Michael Bonaventura

6. **Climate Variability and Food Security in Tanzania: Evidence from Western Bagamoyos**   99
   Paschal Arsein Mugabe

7   The Urban Heat Island Effect in Dutch City Centres: Identifying Relevant Indicators and First Explorations 123
Leyre Echevarría Icaza, F.D. van der Hoeven, and Andy van den Dobbelsteen

8   Planning and Climate Change: A Case Study on the Spatial Plan of the Danube Corridor Through Serbia 161
Tijana Crncevic, Omiljena Dzelebdzic, and Sasa Milijic

9   Programmes of the Republic of Belarus on Climate Change Adaptation: Goals and Results 179
Siarhei Zenchanka

10 A Global Indicator of Climate Change Adaptation in Catalonia ... 191
Ester Agell, Fina Ambatlle, Gabriel Borràs, Gemma Cantos, and Salvador Samitier

Part II Integrating Adaptation Strategies and Educational Approaches

11 Facilitating Climate Change Adaptation on Smallholder Farms Through Farmers' Collective Led On-Farm Adaptive Research: The SAF-BIN Project 205
Romana Roschinsky, Sunil Simon, Pranab Ranjan Choudhury, Augustine Baroi, Manindra Malla, Sukleash George Costa, Valentine Denis Pankaj, Chintan Manandhar, Manfred Aichinger, and Maria Wurzinger

12 Assessing Student Perceptions and Comprehension of Climate Change in Portuguese Higher Education Institutions 221
P.T. Santos, P. Bacelar-Nicolau, M.A. Pardal, L. Bacelar-Nicolau, and U.M. Azeiteiro

13 A Decade of Capacity Building Through Roving Seminars on Agro-Meteorology/-Climatology in Africa, Asia and Latin America: From Agrometeorological Services via Climate Change to Agroforestry and Other Climate-Smart Agricultural Practices ... 237
C. (Kees) J. Stigter

14 West African Farmers' Climate Change Adaptation: From Technological Change Towards Transforming Institutions 253
Daniel Callo-Concha

15 Livelihood Options as Adaptation to Climate Variability Among Households in Rural Southwest Nigeria: Emerging Concerns and Reactions 267
Isaac B. Oluwatayo

| 16 | Climate Change Projections for a Medium-Size Urban Area (Baia Mare Town, Romania): Local Awareness and Adaptation Constraints.. 277 |

Mihaela Sima, Dana Micu, Dan Bălteanu, Carmen Dragotă, and Sorin Mihalache

| 17 | Trends and Issues of Climate Change Education in Japan....... 303 |

Keiko Takahashi, Masahisa Sato, and Yasuaki Hijioka

| 18 | Changes in Attitude Towards Climate Change and Transformative Learning Theory....................................... 321 |

Gherardo Girardi

| 19 | Societal Transformation, Buzzy Perspectives Towards Successful Climate Change Adaptation: An Appeal to Caution............ 353 |

Sabine Tröger

| 20 | Analysis of Climate Change Adaptation Measures Used by Rural Dwellers in the Southeast and Southsouth Zone of Nigeria....... 367 |

C.C. Ifeanyi-Obi

| 21 | Science Field Shops: An Innovative Agricultural Extension Approach for Adaptation to Climate Change, Applied with Farmers in Indonesia....................................... 391 |

C. (Kees) J. Stigter, Yunita T. Winarto, and Muki Wicaksono

Part I
Climate Change Adaptation in Cities and Communities

Chapter 1
Changes in Land Cover Categories within Oti-Kéran-Mandouri (OKM) Complex in Togo (West Africa) between 1987 and 2013

Aniko Polo-Akpisso, Kpérkouma Wala, Soulemane Ouattara, Fousseni Foléga, and Yao Tano

Abstract Oti-Kéran-Mandouri (OKM) is a complex of protected areas with national and international ecological importance. It is located in the flood plain of the Oti River in Togo. Unfortunately, this area is under anthropogenic pressure. In order to enhance biodiversity conservation, this study aims to assess the spatial changes in land cover within OKM. Landsat images from different missions spanning the time steps 1987, 2000 and 2013 were used to produce land cover maps involving six classes. The classification was based on the maximum likelihood algorithm and the change analyses were performed using Land Change Modeler software integrated in Idrisi GIS and Image Processing system. From 1987 to 2013, wetlands, forests and savannahs diminished while cropland and settlements expanded. Considering the overall area of OKM, wetlands decreased from 43.05 % in 1987 to 31.71 % in 2013. Meanwhile, croplands increased from 0.91 % in 1987 to 34.81 % in 2013. Considering their earlier areas in 1987, forests, savannahs and wetlands have experienced an average annual loss of 5.74 %, 3.94 % and 2.02 %, respectively, while croplands increased at an average annual rate of 285.39 %. The main drivers of these changes appear to be the inadequacy of the

A. Polo-Akpisso (✉)
West African Science Service Center for Climate Change and Adapted Land Use, Graduate Study Program on Climate Change and Biodiversity, Université Félix HouphoëtBoigny, Abidjan, 22 BP 461 Abidjan 22, Côte d'Ivoire

Laboratoire de Botanique et Ecologie Végétale, Faculté des Sciences, Université de Lomé, BP 1515 Lomé, Togo

Laboratoire de Zoologie et Biologie Animale, UFR Biosciences, Université Félix HouphoetBoigny, Abidjan, 02 BP 1170 Abidjan 02, Côte d'Ivoire
e-mail: anikopolo@gmail.com; anikopolo@live.com

K. Wala • F. Foléga
Laboratoire de Botanique et Ecologie Végétale, Faculté des Sciences, Université de Lomé, BP 1515 Lomé, Togo

S. Ouattara • Y. Tano
Laboratoire de Zoologie et Biologie Animale, UFR Biosciences, Université Félix HouphoetBoigny, Abidjan, 02 BP 1170 Abidjan 02, Côte d'Ivoire

© Springer International Publishing Switzerland 2016
W. Leal Filho et al. (eds.), *Implementing Climate Change Adaptation in Cities and Communities*, Climate Change Management, DOI 10.1007/978-3-319-28591-7_1

management system and increasing anthropogenic pressures. These are intensified by climate change since adaptive strategies, such as recessional agriculture, play an important role in land cover change. The ongoing process of rehabilitation should be strengthened to enable this protected area to play its roles as Ramsar site, biosphere reserve and priority corridor for the migration of the West African savannah elephant. Data from this study could be used to guide conservation planning, further landscape pattern assessment and land cover modelling in the framework of climate change.

Keywords Land cover change • Recessional agriculture • Conservation • Oti-Kéran-Mandouri • Togo

Introduction

Conservation planning and landscape design are important challenges for ecologists and conservation biologists. One of the clues to overcome these challenges is to gain better understanding of land cover/land use (LCLU) dynamics as well as habitat use by focal animal species. LCLU at any given time form the basis for assessing the pattern of degradation in any ecosystem (Craighead and Convis 2013). Habitat loss and degradation are threats to biodiversity conservation in developing countries where the need to balance conservation and human development is imminent. Habitat degradation due to anthropogenic activities causes biodiversity erosion in African countries (Konaté and Kampmann 2010; Folega et al. 2012). An illustrative example of habitat degradation is the current conversion of species' historical range into small patches of habitat. This is the case of flagship animal species such as the lion, *Panthera leo leo* Linnaeus, 1758 (Henschel et al. 2014) or the savannah elephant, *Loxodonta africana africana* Blumenbach, 1797 (Bouché et al. 2011). These species are categorised as vulnerable species (Blanc 2008; Bauer et al. 2012). The social and economic changes in rural areas across African continent in last decades (Folega et al. 2014b) as well as climate change (Heubes et al. 2012; Diwédiga et al. 2013) are driving the degradation of their natural habitat.

Oti-Kéran-Mandouri (OKM) is a complex of protected areas located in the plain of the Oti River in Togo. The protected areas comprises Oti-Kéran National Park and Oti-Mandouri Wildlife Reserve. This area is an important eco-geographical region for the migration of the West African elephant which unfortunately is under anthropogenic pressure. Kéran National Park and the wildlife reserve of Oti-Mandouri are listed as Ramsar sites (Convention on Wetlands of International Importance) in 1995 and 2008. They are also listed in the top ten priority protected areas currently under rehabilitation process in Togo (UICN 2008). Recent studies in the area show high inter-annual fluctuations with declining tendency of rainfall within the period from 1961 to 2010 (Badjana et al. 2011). These changes are felt by local rural population who implement different indigenous strategies to cope with them. Diwédiga et al. (2013) reported that recessional agriculture is one of the most important climate change adaptation strategies implemented by farmers in this area. Recessional agriculture is an out season cultivation of crop on river sides and wetlands. Crops benefit from soil moisture and organic matter provided by floods. The reasons why farmers implement this activity, as reported by the same study, are the adaptation to shortage period, the compensation to crop lost due to severe delay in rainfall or to flooding and the possibility for double harvest in the same year. However, large part of the wetlands and main rivers are located within the borders of OKM. Previous investigation by the United States Geological Survey (USGS 2013) on LCLU in Togo reported that agricultural expansion is encroaching into Togo's protected areas. However, there is no accurate and detailed data available on such phenomenon for OKM. There is also no data available on the current status of natural habitats within this complex. Up to date land cover data on this area would be a valuable source of information to enhance the management of the protected areas.

Land cover maps at moderate scales enable researchers to characterize spatial-distribution patterns of land cover in order to understand habitat quality. This can be achieved by integrating remote sensing (RS) and geographic information systems (GIS) tools. RS and GIS have revolutionized the scope and power of conservation planning because of their ability to accommodate, integrate, and analyse physical, biological, cultural, social, economic, and political data (Craighead and Convis 2013). They are useful for long term habitat monitoring by enabling the analysis of satellite data provided by Landsat mission archives. In order to promote biodiversity conservation, this study aims to assess the spatial changes in land cover within OKM using remote sensing and GIS. It will provide baseline and detailed data on land cover that will be important for the biological conservation goal of this protected area. These outputs of the study could enhance the management plan design and could be used for modelling LCLU in the framework of climate change for a long-term management program.

Method

Study Area

Oti-Kéran-Mandouri (Fig. 1.1) straddles three districts: Kpendjal, Oti and Kéran. This area is located in the eco-floristic zone I of Togo (Ern 1979). It is characterized by tropical climate with a rainy season from June to October and a dry season between November and May. The annual average rainfall is between 800 and 1000 mm and the temperature vary from 17 to 39 °C during the dry season and from 22 to 34 °C during the rainy season (UICN 2009). The predominant vegetation is Sudanian savannah, with some dry forest patches and gallery forests along rivers (Dimobe et al. 2014; Folega et al. 2014a). This area is a flat plain characterized by

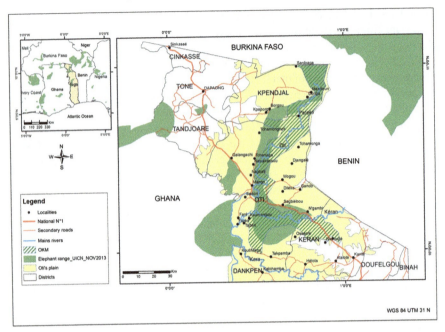

Fig. 1.1 The Oti-Kéran-Mandouri (OKM) study area, with protected areas and district subdivisions discussed in the text

indurate tropical ferruginous soils and swampy tropical ferruginous soils (Dimobe et al. 2014). The large wetlands of the Oti River and its tributaries are important biotopes for migrating birds (Cheke 2001) and large mammals such as the African savannah elephant (Burrill and Douglas-Hamilton 1987; Douglas-Hamilton et al. 1992; Okoumassou et al. 1998). Large portions of this area have been put under protection for biodiversity conservation. Some of these protected areas are Oti-Kéran National Park (9°55′–10°20′ North and 0°25′–1°00′ East) and Oti-Mandouri Wildlife Reserve (10°18′–11° North and 0°30′–0°47′ East). They are part of the most important protected areas in Togo (UICN 2008) and are listed as Ramsar sites (Oti-Kéran in 1997 and Oti-Mandouri in 2007). They have been recently considered as biosphere reserves in 2011. Unfortunately, these protected areas are under continuous anthropogenic pressure that significantly transformed their biotopes into degraded habitats.

Data Acquisition

Landsat images of 30 m spatial resolution were downloaded from the United States Geological Survey via Global Visualization platform (USGS/GloVis). The images were acquired within the time frame of 1987, 2000 and 2013. These time frames

Table 1.1 Landsat image characteristics

Image	Path	Row	Sensor	Date of acquisition	Cloud cover (%)
1	193	053	Landsat 4 Thematic Mapper	1987-10-30	0.00
2	193	052	Landsat 4 Thematic Mapper	1987-10-30	10.00
3	193	053	Landsat 7 Enhanced Thematic Mapper	2000-12-04	0.00
4	193	052	Landsat 7 Enhanced Thematic Mapper	2000-12-30	0.00
5	193	53	Landsat 8 Operational Land Imager	2013-10-29	0.43

were selected because they characterize three important periods in the dynamic of protected areas in Togo as described by Folega et al. (2014b). The first period from colonial to 1990 is marked by strict biodiversity protection; the second period from 1990 to 2000 is marked by anarchic exploitation; and the third period from 2000 characterized by a process of consensual rehabilitation. All the dates of the downloaded images were within the beginning of the dry season to avoid cloud cover and also account for healthy vegetation. The images were already pre-possessed and georeferenced and have not been subjected to further radiometric corrections. Their characteristics are presented below in Table 1.1. Even though the second image (1987) has 10 % of clouds, the study area was not affected.

Image Classification

Landsat 4 scenes (Images 1 and 2) and Landsat 7 scenes (Images 3 and 4) were mosaicked to create a larger image composite. This process was done by matching their radiometric characteristics by equalizing the means and variances of recorded values across the set of images, based on an analysis of comparative values in overlap areas (Eastman 2012). The resulting composites were registered to the Landsat 8 scene with size of 185-km-cross-track-by-180-km-along-track (Lira and Taborda 2014) to enable one scene to cover the entire study area.

The classification scheme was based on the framework recommended by the Intergovernmental Panel on Climate Change (IPCC) for national greenhouse gases emission inventories in Land Use, Land use change and Forestry sector (IPCC 2003). Training sites were defined on the basis of colour infrared composition (Lowry et al. 2005) for six land categories (Table 1.2) distinguished during field campaigns from 11th to 30th November 2013. Based on the defined training sites, a supervised classification using maximum likelihood algorithm was performed on the Landsat the images. Maximum likelihood is one of the most commonly used algorithm in remote sensing image classification (Richards 2013) and therefore the use in this study would facilitate comparison of findings with other studies. The classifications were run with all bands except Band 6 and Band 8 for TM and ETM

Table 1.2 Land categories' definition

N*	Land category	Definition
1	Forest	Vegetation dominated by trees above 7 m of height and with closed canopy. Shrubs or herbaceous may be present
2	Savannah	Vegetation with a continuous grass cover. Trees and shrubs may be present but scattered
3	Wetland	Marshland or areas covered permanently or temporary by water
4	Cropland	All open lands and pastures exploited to grow crops and livestock including traditional agroforestry systems and parklands
5	Settlements	Transportation infrastructure, human settlements of any size, and bare soils
6	Water bodies	Any stretch of water including rivers and water ponds

images. For OLI images, the classification was completed using all bands except Bands 1, 6, 8, 9, 10 and 11. It was performed using Idrisi GIS and Image Processing software.

Accuracy Assessment

The accuracy assessment was done based on Cohen's Kappa (Cohen 1960; Congalton et al. 1983) as well as on user's, producer's and overall accuracies (Liu et al. 2007). The binomial distribution was used to determine the appropriate sample size to obtain unbiased ground reference for a valid statistical testing of the land cover map accuracy. This is recognized as an appropriate mathematical model to use in determining an adequate sample size for accuracy assessment (Hord and Brooner 1976; Hay 1979; Rosenfield and Melley 1980; Fitzpatrick-Lins 1981; Rosenfield 1982). The equation is:

$$N = \frac{z^2 pq}{E^2}$$

where,
   N = Number of samples
   p = Expected or calculated accuracy (in percentage)
   q = 100-p
   E = Allowable error
   Z = Standard normal deviate for the 95 % two-tail confidence level (1.96).

The allowable error E was set to 5 % for an overall expected accuracy of 80 %. The minimum sample size was then calculated and apportioned to the different land cover categories. A minimum threshold of 20 sample points has been set for each land category (van Genderen and Lock 1977).

Table 1.3 Minimum sample size per land category for image 5

Image 5 from October 2013					
Class	Categories	Area (ha)	%Area	Estimated sample	Final sample size
1	Forests	4038.21	2.24	5.50	20
2	Savannahs	33,255.18	18.41	45.29	46
3	Wetlands	57,267.00	31.71	78.01	80
4	Croplands	62,858.07	34.81	85.62	90
5	Settlements	18,454.86	10.22	25.134	27
6	Water bodies	4721.49	2.61	6.43	7
	Total	180,594.80	100	246	270

Reference data have been generated differently for each image. They were based on ground verification data for the recent image (Table 1.3) while generated from the national topographic map established in 1980, the national vegetation map (Afidégnon et al. 2002) and other available maps for the historical images (2000 and 1987).

Ground verification points were generated in ERDAS IMAGINE by applying a stratified random sampling scheme with a window kernel of 3×3 pixels and on the basis of a majority rule. This results in selection of a sample point only if a clear majority threshold of six pixels out of nine in the window belonged to the same class (Maingi et al. 2002). The generation of sample points in this manner ensured that points were extracted from areas of relatively homogenous land cover class.

Change Analysis

Change analyses were performed using Land Change Modeler fully integrated software in the Idrisi GIS and Image Processing system (Eastman 2012). Maps were compared for change detection as following: 1987 and 2000, 2000 and 2013, and 1987 and 2013. Maps were produced with the combination of Land Change Modeler and ArcGIS 10.2.2.

Results

Classification Accuracy

The overall accuracy of the land cover in 1987 is 79.28 %, 2000 is 73.33 %, and the recent image of 2013 is 79.06 %. There is a good agreement of the classification results and reference data with values of Kappa statistics of 0.69, 0.65 and 0.72 as respective overall classification accuracy. However there were low producer's accuracies for earlier land covers (1987 and 2000): settlements and water bodies

Table 1.4 Classification accuracy

Landcover	1987		2000		2013	
	PA (%)	UA (%)	PA (%)	UA (%)	PA (%)	UA (%)
Forests	100	61.11	82.22	78.72	83.87	89.66
Savannahs	82.47	86.02	81.82	60	68	72.34
Wetlands	81.31	87	83.72	70.59	87.10	69.23
Croplands	50	33.33	62.96	85	84.62	91.67
Settlements	26.32	62.50	14.29	100	75	57.14
Water bodies	100	36.36	16.67	100	35.71	83.33
Overall accuracy	79.28 %		73.33 %		79.06 %	
Kappa	0.69		0.65		0.72	
	PA = Producer's accuracy			UA = User's accuracy		

for ETM 2000 (14.29 and 16.67 %) and settlements for TM 1987 (26.32 %). In contrast, user's accuracy was low for cropland and water bodies for TM 1987 (Table 1.4).

Land Cover Dynamics

Wetland was the most represented land cover type. From 1987 to 2013, wetlands, forests and savannahs decreased in areal cover, while cropland and settlements increased (Fig. 1.2). Figure 1.3 shows the change in percentage of the overall area of OKM in each land cover type in 1987, 2000 and 2013. Wetlands decreased in percentage of the overall area from 43.05 % in 1987 to 36.30 % in 2000 and to 31.71 % in 2013. Savannah decreased from 37.72 % in 1987 to 26.34 % in 2000 and to 18.41 % in 2013. Forests increased from 8.84 % in 1987 9.03 % in 2000 but decreased to 2.24 % in 2013. Meanwhile, croplands increased from 0.91 % in 1987 to 13.75 % in 2000 and to 34.81 % in 2013. Settlements decreased from 7.74 % in 1987 to 13.22 % in 2000 and to 10.22 % in 2013. Water bodies decreased from 1.74 % in 1987 to 1.36 % in 2000 but increased to 2.61 % in 2013.

Change Analysis

From 1987 to 2000, the most important changes in land cover were losses in wetland and gains in cropland. Table 1.5 shows the transition in each land category in percentage of the overall area of OKM. Forests were chiefly converted to savannah (3.28 %) and wetlands (2.68 %), and only a minor component was converted to cropland (0.30 %) and settlements (0.18 %). Savannahs were mostly converted to wetlands (14.50 %) and croplands (4.29 %) whereas wetlands were mostly converted to settlements (8.53 %) and savannahs (8.50 %). Some portions of

1 Changes in Land Cover Categories within Oti-Kéran-Mandouri (OKM)... 11

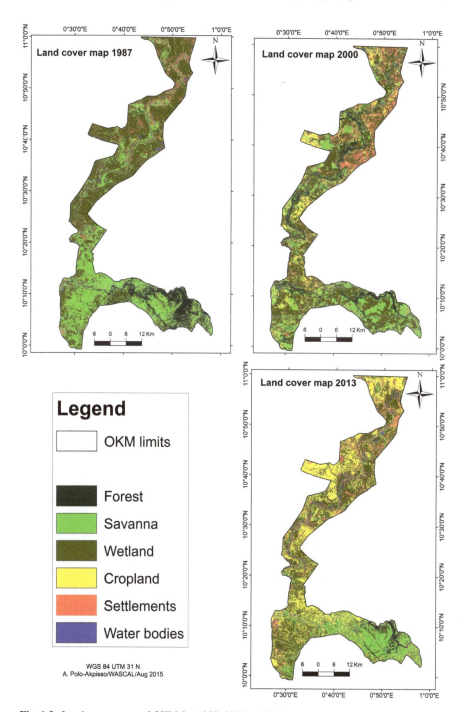

Fig. 1.2 Land cover maps of OKM for 1987, 2000 and 2013

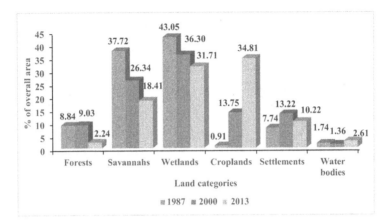

Fig. 1.3 Changes in land categories within OKM complex in Togo from 1987 to 2013

Table 1.5 Transition matrix in percent (%) of overall area from 1987 to 2000 within OKM

From	To	2000					
		Forests	Savannahs	Wetlands	Croplands	Settlements	Water bodies
1987	Forests	2.31	3.28	2.68	0.30	0.18	0.10
	Savannahs	3.29	12.39	14.50	4.29	2.85	0.40
	Wetlands	2.09	8.50	15.92	7.69	8.53	0.32
	Croplands	0.17	0.15	0.30	0.13	0.12	0.04
	Settlements	0.64	1.77	2.62	1.19	1.37	0.16
	Water bodies	0.53	0.27	0.28	0.14	0.18	0.34

croplands were also converted to wetlands (0.30 %) and forests (0.17 %). Settlements were mostly converted to wetlands (2.62 %) and savannahs (1.77 %). Much less change was observed for water bodies, and this largely comprised conversion to forests (0.53 %) and wetlands (0.28 %).

From 2000 to 2013, croplands and wetlands remained land categories that changed most and wetlands remained the land category where changes have been more important. The transition in each land category in percentage of overall area of OKM is shown in Table 1.6. Forests were mostly converted in wetlands (2.02 %) and in croplands (1.95 %). Savannahs were mostly converted in wetlands (7.47 %) and in croplands (7.40 %) while 11.74 % and 6.92 % of wetlands were converted in croplands and in savannahs respectively. Part of croplands were also converted in others land categories: 3.70 % were converted in wetlands, 1.51 % in settlements

Table 1.6 Transition matrix in percent (%) of overall area from 2000 to 2013 within OKM

From	To	2013					
		Forests	Savannahs	Wetlands	Croplands	Settlements	Water bodies
2000	Forests	1.05	1.79	2.02	1.95	1.30	0.92
	Savannahs	0.67	8.13	7.47	7.40	2.21	0.45
	Wetlands	0.46	6.92	13.78	11.74	3.04	0.37
	Croplands	0.02	0.81	3.70	7.57	1.51	0.15
	Settlements	0.02	0.69	4.50	5.91	1.96	0.14
	Water bodies	0.02	0.07	0.23	0.24	0.20	0.60

and 0.81 % in savannahs. Water bodies were mostly converted in croplands (0.24 %) and in wetlands (0.23 %).

The overall change between 1987 and 2013 showed high variation in wetlands and croplands. Figure 1.4 shows land cover changes from 1987 to 2013 and Table 1.7 reports the transition in each land category. Wetlands were predominantly converted to croplands (16.84 %) and to settlements (6.19 %). There was also transition of savannahs to wetlands (11.33 %) and croplands (12.80 %). Forests were converted to savannahs (4.02 %) and croplands (1.71 %), and less so to wetlands (0.83 %) and water bodies (0.05 %). Croplands were also converted to wetlands (0.28 %), settlements (0.14 %), water bodies (0.08 %), savannahs (0.06 %) and forests (0.01 %). Settlements were mostly converted to croplands (2.85 %) and wetlands (2.81 %). Water bodies were converted to settlements (0.28 %) and croplands (0.27 %) to a greater extent than they were converted to wetlands (0.18 %), savannah (0.06 %) and forests (0.01 %).

The net change in each land category i.e., the result of taking the earlier land cover areas, adding the gains and then subtracting the losses (Eastman 2012) is reported in Table 1.8. Croplands and settlements increased at an average of 108.13 % and 5.45 % respectively, of the proportion of their own area from 1987 to 2000. From 2000 to 2013, croplands gained from all other land categories and continue then to increase but at a rate of 11.77 % per year of their own area. Forests and savannahs decreased at an annual average rate by 5.79 % and 2.32 %, respectively of their own area. Globally from 1987 to 2013, forests, savannahs and wetlands have decreased at an average annual percentage of their own area by 5.74 %, 3.94 % and 2.02 %, respectively, while croplands increased at an average annual rate of 285.39 %.

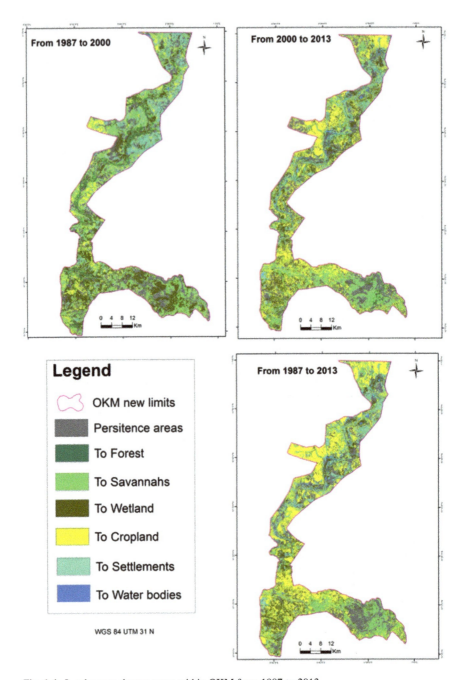

Fig. 1.4 Land cover change maps within OKM from 1987 to 2013

Table 1.7 Transition matrix in percent (%) of overall area from 1987 to 2013 within OKM

From	To	2013					
		Forests	Savannahs	Wetlands	Croplands	Settlements	Water bodies
1987	Forests	2.00	4.02	0.83	1.71	0.24	0.05
	Savannahs	0.16	10.67	11.33	12.80	2.23	0.54
	Wetlands	0.04	3.03	16.29	16.84	6.19	0.66
	Croplands	0.01	0.06	0.28	0.34	0.14	0.08
	Settlements	0.02	0.57	2.81	2.85	1.15	0.35
	Water bodies	0.01	0.06	0.18	0.27	0.28	0.93

Discussion

Image Classification and Accuracy Assessment

The overall accuracy assessment values for all the classified images reflect the confusion between some land categories, especially settlements and croplands. This confusion might explain the poor producer's accuracy for both classes. The similar confusions were previously reported by Folega et al. (2014b). The study area is characterized by a cultural landscape marked by traditional agroforestry system wherein farmlands and crops are associated with parklands of multipurpose trees (Wala et al. 2005; Folega et al. 2011; Kebenzikato et al. 2014; Padakale et al. 2015). This cultural landscape is the source of the misclassification of settlements and croplands and even the source of confusion with savannahs (Badjana et al. 2014b). Misclassification in water bodies is due to the low resolution of Landsat imagery limiting the distinction between water bodies with surrounding vegetation.

Land Use and Land Cover Change Analysis

Wetlands are the most important land category found within OKM. This is well in line with the classification of OKM as a Ramsar site (Cheke 2001; Adjonou 2012). It is also the land category that is the most converted, particularly to croplands. There is high pressure on wetlands because they are systematically converted into large paddy fields. Agriculture is then the main driver of change in Landcover and Landuse dynamics (LCLU). It is a shifting cultivation such as elsewhere in Togo and in West Africa. This result is in line with previous studies in tropical regions and particularly in West Africa (Lambin et al. 2003; Houessou et al. 2013; Wood et al. 2004). Other studies, concluded that agriculture, population growth and indirect effects of climate change remain the main factors of land change and the

Table 1.8 Net change in each land category within OKM

	1987–2000				2000–2013				1987–2013			
	Total net change (ha)	Average net change per year (ha)	Annual net change in % of earlier area		Total net change (ha)	Average net change per year (ha)	Annual net change in % of earlier area		Total net change (ha)	Average net change per year (ha)	Annual net change in % of earlier area	
Forests	336	25.85	0.16		−12,269	−943.77	−5.79		−11,923	−917.15	−5.74	
Savannahs	−20,548	−1580.62	−2.32		−14,321	−1101.62	−2.32		−34,869	−2682.23	−3.94	
Wetlands	−12,186	−937.38	−1.21		−8285	−637.31	−0.97		−20,472	−1574.77	−2.03	
Croplands	23,190	1783.85	108.13		38,019	2924.54	11.77		61,209	4708.38	285.39	
Settlements	9894	761.08	5.45		−5415	−416.54	−1.75		4479	344.54	2.47	
Water bodies	−685	−52.69	−1.68		2272	174.77	7.14		1586	122.00	3.89	

main threat to biodiversity conservation (Landmann et al. 2010; Heubes et al. 2012). Changes in populations' livelihood through the implementation of indigenous adaptation strategies (Badjana et al. 2011; Diwédiga et al. 2013; Ojoyi and Mwenge Kahinda 2015) could be considered as the indirect effects of climate change. The implementation of recessional agriculture as an adaptive strategy in the study area shows that climate change and LCLU change are tightly linked.

Increase in settlements could be related to the annual rate of population growth in the study area. Actually, OKM is within a region where the population growth rate is the highest in Togo (3.18 %) (RGPH 2011). Moreover, Kpendjal, Oti and Keran districts are the poorest in Togo (DSRP-C 2009). Therefore, there is high pressure on natural resources by the local population. The intrinsic relationship between natural resource degradation and poverty has been shown by many studies (Konaté and Linsenmair 2010; Dimobe et al. 2014). There is a particular problem of natural resource management that increases the pressure on protected areas in Togo after 1990. As described by Folega et al. (2014b), the period from 1990 to 2000 was marked by illegal and anarchic exploitation of protected resources by bordering human populations. This situation resulted from the political, economic and social troubles of 1990 mainly due to uncontrolled democratic opening process and the semi-military and repressive management system implemented from colonial period to 1990. These illegal and anarchic anthropogenic activities led to high land conversion. This could explain the increase of croplands from 1987 to 2000. This increment is higher than the reported agricultural expansion on the national scale for a period of 25 years from 1975 to 2000 by USGS (2013). While croplands increased from 0.91 % in 1987 to 34.81 % in 2013 of the total area of OKM, they have been reported to increase from 12.1 % of the country's land area in 1975 to 21.8 % in 2000.

As shown by transition maps, there is a core area of remaining natural ecosystems within OKM in Keran district. This area has been already mentioned by UICN (2008) as the most conserved area. Periodic visits to this area by savannah elephant coming from bordering countries have been mentioned in several studies (Okoumassou et al. 2004; UICN 2008). This area requires a particular conservation attention. Its remaining natural habitats are still under threats from some anthropogenic activities such as wood cutting (Adjonou et al. 2009) and charcoal production (Kokou et al. 2009). In the centre of Benin, Arouna et al. (2011) found also that charcoal production represented the main activity leading to land cover change. Moreover, there are also threats from transhumance on which no accurate data are currently available. There are in the order of a 1000 heads of cattle pasturing through this area each year damaging crops and invading protected areas. Usually, foraging plants are cut down for cattle and the savannah is burnt without any respect to the fire calendar established for this area.

Considering the net change for each land category (Table 1.8), forests increased by 2.08 % from 1987 to 2000 and decreased by 75.27 % from 2000 to 2013. The increase of forests from 1987 to 2000 could be explained by natural regeneration since forests are located along rivers and in the well protected core area of the former Oti-Keran national park (southern part of our study area). These areas were less pressured when social claims started from 1990. However, the decrease of forests from 2000 to 2013 is the result of the combined effects of increasing anthropogenic activities (wood cutting and charcoal production) and climate change. Recent studies demonstrated that climate change effects in Northern Togo are expressed by inter-annual fluctuations with declining tendency of rainfall and increasing tendency of temperature on the period from 1961 to 2010 (Badjana et al. 2011, 2014a). The deficit in rainfall could weaken plant resilience to fire events. Furthermore, one of the most adaptive strategies to climate change implemented in this are by local population is recessional agriculture. This activity consists of crop cultivation during the dry season on river banks and in wetlands where crop could benefit from flooding and soil moisture. This practice involves the destruction of gallery forest (Diwédiga et al. 2013; Badjana et al. 2014b). Recessional agriculture is categorized as a wetland agriculture and considered as one of the largest overall agriculture adaptations (Menotti and O'Sullivan 2013). It is described as being highly productive in many tropical regions (Whitmore and Turner 2001). The decrease of forests from 2000 to 2013 reflected the ineffectiveness of the management system of this complex of protected areas and the weakness of the current consensual rehabilitation process.

Conclusion

This study provides detailed analysis on the dynamics of Land cover of OKM. There has been important changes in land cover within the protected area from 1987 to 2013. Forests, savannahs and wetlands were converted into human transformed landscapes. Wetlands were the most converted land category. Croplands and settlements have increased with a high magnitude especially from 1987 to 2000. Drivers of all these changes appear to be the ineffectiveness of management system and increasing anthropogenic pressure which is emphasized by climate change effects since adaptive strategy such as recessional agriculture leads to gallery forest degradation and wetland conversion in croplands. These changes threaten biodiversity conservation role of the protected area and even its status not only as Ramsar site but also as priority corridor for the migration of the West African savannah elephant. Results from this study could be used to guide conservation planning in the current process of protected areas requalification for which OKM has been listed as one of the ten top priority protected areas to be requalified. Further investigations will focus on landscape patterns to assess the connectivity of remaining natural land patches since isolation would disrupt the connectivity of the protected areas network for large mammal migration.

Acknowledgement This study was funded by the German Federal Ministry of Education and Research (BMBF) through the West African Science Service Centre on Climate Change and Adapted Land Use (WASCAL) initiative. We acknowledge all the anonymous reviewers who added a great value to the first version of this manuscript.

References

Adjonou K (2012) Rapport De Collecte Des Données Nationales—Togo. PARCC-UNEP-UICN-GEF, Lomé

Adjonou K, Bellefontaine R, Kokou K (2009) Les Forêts Claires Du Parc National Oti-Ke′Ran Au Nord-Togo: Structure, Dynamique Et Impacts Des Modifications Climatiques Récentes. Secheresse 20:1–10

Afidégnon D, Carayon J-L, Fromard F, Lacaze D, Guelly KA, Kokou K, Woegan YA, Batawila K, Blasco F, Akpagana K (2002) Carte De Végétation De La Végétation Du Togo. L.B.E.V., L.E. T., Lomé, Toulouse

Arouna O, Toko I, Djogbénou CP, Sinsin B (2011) Comparative analysis of local populations' perceptions of socioconomic determinants of vegetation degradation in Sudano-Guinean Area in Benin (West Africa). Int J Biodiv Conserv 3:327–337

Badjana HM, Batawila K, Wala K, Akpagana K (2011) Évolution Des Paramètres Climatiques Dans La Plaine De L'oti (Nord-Togo): Analyse Statistique, Perceptions Locales Et Mesures Endogènes D'adaptation. Afr Sociol Rev 15:77–95

Badjana HM, Hounkpè K, Wala K, Batawila K, Akpagana K, Edjamé KS (2014a) Analyse De La Variabilité Temporelle Et Spatiale Des Séries Climatiques Du Nord Du Togo Entre 1960 Et 2010. Eur Sci J 10:257–272

Badjana HM, Selsam P, Wala K, Flügel W-A, Fink M, Urban M, Helmschrot J, Afouda A, Akpagana K (2014b) Assessment of land-cover changes in a sub-catchment of the Oti Basin (West Africa): a case study of the Kara River Basin. Zbl Geol Paläont Teil I 1:151–170

Bauer H, Nowell K, Packer C (2012) Panthera leo. The IUCN red list of threatened species. Version 2014.3. www.iucnredlist.org. Accessed 27 Feb 2015

Blanc J (2008) Loxodonta Africana. In: IUCN (ed) IUCN Red list of threatened species

Bouché P, Douglas-Hamilton I, Wittemyer G, Nianogo AJ, Doucet J-L, Lejeune P, Vermeulen C (2011) Will elephants soon disappear from West African Savannahs? PLoS One 6:e20619

Burrill A, Douglas-Hamilton I (1987) African elephant database project: final report. UNEP/GRID, Nairobi

Cheke RA (2001) Important bird areas in Africa and Associated Islands: priority sites for conservation. In: Fishpool LDC, Evans MI (eds). Pisces Publications and BirdLife International, Newbury and Cambridge

Cohen I (1960) A coefficient of agreement of nominal scales. Educ Psychol Meas 20:37–46

Congalton RG, Oderwald R, Mead R (1983) Assessing landsat classification accuracy using discrete multivariate analysis statistical techniques. Photogramm Eng Remote Sens 49:1671–1678

Craighead FL, Convis CLJ (2013) Conservation planning: shaping the future. ESRI Press, Redlands, CA

Dimobe K, Wala K, Dourma M, Kiki M, Woegan Y, Folega F, Batawila K, Akpagana K (2014) Disturbance and population structure of plant communities in the wildlife reserve of Oti-Mandouri in Togo (West Africa). Ann Res Rev Biol 4:2501–2516

Diwédiga B, Batawila K, Wala K, Hounkpè K, Gbogbo AK, Akpavi S, Tatoni T, Akpagana K (2013) Exploitation Agricole Des Berges: Une Strategie D'adaptation Aux Changements Climatiques Destructrice Des Forets Galleries Dans La Plaine De L'oti. Afr Sociol Rev/Revue Africaine de Sociologie 16:77–99

Douglas-Hamilton I, Michelmore F, Inamdar I (1992) African elephant database. GEMS/GRID/UNEP, Nairobi

DSRP-C (2009) Document complet de Stratégie de Réduction de la Pauvreté 2009–2011 (DSRP-C), Version finale. 117 p

Eastman JR (2012) Idrisi Selva manual. Clark University, Worcester

Ern H (1979) Die Vegetation Togos. Gliederung, Gefährdung, Erhaltung. Willdenowia 9:295–312

Fitzpatrick-Lins K (1981) Comparison of sampling procedures and data analysis for a land use and land cover maps. Photogramm Eng Remote Sens 47:343–351

Folega F, Samake G, Zhang C-y, Zhao X-h, Wala K, Batawila K, Akpagana K (2011) Evaluation of agroforestry species in potential fallows of areas gazetted as protected areas in North-Togo. Afr J Agric Res 6:2828–2834

Folega F, Dourma M, Wala K, Batawila K, Zhang CY, Zhao XH, Koffi A (2012) Assessment and impact of anthropogenic disturbances in protected areas of Northern Togo. Forestry Stud China 14:216–223

Folega F, Dourma M, Wala K, Batawila K, Xiuhai Z, Chunyu Z, Akpagana K (2014a) Basic overview of Riparian Forest in Sudanian Savanna ecosystem: case study of Togo. Rev Écol (Terre Vie) 69:24–38

Folega F, Zhang CY, Zhao XH, Wala K, Batawila K, Huang HG, Dourma M, Akpagana K (2014b) Satellite monitoring of land-use and land-cover changes in Northern Togo protected areas. J Forestry Res 25:385–392

Hay AM (1979) Sampling designs to test land use map accuracy. Photogramm Eng Remote Sens 45:529–533

Henschel P, Coad L, Burton C, Chataigner B, Dunn A, MacDonald D, Saidu Y, Hunter LTB (2014) The lion in West Africa is critically endangered. PLoS One 9:e83500

Heubes J, Schmidt M, Stuch B, García Márquez JR, Wittig R, Zizka G, Thiombiano A, Sinsin B, Shaldach R, Hahn K (2012) The projected impact of climate and land use change on plant diversity: an example from West Africa. J Arid Environ 96:48–54

Hord RM, Brooner W (1976) Land use map accuracy criteria. Photogramm Eng Remote Sens 42:671–677

Houessou LG, Teka O, Imorou IT, Lykke AM, Sinsin B (2013) Land use and land-cover change at "W" biosphere reserve and its surroundings areas in Benin Republic (West Africa). Environ Nat Resour Res 3:87–100

IPCC (2003) Good practice guidance for land use, land-use change and forestry. In: Penman J, Gytarsky M, Hiraishi T, Krug T, Kruger D, Pipatti R, Buendia L, Miwa K, Ngara T, Tanabe K, Wagner F (eds) IPCC National Greenhouse Gas Inventories Programme. Intergovernmental Panel on Climate Change, Geneva

Kebenzikato AB, Wala K, Dourma M, Atakpama W, Dimobe K, Pereki H, Batawila K, Akpagana K (2014) Distribution Et Structure Des Parcs À Adansonia Digitata L. (Baobab) Au Togo (Afrique De L'ouest). Afrique SCIENCE 10:434–449

Kokou K, Nuto Y, Atsri H (2009) Impact of charcoal production on woody plant species in West Africa: a case study in Togo. Sci Res Essays 4:881–893

Konaté S, Kampmann D (2010) Biodiversity Atlas of West Africa, volume Iii: Côte D'ivoire. Abidjan, Frankfurt/Main

Konaté S, Linsenmair KE (2010) Biological diversity of West Africa: importance, threats and valorisation. In: Konaté S, Kampmann D (eds) Biodiversity Atlas of West Africa Volume Iii: Côte D'ivoire. Abidjan, Frankfurt/Main

Lambin EF, Geist HJ, Lepers E (2003) Dynamics of land-use and land-cover change in tropical regions. Ann Rev Environ Resour 28:205–241

Landmann T, Machwitz M, Schmidt M, Dech S, Vlek P (2010) Land cover change in West Africa as observed by satellite remote sensing. In: Konaté S, Kampmann D (eds) Biodiversity Atlas of West Africa: Côte D'ivoire. Abidjan, Frankfurt/Main

Lira C, Taborda R (2014) Advances in applied remote sensing to coastal environments using free satellite imagery. In: Finkl CW, Makowski C (eds) Remote sensing and modeling: advances in Coastal and Marine resources. Springer International Publishing, New York, NY

Liu C, Frazier P, Kumar L (2007) Comparative assessment of the measures of thematic classification accuracy. Remote Sens Environ 107:606–616

Lowry JHJ, Ramsey RD, Boykin K, Bradford D, Comer P, Falzarano S, Kepner W, Kirby J, Langs L, Prior-Magee J, Manis G, O'Brien L, Sajwaj T, Thomas KA, Rieth W, Schrader S, Schrupp D, Schulz K, Thompson B, Velasquez C, Wallace C, Waller E, Wolk B (2005) Southwest regional gap analysis project: final report on land cover mapping methods. RS/GIS Laboratory, Utah State University, Logan

Maingi JK, Marsh SE, Kepner WG, Edmonds CM (2002) An accuracy assessment of 1992 landsat-mss derived land cover for the Upper San Pedro Watershed (U.S./Mexico). United States Environmental Protection Agency, Washington, DC

Menotti F, O'Sullivan A (2013) The Oxford handbook of wetland archaeology. Oxford University Press, Oxford

Ojoyi MM, Mwenge Kahinda J-M (2015) An analysis of climatic impacts and adaptation strategies in Tanzania. Int J Clim Change Strategies Manage 7:97–115

Okoumassou K, Barnes RFW, Sam MK (1998) The distribution of elephants in North-Eastern Ghana and Northern Togo. Pachyderm 26:52–60

Okoumassou K, Durlot S, Akpamou K, Segniagbeto H (2004) Impacts Humains Sur Les Aires De Distribution Et Couloirs De Migration Des Éléphants Au Togo. Pachyderm 36:69–79

Padakale E, Atakpama W, Dourma M, Dimobe K, Wala K, Guelly KA, Akpagana K (2015) Woody species diversity and structure of Parkia Biglobosa Jacq. Dong Parklands in the Sudanian Zone of Togo (West Africa). Ann Res Rev Biol 6:103–114

RGPH (2011) Recensement Général De La Population Et De L'habitat: Résultats Définitifs. In: c. d. l. P. Ministère auprès du Président de la République, du Développement et de l'Aménagement du Territoire (ed). DGSCN

Richards JA (2013) Supervised classification techniques. In: Remote sensing digital image analysis: an introduction, 5th edn. Springer, New York, NY

Rosenfield GH (1982) Sample design for estimating change in land use and land cover. Photogramm Eng Remote Sens 48:793–801

Rosenfield GH, Melley ML (1980) Applications of statistics to thematic mapping. Photogramm Eng Remote Sens 48:1287–1294

UICN (2008) Evaluation De L'efficacité De La Gestion Des Aires Protégées: Aires Protégées Du Togo. Programme Afrique Centrale et Occidentale (PACO)

UICN (2009) Baseline study B: Etat Actuel De La Recherche Et De La Compréhension Des Liens Entre Le Changement Climatique, Les Aires Protégées Et Les Communautés, Projet: Evolution Des Systèmes D'aires Protégées Au Regard Des Conditions Climatiques, Institutionnelles, Sociales, Et Économiques En Afrique De L'ouest, Rapport Final. GEF/UNEP/WCMC/UICN

USGS (2013) West Africa land use and land cover trends project. http://lca.usgs.gov/lca/africalulc/results.php#togo_lulc. Accessed 27 Feb 2015

van Genderen JL, Lock BF (1977) Testing land use map accuracy. Photogramm Eng Remote Sens 43:1135–1137

Wala K, Sinsin B, Guelly KA, Kokou K, Akpagana K (2005) Typologie Et Structure Des Parcs Agroforestiers Dans La Préfecture De Doufelgou (Togo). Science et changements planétaires/Sécheresse 16:209–216

Whitmore TM, Turner BL (2001) Cultivated landscapes of Middle America on the eve of conquest. Courier Corporation, New York, NY

Wood EC, Tappan GG, Hadj A (2004) Understanding the drivers of agricultural land use change in South-Central Senegal. J Arid Environ 59:565–582

Chapter 2
Exposure of Rural Communities to Climate Variability and Change: Case Studies from Argentina, Colombia and Canada

David Sauchyn, Jorge Julian Velez Upegui, Mariano Masiokas, Olga Ocampo, Leandro Cara, and Ricardo Villalba

Abstract This paper presents results from studies of exposure to climate change and extreme events in the Mendoza River Basin in western Argentina, the Chinchiná River basin in the Colombian Andes, and the Oldman River basin and Swift Current Creek watershed in the Canadian Prairies. These case studies are a major component of an international research project: "Vulnerability and Adaptation to Climate Extremes in the Americas" (VACEA). This project is very much interdisciplinary; with social and natural science providing context and direction for research in the other realm of scholarship, producing insights that very likely would not arise from a more narrow disciplinary perspective. A large number of interviews with local actors revealed that agricultural producers and local officials recognize their high degree of exposure and sensitivity to climate variability and extreme weather events, although they generally do not associate this with climate change. Case studies of exposure demonstrate that the perceptions of the local actors are consistent with the nature of the regional hydroclimatic regimes. In all four river basins, climate variability between years and decades masks any regional expression of global climate change. These modes of periodic variability dominate the paleoclimate of past centuries and the recorded hydroclimate of recent decades. The exposure variables examined in this paper, indices of stream flow, snowpack, water excess and deficit, vary in coherence with the characteristic frequencies of large-scale ocean–atmosphere circulation patterns, specifically the ENSO and PDO. Projections of the future states of these variables require the use of climate

D. Sauchyn (✉)
Prairie Adaptation Research Collaborative, University of Regina, Regina, SK, Canada
e-mail: sauchyn@uregina.ca

J.J. Velez Upegui
Department of Civil Engineering, Universidad Nacional de Colombia, Manizales, Colombia

M. Masiokas • L. Cara • R. Villalba
Instituto Argentino de Nivología, Glaciología y Ciencias Ambientales (IANIGLA), CCT-CONICET, Mendoza, Argentina

O. Ocampo
Universidad Autonoma de Manizales, Manizales, Colombia

models that are able to simulate the internal variability of the climate system and the teleconnections between ocean–atmosphere oscillations and regional hydroclimate.

Keywords Rural agricultural communities • Exposure • Climate change • Climate variability • Canada • Colombia • Argentina

Introduction

Vulnerability and adaption to climate change, and the related concepts, are defined by the IPCC (Agard and Schipper 2014) in terms of the extent to which a social and natural system is or could be adversely or beneficially affected:

- **Adaptation**: "The process of adjustment to actual or expected climate and its *effects*."
- **Adaptive capacity**: "The ability of systems [etc.] to adjust to potential *damage*, to take advantage of opportunities, or to respond to consequences."
- **Exposure**: "The presence of people [etc.] in places that could be adversely *affected*."
- **Sensitivity**: "The degree to which a system or species is *affected*, either adversely or beneficially, by climate variability or change."
- **Vulnerability**: "The propensity or predisposition to be adversely *affected*."

These definitions imply that "a predisposition to be adversely affected" (i.e., vulnerability), requires, in the first place, "The presence of people ... in places that could be adversely affected" (i.e., exposure). Secondarily, a necessary condition is that social or natural systems must have either some sensitivity or lack adaptive capacity, or both. In general, rural communities are among the most vulnerable places because they tend to be on the social and economic margins of society and the typical livelihood, agriculture, is highly exposed and sensitive to climate variability and extremes (Hales et al. 2014). Rural livelihoods and social structures are vulnerable to the impacts of climate change on natural resources, including shifts in supplies of water and ecological goods and services. Relative to urban areas, rural communities tend to have less capacity for responding to and preparing for climate changes, as a function of their physical isolation, lower economic diversity, aging population, and less access to formal institutions infrastructure, health care and emergency response systems (Hales et al. 2014).

This paper examines the exposure of rural agricultural communities in Argentina, Canada and Colombia to climate variability and change. These case studies are a key component of the major interdisciplinary research project "Vulnerability and Adaptation to Climate Extremes in the Americas" (VACEA).[1] The goal of the VACEA project is to improve the understanding of the vulnerability of rural agricultural and indigenous communities to shifts in climate variability and to the

---

[1] http://www.parc.ca/vacea/

frequency and intensity of extreme climate events, and to engage governance institutions in enhancing the adaptive capacity of these communities. This paper presents results of our studies of exposure to climate change and extreme events in the Mendoza River Basin in western Argentina, the Chinchiná River basin in the Colombian Andes, and the Oldman Riverbasin and Swift Current Creek watershed in the Canadian Prairies. Elsewhere in this volume, related papers (Diaz, Mussetta et al. Hurlbert et al. Marchildon et al.) present the results of research on the sensitivity and adaptive capacity of rural communities in these river basins. Our evaluation of exposure in this paper is based on the statistical analysis of trends, variability and extremes in time series of climate and water variables. Whereas much of these data are instrumental observations from monitoring networks, we also analyze output from climate models and paleoclimate data inferred from climate proxies such as tree rings.

The research design of the VACEA project follows a vulnerability assessment model (Fig. 2.1) and its associated community-oriented, participatory methodologies. This model frames vulnerability as a function of exposure to climate hazards and their impacts, and the social conditions that determine sensitivity and adaptive capacity (Smit and Wandel 2006), providing a consistent basis for interdisciplinary and comparative research. This "bottom-up" approach incorporates the assessment of vulnerability according to how actors perceive their exposures, sensitivities, and adaptive capacity in the context of other stressors and changes (for example, drought will exaggerate vulnerability to falling international market prices). While the perspective of local actors is critical for evaluating the social variables,

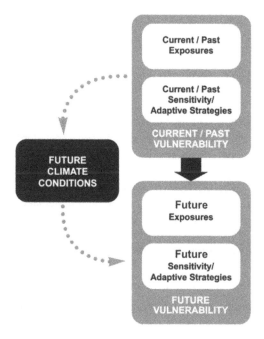

Fig. 2.1 A conceptual model for the assessment of vulnerability to climate change

it also provides important context for the study of physical exposure to regional climate variability and extremes and to impacts on the agro-ecosystems and environmental services that support rural populations.

The Evaluation of Exposure in a Social Science Context

The VACEA project was designed to undertake climate change science in a social science context. The advantage of this bottom-up, community-based approach is a greater relevance, applicability and interdisciplinarity of the science. Alternatively, the historical and future hydroclimate of any study area could be characterized by processing climatic, ecological and hydrologic data, accessing observations and model outputs available from online repositories. Without a sampling of the human population, however, researchers run the risk of investigating environmental changes that are of little or no consequence from the perspective of the local actors; "relying on hazard/impact modelling alone can lead to entirely erroneous conclusions about the vulnerability of rural communities, with potential to significantly misdirect policy intervention" (Nelson et al. 2010: 18). Therefore, the case studies of exposure reported here were very much informed by community vulnerability assessments carried out during the initial stages (2011–13) of the VACEA project. Hundreds of local actors, mainly agricultural producers and local officials, were interviewed using a common methodology, a semi-structured questionnaire with mostly open-ended interview questions.

The social surveys have focused the VACEA project on the weather events and climate changes of greatest concern to the residents of the rural communities. As an example, in the Swift Current Creek watershed in southwestern Saskatchewan (Canada) survey participants identified

- The stresses caused by periodic drought, including crop damage resulting from a lack of soil moisture, poor surface water quality, and a slow institutional response.
- Heightened sensitivity to water deficits where there is less access to surface and groundwater or irrigation is underdeveloped, or because the impacts of drought are exacerbated by a growing cost/price squeeze, a shift in production to single commodities, and increasing farm size and dependence on technology, which tend to offset the benefits of traditional soil and water conservation practices.
- The impacts of excess precipitation: damage to transportation and irrigation infrastructure, delayed seeding and harvest, and changes to water quality, which also are linked to the intensification and industrialization of agriculture and other industrial development (e.g., oil and gas).

Some of these impacts and sensitivities are unique to the physical geography and agricultural production in the Canadian prairies; but most are shared with the river basins in South America. For example, climate extremes impact water quality in the Chinchiná River basin. In the Mendoza River basin, access to water is an issue:

producers located upstream, or nearer to the sources of water, tend to be less sensitive to drought than those downstream of urban centers or at the far end of the irrigation network.

A common theme emerging from the community vulnerability assessments is that 'long-term' climate changes and their affects are not obvious to most agricultural producers. They experience weather not climate. Their adaptive capacity is tuned to seasonal and interannual variability. The main threat posed by climate change is from extreme and unexpected weather, which is more often viewed as natural climatic variability rather than an indication of climate change. This distinction between climate change, a monotonic trend, and climate variability, short-term periodicity, is somewhat arbitrary given that these concepts describe statistical properties of a single climate system. However, it is a meaningful distinction in terms of climate impacts and adaptation strategies (Sauchyn et al. 2015). Observable and projected trends in climate variables, notably the consistent rise in global mean surface temperature, can be attributed directly to the increased concentration of anthropogenic greenhouse gases (IPCC 2013). Furthermore, a large number of studies have documented changes in biological and physical systems that are consistent with global warming (IPCC 2014). Whereas, in absence of anthropogenic global warming, these trending changes may not exist, climate is inherently variable and extreme events occur even in the times of relatively stable climate, although there is substantial scientific evidence suggesting that shifts in amplitude and frequency are occurring in response to warming of the oceans and atmosphere (IPCC 2012; Trenberth 2012).

In the literature on climate change impacts, vulnerability and adaptation, climate variability and extremes generally are viewed as a bigger problem than changes in average conditions (Katz and Brown 1992), although referring to "organizations that specialize in hazard issues" Orlove (2009: 160) suggested "this emphasis directs attention towards short-term acute problems of moderate importance; and away from long-term chronic problems of greater importance, particularly water availability". The relative importance of "short-term acute" versus "long-term chronic" problems would depend very much on context, and specifically the severity of climate events relative to the local capacity to cope. The case study areas considered here are interior river basins characterized by large inter-annual and decadal climatic variability. Despite a history of adaptation to this variable climate, extreme events (floods, droughts and storms) in these river basins have been among the mostly damaging and costly natural events in their respective countries. In these regions, extreme weather has not been of "moderate importance". The problematic climatic changes in these rural watersheds are not a monotonic rise of temperature, or changes in the mean states of other climate variables; it is the prospect of climate variability and extremes that are outside the experience and adaptive capacity of the rural communities.

Case Studies of Exposure

Despite the tens of thousands of kilometres that separate the four case study river basins (Table 2.1), and a difference in agricultural commodities, there is much similarity in terms of exposure of the rural communities to hydroclimatic variability and extremes. Three of the four watersheds (Chinchiná, Mendoza, Oldman) have mountain headwaters shedding snowmelt and rainfall runoff. The fourth stream, Swift Current Creek, also is fed primarily with snowmelt waters although from a prairie upland (Cypress Hills) as opposed to the Cordillera. Droughts and floods are serious climate hazards in these basins, threatening water supply for human consumption and agricultural production. In all four regions, irrigation is an important agricultural practice and adaptation to permanently or seasonally dry conditions. The communities in the Mendoza River basin have been called "hydraulic societies" dependent entirely on the diversion of water from the Rio Mendoza; 98 % of the population of the Province of Mendoza resides in the oases that represent only 3 % of the total surface area. Southern Alberta has more than 60 % of Canada's irrigated area and most of this is located in the Oldman River Basin. These dry river basins also suffer from periodic inundation from the flooding of mountain rivers. Flooding and intense rain also is a serious hazard in the Chinchiná River Basin, and even though it is located in the tropical Colombian Andes, fruit crops are irrigated to overcome seasonal moisture deficits. Thus there are sufficient similarities among the chosen river basins to enable a multi-national comparative study of the human and environmental dimensions of the impacts and adaptive responses to short-term climate variability and extreme events.

Climate Change

Because anthropogenic climate change presents a new distribution of weather statistics, a shift in means and extremes, the regional consequences of change in the earth's energy balance can be understood only from the analysis of output from multiple runs of climate models and different greenhouse gas emission scenarios. In

Table 2.1 Geographical characteristics of the case study river basins

River Basin	Country	Region or province	Size (km^2)	Agricultural production
Chinchiná	Colombia	Caldas	1052	Coffee, fruits, maize, cattle
Mendoza	Argentina	Mendoza	17,821	Grapes, fruits, cattle, horticulture, goats
Oldman	Canada	Alberta	26,700	Grains, pulses, forage, vegetables, cattle
Swift Current	Canada	Saskatchewan	5592	Grains, pulses, forage, cattle

each country, climate change scenarios were derived from global and regional models that are able to simulate important aspects of the regional climate, including the spectral and geographic characteristics of teleconnection patterns (e.g., Lapp et al. 2011) and historical climatic variability. Thus a model that performs well for the Chinchiná River basin in Colombia is not the best model for projecting future climate in western Canada or west-central Argentina. Despite this difference in the source of model projections, there are important common tendencies, including rising minimum temperatures resulting in a decline of the extent of snow and glacier ice and an earlier onset of annual snowmelt.

The case of the Chinchiná River basin is used to illustrate the projection of climate changes and their impact on water supplies. The projections in Fig. 2.2 are expressed as anomalies; the difference in mean annual temperature (°C) and precipitation (%) between 2010–2039 and 1981–2010. The source of these data is six global climate models from the CMIP3 archive built for the IPCC Fourth Assessment Report (Meehl et al. 2007). The temperature anomalies are all positive with greater warming at mid elevations. There is a large range of precipitation anomalies, especially at higher elevations, but with a median projection of little change.

In Fig. 2.3, which illustrates the impact of the projected climate changes on the Chinchiná River, there is little change in mean monthly flows. The projected mean flow rates follow the historical bimodal distribution. On average streamflow is reduced by about 1 %. This compares to an estimated error of around 5 % in the hydrological model.

Climate Variability

Earlier in this paper, we reported that the agricultural producers interviewed in the four case study areas made little or no reference to the threats or opportunities presented by long-term trends in climate variables. Rather their focus and concerns

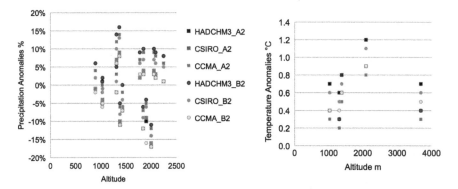

Fig. 2.2 CMIP3 (IPCC AR4) climate change scenarios for the Chinchiná River basin: precipitation (%) and temperature (°C) anomalies, 2010–2039 versus 1981–2010 (Colour figure online)

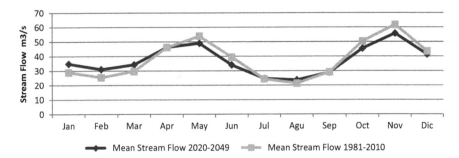

Fig. 2.3 A projection of the future flow of the Chinchiná River based on a historical run (1981–2010) of the Tetis hydrological model and a future (2020–2049) simulation, created by driving the Tetis model with output from the climate model NCC_NORESM1 (RCP 8.5)

relate to recent weather events, which they perceive as unusually severe, but not necessarily an indication of climate change. They tend to regard extreme weather events as typical of their variable climate but nonetheless stressful. The clear message for the natural scientists on the VACEA project was that, if our research is to have relevance for the local administrations and agricultural economy, we should address this dominant concern: exposure and sensitivity to climate variability and extreme events. At the same time, adaptation planning and policy making at higher levels would benefit from new climate change projections and research on how weather is changing as a consequence of the anthropogenic warming of the oceans and atmosphere.

The histories of the studied communities and local economies are punctuated by the impacts of flooding and drought. Therefore considerable scientific effort was applied to characterizing and analyzing these hydrologic extremes, mostly by deriving, from instrumental weather and water records, maps and time series of commonly used indices of drought and excess water, such as the Palmer Drought Severity Index, Climate Moisture Index and Standardized Evapotranspiration Precipitation Index (SPEI). For example, in Canada we found that that the SPEI is an effective index of moisture variability in terms of the impact on agricultural production as illustrated in Fig. 2.4, a plot of growing season (May to August) SPEI and spring wheat yields for the Swift Current Creek watershed over the period 1956–2012 (From Wittrock et al. 2014). Both variables are dimensionless. The crop yield data were processed to remove a positive trend that reflects technological innovation and better farming practices over the past six decades, and in general progressive adaptation to a dry climate and relatively short growing season. In Fig. 2.4, negative departures in wheat yields occur in years with a negative water balance; they are lowest in the second of two or more dry years (i.e., 1961 and 1985). Higher yields (positive departures from the long-term trend) are associated with a positive SPEI, however, excessive moisture can suppress yields, such as in the most recent 4 years, by delaying the seeding and germination of annual crops.

The risks and impacts associated with extreme water levels and sustained dry spells, as illustrated in Fig. 2.4 and reported by the local actors, focused our research

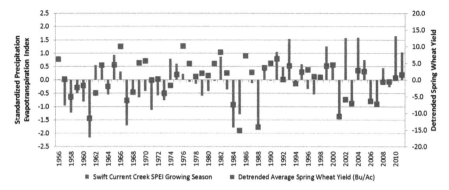

Fig. 2.4 Time series (1956–2011) of dimensionless growing season SPEI and detrended spring wheat yields, Swift Current Creek watershed, Saskatchewan, Canada. From: Wittrock et al. (2014) (Colour figure online)

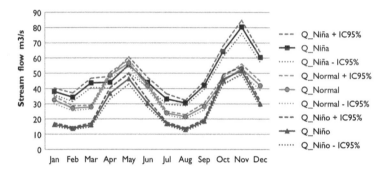

Fig. 2.5 The annual water balance (1981–2010) of the Chinchiná River Basin for La Niña (*blue*), El Niño (*red*) and normal (*green*) weather conditions. The three simulations, plotted with 95 % confidence intervals (IC), were run using a lumped hydrological model (Colour figure online)

activity on the causes, frequency, severity and duration of excessive moisture and water scarcity, and particularly variability at annual and decadal time scales. Furthermore, climate model projections (e.g., Kharin et al. 2007; Tebaldi et al. 2006; IPCC 2013) converge on a scenario of amplified hydroclimatic variability with twenty-first century greenhouse warming. In all four of our case study watersheds, the hydroclimatic variability is linked to the correlation over space and time between the regional hydroclimatic regimes and large-scale patterns of coupled ocean–atmosphere circulation, specifically the El Niño-Southern Oscillation (ENSO), and Pacific Decadal Oscillation (PDO) Masiokas et al. 2006, 2010; Poveda and Mesa 1997; Shabbar et al. 2011; Villalba et al. 2011).

The strong influence of the ENSO is observed across the Colombian Andes (Poveda et al. 2011) including the Chinchiná River basin. Figure 2.5 shows the results of a simulation of the annual water balance (1981–2010) using a lumped hydrological model driven with monthly weather conditions during El Niño, La Niña and normal years. While flows are similar during the early wet season (AMJ) in La Niña and normal years, and during the later wet season (ON) in normal and El Niño years, there is otherwise a significant difference according to the ENSO phase. El Niño produces abnormally low flows during the two dry seasons (JFM and JAS), and there are large positive anomalies under La Niña weather conditions, especially during the late wet season (OND). Overall, El Niño is associated with a 25 % reduction in river discharge, while there is an increase of 27 % in La Niña years.

The influence of Pacific Ocean sea surface temperature (SST) is also very evident in the hydroclimate of the Andes of west-central Argentina (location of the Mendoza River basin), as demonstrated by the plots in Fig. 2.6 of regional snowpack and stream flow records for the period 1909–2010 (Masiokas et al. 2010, 2012). A slightly negative but non-significant trend in regional river discharges is obscured by the large inter-annual and decadal scale variability. Using an objective and relatively simple method of detecting the driest and wettest intervals in these time series, Masiokas et al. (2012) identified statistically significant regime shifts in 1945, when mean water levels dropped 32 %, and in 1977, when they increased by 29 %. Also plotted in Fig. 2.6 is the July–June PDO index of sea surface

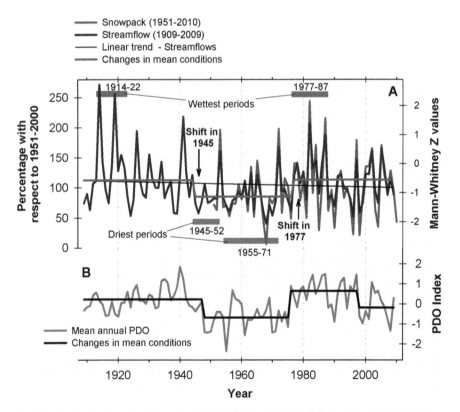

Fig. 2.6 (a) Regional snowpack (*red*) and streamflow (*blue*) records from west-central Argentina and adjacent Chilean Andes. Statistically significant shifts in mean flow conditions (*green line*) occurred in 1945 and 1977; (b) a 1909–2009 time series of the mean July to June PDO index (*gray*) and shifts in mean levels (*black*). *Source*: Masiokas et al. (2010, 2012) (Colour figure online)

temperatures in the extra-tropical, northern hemisphere Pacific Ocean. Decadal-scale variability dominates the PDO index, with definite shifts in the mean value every few decades. Clearly evident in Fig. 2.6 are the coinciding patterns of decadal-scale variations in regional streamflow and snowpack and the temporal pattern of the PDO.

The teleconnections between Pacific Ocean climate oscillations and regional hydroclimate, demonstrated above for the Andes, extend throughout the Cordillera of the western Americas (Villalba et al. 2011). Whereas a 30-year record of hydroclimate, such as in Fig. 2.5, is sufficient to capture the short-term affects of ENSO, the detection of lower-frequency (decadal) periodicity requires longer hydroclimatic time series, such as in Fig. 2.6. Even longer records of annual hydroclimate, inferred from the growth of long-lived trees, are available from the central and southern Andes of Argentina and the Canadian Rocky Mountains. In these temperate mid-latitudes (30–60°), tree growth is confined to a distinct growing season and tends to be limited by available soil moisture because at these latitudes the eastern slopes of the cordillera are dry. Using moisture-sensitive tree-ring chronologies, and the well-established methods and principles of dendrochronology (Hughes et al. 2011), researchers associated with the VACEA project have reconstructed annual streamflow for the past six to eight centuries.

The time series in Fig. 2.7, from west-central Argentina and the adjacent Chilean Andes, includes snowpack reconstructed back to 1866 using rainfall data, and

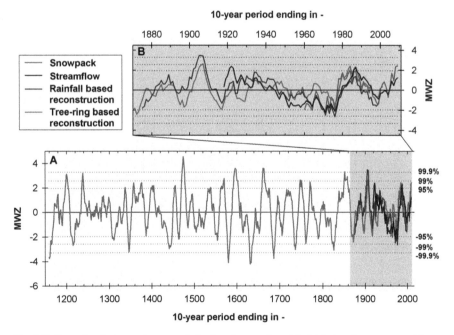

Fig. 2.7 Records of measured snowpack (*red*) and streamflow (*dark blue*) and reconstructed from rainfall data (since 1866, *light blue*) and from tree-rings (1150–2001, *green*). The data were standardized using Mann–Whitney Z statistics (MWZ) for 10-year intervals. All four curves display coherent patterns of decadal variation. *Horizontal dotted lines* indicate statistically wetter or drier conditions (Masiokas et al. 2012) (Colour figure online)

annual streamflow extended back to 1150 using tree rings. All data were standardized and averaged over 10-year intervals (Masiokas et al. 2012). These instrumental and proxy time series display coherent patterns of decadal variation, providing important context for the interpretation of trends and variability in recent observations of climate and water variables, and model projections of future hydroclimate.

A similar tree-ring reconstruction of annual streamflow for the Oldman River in western Canada was developed for this paper using methods described in Sauchyn et al. (2014). This proxy hydrometric time series, spanning 1377–2010, is plotted in Fig. 2.8. It reveals strong decadal-scale variability linked to SST oscillations in the northern Pacific Ocean, the primary source of precipitation over northwestern North America, particularly in winter. Figure 2.9 gives the results of a spectral analysis of this proxy streamflow record. The dominant modes of variability correspond to periodicities of approximately 2–4, 13 and 60 years. The highest and lowest of these frequencies are characteristic of the ENSO and PDO, providing further evidence of the strong teleconnections between these ocean–atmosphere oscillations and the hydroclimate of the case study river basins.

This teleconnection between the PDO and regional hydroclimate has important implications for the timing and intensity of extreme events, as highlighted by Fig. 2.10. Gurrapu et al. (2016) stratified the peak annual flow series for the Crowsnest River, a tributary of the Oldman River, into years of negative versus positive PDO. The corresponding flood frequency curves show a distinct separation; flood risk is relatively higher in the cool (negative) phase. At very high flows, with a return period of 10–50 years, there is some overlap but only in the 95 % confidence limits. Thus the probability of exposure to excess water (and conversely

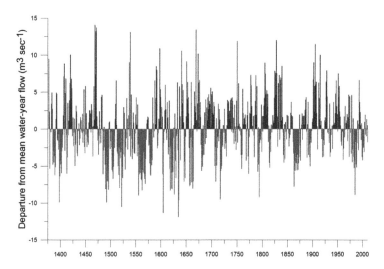

Fig. 2.8 A tree-ring reconstruction of the water-year (October–September) flow of the Oldman River from 1377 to 2010. These 12-month inferred flows are plotted as positive (*blue*) and negative (*red*) departures from the mean water-year flow for the reconstruction period. See Sauchyn et al. (2014) for a description of the methodology (Colour figure online)

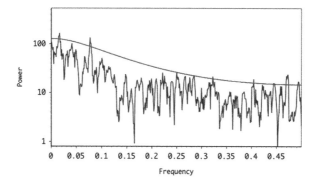

Fig. 2.9 The results of a multi-taper method (MTM) spectral analysis of the Oldman River reconstruction shown in Fig. 2.8. The most powerful frequencies, exceeding a significant level of 95 % (*red curve*), represent periodicity at approximately 2–4, 13 and 60 years (Colour figure online)

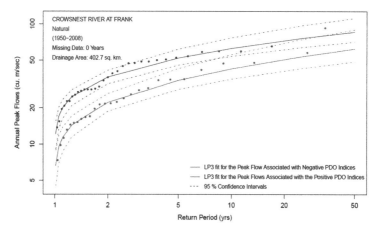

Fig. 2.10 Exceedence curves for the peak annual flow of the Crowsnest River, Alberta (Canada), during negative and positive phases of the PDO (From: Gurrapu et al. 2016) (Colour figure online)

drought) varies between decades, depending on the phase of the PDO and the state of ENSO (Fig. 2.5).

Conclusions

Climate change will increasingly impact the livelihoods of rural people, who are often disproportionately more vulnerable, given their dependency on natural resources and exposure to other stressors. The way that residents of agricultural communities perceive climate change was a key element and driver of research on the VACEA international research project. This critical information was the basis for not only an evaluation of sensitivity and adaptive capacity, but also, as reported in this paper, case studies of exposure to the most locally-relevant aspects of climate variability and change. The VACEA approach to interdisciplinary climate change research, whereby social and natural science provide context and direction for research in the other academic realm, has produced insights that very likely would not arise from a more narrow disciplinary perspective. The community vulnerability assessments (CVA) revealed that local actors were either not cognizant or not concerned about 'long-term' trends in climate variables. From the perspective of the science and impacts of global warming, as reported by the IPCC (2014), it would be easy to conclude that these rural farmers and administrators are naïve or in denial. However, five years of study of the weather, climate, hydrology, economy and sociology of communities in four watersheds suggests otherwise. Local agricultural producers observe and experience short-term variability, in relation to their seasonal to annual planning horizon and the impacts of wet/dry cycles and extremes. Given the degree of variability in the hydroclimate of the case study river basins, and the sensitivity of the rural livelihoods to this variability and weather extremes, the perception of the local actors conforms to their experiences and the nature of the regional hydroclimate.

In the case study river basins, climate variability between years and decades dominates the paleoclimate of past centuries, and the recorded hydroclimate of recent decades, masking the regional expression of global climate change. Exposure variables examined in this paper, indices of stream flow, snowpack, water excess and deficit, vary in coherence with the characteristic frequencies of ocean–atmosphere circulation patterns, specifically the ENSO and PDO. This finding suggests that future projections of these variables should be derived from climate models that are able to simulate the internal variability of the climate system and the teleconnections between ocean–atmosphere oscillations and regional hydroclimate. Recent climate modelling studies of the detection and predictability of anthropogenic climate change have concluded that much of the discrepancy among climate projections at regional scales is irreducible owing to the internal variability of the climate system (Knutti and Sedláček 2012), and therefore natural climate variability poses inherent limits to climate predictability (Deser et al. 2012).

Our findings, on the spatial and temporal patterns of stream flow and related variables, provide important perspective for understanding of vulnerabilities to climate variability and change in rural agricultural communities. They place the hydrologic extremes experienced by the communities in a long-term and natural science context. For example, the relative magnitude of extreme weather, such as the droughts of 1999–2002 and 2004–2005 in the Oldman and Mendoza River basins, respectively, locally perceived as severe, can be revaluated in the light of the regional variability and forcing of climate and hydrology over the past decades and centuries. Extreme dry years are not as uncommon as believed by local actors, especially as they become removed in time or space from the impacts of drought. Raising awareness of the possibility of severe and prolonged drought, and conversely high water levels that exceed recent extremes, should encourage appropriate adaptation measures. The important implication of our case studies of rural communities in three countries is that evidence-based adaptation policy and planning requires deep interdisciplinarity whereby research in the natural sciences depends on social science for the key variables and relevant scales and the resulting studies of exposure inform the evaluation of community adaptive capacity and governance. These lessons are transferrable to regions where rural comminutes are exposed and sensitive to a range of environmental fluctuations and stresses from extreme weather events to inter-annual and -decadal variability and long-term changes.

Acknowledgements The VACEA project is funded by the Government of Canada under the International Research Initiative on Adaptation to Climate Change (IRIACC) program. The funding agencies are the International Development Research Centre (IDRC), the Natural Sciences and Engineering Research Council (NSERC), and the Social Sciences and Humanities Research Council (SSHRC).

References

Agard J, Schipper L (eds) (2014) IPCC WGII AR5 glossary. http://ipcc-wg2.gov/AR5/images/uploads/WGIIAR5-Glossary_FGD.pdf. Accessed 08 Apr 2015

Deser C, Knutti R, Solomon S, Phillips AS (2012) Communication of the role of natural variability in future North American climate. Nat Clim Change 2:775–779

Gurrapu S, St-Jacques J-M, Sauchyn DJ, Hodder KR (2016) The influence of the PDO and ENSO on the annual flood frequency of Southwestern Canadian Prairie Rivers. J Am Water Resour Assoc (in press)

Hales D, Hohenstein W, Bidwell MD, Landry C, McGranahan D, Molnar J, Morton LW, Vasquez M, Jadin J (2014) Ch. 14: Rural communities. In: Melillo JM, Terese Richmond TC, Yohe GW (eds) Climate change impacts in the United States: the Third National Climate Assessment. U.S. Global Change Research Program, pp 333–349. doi:10.7930/J01Z429C

Hughes MK, Swetnam TW, Diaz HF (eds) (2011) Dendroclimatology: progress and prospects. Springer Science, Dordrecht, p 365

IPCC (2012) Managing the risks of extreme events and disasters to advance climate change adaptation. In: Field CB, Barros V, Stocker TF, Qin D, Dokken DJ, Ebi KL, Mastrandrea MD, Mach KJ, Plattner G-K, Allen SK, Tignor M, Midgley PM (eds) A special report of Working Groups I and II of the Intergovernmental Panel on Climate Change. Cambridge University Press, Cambridge

IPCC (2013) Climate change 2013: the physical science, contribution of Working Group I to the fifth assessment report of the Intergovernmental Panel on Climate Change. Cambridge University Press, Cambridge

IPCC (2014) Climate change 2013: impacts, vulnerability and adaptation, contribution of Working Group II to the fifth assessment report of the Intergovernmental Panel on Climate Change. Cambridge University Press, Cambridge

Katz RW, Brown BG (1992) Extreme events in a changing climate: variability is more important than averages. Clim Change 21:289–302

Kharin VV, Zwiers FW, Zhang X, Hegrel GC (2007) Changes in temperature and precipitation extremes in the IPCC ensemble of global coupled model simulations. J Climate 20:1419–1444

Knutti R, Sedláček J (2012) Robustness and uncertainties in the new CMIP5 climate model projections. Nat Clim Change. Published online: 28 October 2012. doi:10.1038/Nclimate1716

Lapp SL, St. Jacques JM, Barrow EM, Sauchyn DJ (2011) GCM projections for the Pacific decadal oscillation under greenhouse forcing for the early 21st century. Int J Climatol. doi:10.1002/joc.2364

Masiokas MH, Villalba R, Luckman BH, Le Quesne C, Aravena JC (2006) Snowpack variations in the central Andes of Argentina and Chile, 1951–2005: large-scale atmospheric influences and implications for water resources in the region. J Clim 19:6334–6352

Masiokas MH, Villalba R, Luckman B, Mauget S (2010) Intra- to multidecadal variations of snowpack and streamflow records in the Andes of Chile and Argentina between 30° and 37°S. J Hydrometeorol 11:822–831

Masiokas MH, Villalba R, Christie DA et al (2012) Snowpack variations since AD 1150 in the Andes of Chile and Argentina (30°–37°S) inferred from rainfall, tree-ring and documentary records. J Geophys Res 117:D05112. doi:10.1029/2011JD016748

Meehl GA, Covey C, Delworth T, Latif M, McAvaney B, Mitchell JFB, Stouffer RJ, Taylor KE (2007) The WCRP CMIP3 multimodel dataset, a new era in climate change research. Bull Am Meteorol Soc 88:1383–1394

Nelson R, Kokic P, Crimp S, Martin P, Meinke H, Howden SM, de Voil P, Nidumolu U (2010) The vulnerability of Australian rural communities to climate variability and change: Part II—integrating impacts with adaptive capacity. Environ Sci Policy 13:18–27

Orlove B (2009) The past, the present and some possible futures of adaptation. In: Adger N et al (eds) Adapting to climate change: thresholds, values and governance. Cambridge University Press, Cambridge, pp 131–163

Poveda G, Mesa OJ (1997) Feedbacks between hydrological processes in tropical South America and large-scale oceanic atmospheric phenomena. J Clim 10:2690–2702

Poveda G, Álvarez DM, Rueda ÓA (2011) Hydro-climatic variability over the Andes of Colombia associated with ENSO: a review of climatic processes and their impact on one of the Earth's most important biodiversity hotspots. Clim Dyn. doi:10.1007/s00382-010-0931-y

Sauchyn DJ, Vanstone J, St. Jacques J-M, Sauchyn R (2014) Dendrohydrology in Western Canada and applications to water resource management. J Hydrol. http://dx.doi.org/10.1016/j.jhydrol.2014.11.049

Sauchyn DJ, Bonsal B, Kienzle SW, St Jacques J-M, Vanstone J, Wheaton E (2015) Adaptation according to mode of climate variability: a case study from Canada's Western interior. In: Leal W (ed) Handbook of climate change adaptation. Springer Science, Berlin Heidelberg

Shabbar A, Bonsal BR, Szeto K (2011) Atmospheric and oceanic variability associated with growing season droughts and pluvials on the Canadian Prairies. Atmosphere-Ocean. doi:10.1080/07055900.2011.564908

Smit B, Wandel J (2006) Adaptation, adaptive capacity and vulnerability. Glob Environ Change 16:282–292

Tebaldi C, Hayhoe K, Arblaster JM, Mehl GA (2006) Going to the extremes: an intercomparison of model-simulated historical and future changes in extreme events. Clim Change 79:185–211

Trenberth KE (2012) Framing the way to relate climate extremes to climate change. Clim Change 115:283–290. doi:10.1007/s10584-012-0441-5

Villalba R, 30 others (2011) Dendroclimatology from regional to continental scales: understanding regional processes to reconstruct large-scale climatic variations across the Western Americas. In: Hughes MK, Swetnam TW, Diaz HF (eds) Dendroclimatology: progress and prospects. Springer Science, Dordrecht, pp 175–230

Wittrock V, Wheaton E, Bonsal B, Vanstone J (2014) Connecting climate crop yields: case studies of the swift current Creek and Oldman River Watersheds. Prepared for the vulnerability and adaptation to climate extremes in the Americas (VACEA) of the Prairie Adaptation Research Collaborative, Regina, SK. SRC # 13224-2E13. Saskatchewan Research Council, Saskatoon

Chapter 3
Small Scale Rain- and Floodwater Harvesting for Horticulture in Central-Northern Namibia for Livelihood Improvement and as an Adaptation Strategy to Climate Change

Alexander Jokisch, Wilhelm Urban, and Thomas Kluge

Abstract This paper presents research results of the German-Namibian joint research project CuveWaters in which different technologies for small-scale rain- and floodwater harvesting were introduced as pilot plants in central-northern Namibia as part of a broader Integrated Water Resources Management (IWRM) approach. Central-northern Namibia has semi-arid climate conditions with clearly distinctive dry and wet seasons. Rain- and floodwater harvesting for irrigation purposes are intended to increase resilience in agricultural production by building buffers for interseasonal dry spells and to make irrigation farming possible during the dry season. This is intended to improve availability of vegetables in rural parts of Namibia and to derive income on local markets. Besides raising temperatures, climate change in sub-Saharan Africa is predicted to increase rainfall variability. Therefore, these adaptations also present a precondition for adapting to future climate change. Within the project, different organisational approaches such as harvesting of rainwater at the household as well as at the communal level were tested, as well as different locally available tank construction materials. All technologies were developed in cooperation with the local communities and framed by capacity development measures which yielded very good results and enabled a diffusion of the technology in the region. Based on 5 years of research the construction of ferrocement tanks on the household level and ponds covered with shade nets on the communal level can be recommended. Due to high evaporation rates all gardens irrigated with harvested rainwater were equipped with water saving drip irrigation systems.

A. Jokisch (✉) • W. Urban
Department of Civil and Environmental Engineering, Institute IWAR, Chair of Water Supply and Ground Water Protection, Technische Universität Darmstadt, Franziska-Braun-Str. 7, 64287 Darmstadt, Germany
e-mail: alex.jokisch@gmx.de

T. Kluge
Institute for Social-Ecological Research (ISOE), Hamburger Allee 45, 60486 Frankfurt/Main, Germany

In combination with capacity development focusing on water management this enabled the users to irrigate their gardens throughout the dry season.

Keywords Central-northern Namibia • Rainwater harvesting • Floodwater harvesting • Climate change adaptation

Introduction

Today, population growth and economic development are putting increasing pressure on natural resources such as land and water (Foley et al. 2011). While worldwide more and more people are living in cities, it is predicted that Africa's rural population depending on agriculture as their main business will grow at least until 2050 (de Fraiture et al. 2010). To achieve sustainable development, integrated approaches are necessary that combine the sustainable use of limited water resources and at the same time increase water availability and agricultural productivity. Compared to other parts of the world, agriculture in sub-Saharan Africa is still characterised by very small scale farming that is very much affected by climate variability such as droughts and floods, as well as interseasonal dry spells. In this regard Namibia, located in south-western Africa, is no exception.

Namibia is often referred to as the driest country in sub-Saharan Africa (Kluge et al. 2008) and large parts of the country are covered by desert (Mendelsohn et al. 2002). While Namibia as a whole is one of the countries with the lowest population density worldwide, almost half of the Namibian population (approximately 850,000 people) is concentrated in the central-northern parts of the country (Namibia Statistics Agency 2011) which comprises only 10 % of the country's surface area (Mendelsohn et al. 2002). Central-northern Namibia is characterised by a highly variable climate typical for semi-arid regions all around the world. During the rainy season, which lasts from October to March, a mean annual rainfall of 470 mm is recorded (Sturm et al. 2009) while during the rest of the year there is almost no precipitation. The region is further characterised by a very flat topography and the so-called Oshana water system, draining the whole region from north to south towards the Etosha Salt Pan. The Oshanas are very shallow ephemeral river streams, together making up the Cuvelai River that has its headwaters in the Encoco Mountains of southern Angola between the catchments of the much larger Kunene and Okavango rivers (Mendelsohn et al. 2000). In some years the Oshanas carry so much water that the region experiences a flood, where these floods are known locally as Efundja. In other years, the Oshanas carry a smaller volume of water that mainly originates from local rainfall, but hardly infiltrates due to the local soil conditions. Years with little rainfall and only local surface water flows are also common.

Most people in the region practice subsistence farming with the main staple food being pearl millet (*Pennisetum glaucum*), locally known as Mahangu. Most also keep livestock such as cattle and goats. Besides subsistence farming, employment opportunities in the region are very scarce and thus unemployment rates are very high. Compared to most other parts of the country, the central-northern regions present favourable conditions for rain fed farming but agricultural production is limited by climate variability, very low soil fertility and limited knowledge about improved agricultural technologies (FAO 2009) such as irrigation and conservation agriculture as well as rainwater harvesting.

Climate Change in Central-Northern Namibia

While greenhouse gas emissions are mainly produced in developed countries of the northern hemisphere, those most severely affected by the consequences of climate change are expected to be the developing countries in the southern hemisphere, especially in sub-Saharan Africa (IPCC 2014). According to IPCC climate change projections for Africa "mean annual temperature rise over Africa, relative to the late twentieth century mean annual temperature, is *likely* to exceed 2 °C in the Special Report on Emissions Scenarios (SRES) A1B and A2 scenarios by the end of this century" (Niang et al. 2014). Furthermore, "African ecosystems are already being affected by climate change, and future impacts are expected to be substantial" (Niang et al. 2014). According to Reid et al. (2007), the most influential consequence for Namibia besides rising temperature will be more variable rainfall, leading to a decrease in consecutive wet days. Further consequences will be a later start of the rainy season and an earlier cessation leading to an overall shorter rainy season and an longer dry season (Dirkx et al. 2008). Namibia's economy is heavily dependent on agriculture, especially as the only employment opportunity for large parts of the population, particularly in rural and remote parts of the country. As with other semi-arid areas around the world, this type of agriculture is highly susceptible to climate variability and change (Morton 2007; Zeidler et al. 2010).

Rainwater and Floodwater Harvesting as Climate Change Adaptation

In the face of climate change, especially with changes such as increasing rainfall variability, more frequent intensive rainfall events, longer and more frequent dry spells during the rainy season and a shorter rainy season, several authors have highlighted the importance of countermeasures such as water harvesting, the construction of dams and storage of water underground (Barron 2009; Niang et al. 2014; Zeidler et al. 2010). Nevertheless the authors of the latest "IPCC Climate Change Report, Chapter 22 Africa" point out that especially for southern Africa data availability is very limited and future adaptation is complicated due to uncertainties about the extent of change (Niang et al. 2014). In this regard Grey and Sadoff (2007) highlight that in Africa the unmanaged degree of climate variability is much higher than predicted climate changes, thus making it more important for Namibia (as for other countries in sub-Saharan Africa) to better adapt to current variability and recognise this as a precondition for future climate changes. One of the most promising measures to adapt to current and future climate and water variability is storage of water to build buffers for droughts and dry spells (Tuinhof et al. 2012).

Water storage increases the availability of and access to water, thus increasing adaptive capacity, agricultural production, and water security, and thereby decreasing the climate vulnerability of societies (Mc Cartney and Smakhtin 2010; see Fig. 3.1). Other authors have also found that water harvesting can make a useful contribution to agricultural productivity both under current climate conditions as well as in future (Boelee 2013; Critchley and Siegert 1991; Ngigi 2009). The aim of water storage intervention in future would thus be to increase resilience to climate variability. Resilience is defined here as the degree to which a system can cope with changes while still remaining in the same trajectory, so that it keeps more or less the same functions, feedbacks and identity (Walker and Salt 2006). In this regards Enfors and Gordon (2007) defined a resilient dryland agro-ecosystem, as a system "that maintains its capacity to generate food and other vital ecosystem services over time, by absorbing or adapting to disturbances such as dry-spells, droughts and floods."

Namibia has a long tradition of water storage especially in the semi-arid farmland in the interior of the country where extensive cattle and sheep farming is practiced. Here early European settlers constructed hundreds of sand dams where water was stored in the riverbeds of ephemeral river streams (Stengel 1963). While sand storage dams on the farmland in the interior of the country are still in use today, water storage approaches in the densely settled northern regions of the country were not adopted until much later. This is partly due to the different topography of the region, since the storage in local sand is impossible due to high salt content while the construction of other types of dams is restricted by the flat topography. Nevertheless, first attempts to construct larger dams were made after severe famines in the early 1930s (Mendelsohn et al. 2000). After 1950, a large scale investment was made and altogether 320 earth dams with a capacity of 30,000 m^3 each and a further 65 pump storage dams were constructed, where some dams were also equipped with purification plants to serve as a source of drinking water (Stengel 1963; Mendelsohn et al. 2000). Besides supplying hospitals, mission stations and schools with water, large scale irrigation plots were implemented next to these dams, with the stored water used for irrigation. Today, most

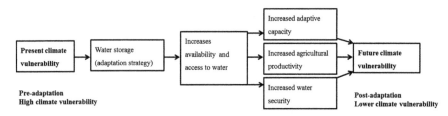

Fig. 3.1 Water storage as adaptation strategy to reduce climate change vulnerability (adapted from Mc Cartney and Smakhtin 2010)

of these earth dams are characterised by low water quality and high salt contents, and most pump storage dams are in very poor condition with purification equipment either broken or stolen (Drießen and Jokisch 2011). The main problem of water storage in these large earth and pump storage dams was the high evaporation rates in the region, which reach up to 1960 mm/year (Mendelsohn et al. 2002). Several attempts to avoid evaporation by constructing roofs failed due to the size of the dams. Another problem has been quality degradation due to the high salt content of the local soil that was used for dam construction as well as water contamination due to free roaming livestock (Drießen and Jokisch 2011). No major investments have been made in this kind of water storage infrastructure since the 1970s. This is partly because after 1970 the government promoted the construction of a large scale water pipeline system fed by water from the Kunene River. Water for the pipeline system is abstracted in the Callueque Dam on the Angolan side of the border and transported into the region via a 150 km open canal and nowadays supporting most towns and villages (Bittner Water Consult 2006). Due to recent population growth this water pipeline system is now at its carrying capacity. Furthermore, due to the implementation of several water purification plants in the major towns of the region, it is now only intended to provide good quality drinking water. Therefore, if agricultural production is to be increased, and water made available for irrigation, other solutions have to be found.

The CuveWaters Project

The research and development project "CuveWaters—Integrated Water Resources Management in Namibia" identified water as the key resource for sustainable development in the region and has been working in central-northern Namibia since 2006. CuveWaters is a joint research project of German and Namibian partners sponsored by the German Federal Ministry for Education and Research (BMBF). The CuveWaters approach for Integrated Water Resources Management (IWRM) is based on a multi-resource mix in which different technologies are providing water of different quality and quantity for different purposes. All technologies were developed together with Namibian partners and in cooperation with the local communities following a demand responsive approach (Deffner and Mazambani 2010) which is characterised by several workshops, stakeholder meetings on local, regional and national level, as well as social-cultural surveys in the respective villages where technologies were meant to be implemented.

All technologies were implemented in the study region as pilot plants to test the technologies within the local setting (see Fig. 3.2). Solar driven groundwater desalination plants were implemented in the villages of Amarika and Akutsima to provide these remote areas with a reliable source of drinking water. In the fast growing town Outapi, a pilot plant for sanitation and water-reuse was constructed where the water can be reused for the irrigation of vegetables. Furthermore two technologies for water storage were selected and adapted to the local context in consultation with Namibian experts and local communities. Rainwater harvesting

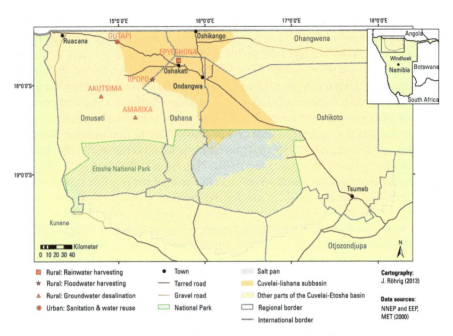

Fig. 3.2 Location of CuveWaters pilot plants in central-northern Namibia

as developed within the project's context can be defined as the collection of rainwater from impermeable surfaces such as roofs, roads or concrete. The technology has already been tested in other regions of Africa where a huge potential for it as an alternative water supply option and as measure against climate change was found (Gould and Nissen-Petersen 2003; Mwenge Kahinda and Taigbenu 2011). In the village Epyeshona, close to the region's biggest town Oshakati, different pilot plants for rainwater harvesting were implemented at the household and communal level. Rainwater harvesting within the project was only intended for the provision of irrigation water or for domestic purposes such as cooking and washing. Household rainwater harvesting tanks of different materials were constructed at three households in the village. Furthermore, a ground catchment was constructed where rainwater is harvested in an underground tank that supplies the common garden area of five households with irrigation water. The second water storage option floodwater harvesting here refers to the harvesting and storage of water from the Oshanas, the local ephemeral river streams. In all cases, water is collected during the rainy season and can be used during the rainy season to bridge dry spells or to irrigate gardens during the dry season. The pilot plant for the storage of Oshana floodwater was constructed in Iipopo village, where the plant and attached gardens were managed by ten people from the local community. All constructions were flanked by capacity development measures, where several people in each village were trained in construction (Zimmermann et al. 2012), and local people received training in all aspects of sustainable water use, irrigation and horticulture.

3 Small Scale Rain- and Floodwater Harvesting for Horticulture in Central... 45

Fig. 3.3 CuveWaters rainwater harvesting pilot plants (from *top left* clockwise: communal approach with greenhouse and outside garden, communal approach ground catchment with underground tank, household approach ferrocement tank with garden, spinach inside household greenhouse) (Liehr et al. 2015)

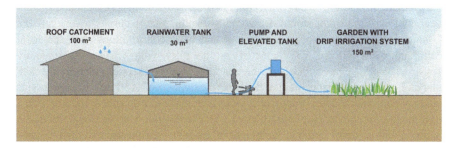

Fig. 3.4 Concept for rainwater harvesting in the household approach (Jokisch et al. 2015)

Different rainwater harvesting approaches adapted from other parts of the world were tested over the course of 5 years of monitoring and research. In the pilot village Epyeshona two different organisational approaches were tested, the household and the communal approach. Three different rainwater harvesting tank technologies using different tank construction materials were tested at the household level (Figs. 3.3 and 3.4). At different households within the pilot village one polyethylene tank, one ferrocement tank and one brick tank were constructed in order to compare the different materials under local conditions. For all three technologies the materials were purchased locally and local staff was trained in the construction of the tanks (Zimmermann et al. 2012). Rainwater is harvested on the roofs of private buildings and stored in tanks of 30 m³ each. The water can then be pumped to elevated tanks using pedal pumps and used in water saving drip irrigation system supplying household gardens. Garden products such as tomatoes, green pepper and onions are intended to be used for household consumption or for selling on local markets.

In the case of the communal approach, five households share the water storage tanks and the attached gardens. The pilot plant consists of an underground tank made of bricks and ferrocement (120 m^3) and a pond (80 m^3) made of dam liner. Rainwater is mainly collected on a purpose-made ground catchment constructed of concrete slabs. Both reservoirs have roofs made of shade net to avoid evaporation as well as pollution of the water stored. The water can then be pumped to an elevated tank when it is needed either inside the jointly used greenhouse or in the outside gardens where each participating household has its own beds for vegetable production. Both the greenhouse and the outside gardens are equipped with water saving drip irrigation (see Figs. 3.3 and 3.5).

While for rainwater harvesting the collection area for rainwater, and thus the overall amount of water that can be harvested, is limited to already existing surfaces such as roofs or roads or by limited space for purpose made catchments, the amount of floodwater that could be harvested from the local Oshana water system is much higher. Therefore CuveWaters developed a pilot plant for the harvesting of this floodwater based on technologies already applied for rainwater harvesting. Based on suggestions made by Namibian experts, the pilot plant was designed to be much smaller than earlier attempts for floodwater storage in the region (Stengel 1963), thus making it possible to construct roofs in order to avoid evaporation and quality degradation of the water stored. Based on a satellite picture survey a very reliable Oshana in the village Iipopo was found and after consultations with the local community the village was selected as the pilot village. The subsequently constructed pilot plant consists of three storage reservoirs, one underground tank made of bricks and ferrocement and a concrete dome roof (130 m^3) and two ponds made of dam liner (135 m^3 each), one with a roof made of iron sheets and one with a roof made of shade net (see Figs. 3.6 and 3.7). Floodwater from the local Oshana is pumped with a motor pump at the height of the rainy season and can then be used for irrigation during the dry season. To bridge dry spells during the rainy season water is taken directly from the Oshana. The pilot plant further consists of a greenhouse (176 m^2) and an open garden area (1200 m^2) both equipped with a

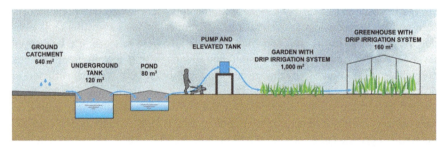

Fig. 3.5 Concept for rainwater harvesting in the communal approach (Jokisch et al. 2015)

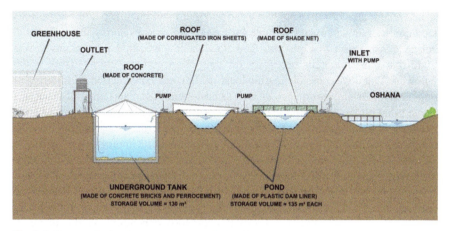

Fig. 3.6 Concept for floodwater harvesting (C. Treskatis)

Fig. 3.7 Floodwater harvesting pilot plant (from *top left* clockwise: covered ponds, greenhouse, crop from outside gardens, underground tank and outside gardens) (Liehr et al. 2015)

water saving drip irrigation system (Jokisch et al. 2014). The pilot plant is managed by ten households from the local community that were selected based on the demand-responsive approach (Deffner and Mazambani 2010).

Results

In the case of rainwater harvesting the tanks and related infrastructure were constructed between 2009 and 2011, and subsequently monitored over a period of 4–6 years. It was found that in years with normal or slightly below normal rainfall sufficient amounts of rain were harvested and that in all but one out of five seasons good quality rainwater was available for irrigation purposes during the dry season. During construction of both rain- and floodwater harvesting pilot plants it was found that most materials necessary for tank construction were available in the region and that it was possible to train local staff in tank construction. Monitoring of long term material performance as well as cost–benefit and sustainability analyses showed that on the household level ferrocement tanks, and on the communal level greenhouses with ponds covered by shade net are the most appropriate options. Compared to the other materials used, the ferrocement technology was associated with the lowest maintenance costs and was coping best with the harsh environmental conditions of the project region such as high salt contents of the local soil as well as very high rates of solar radiation which was especially affecting the polyethylene tank negatively. Investment costs for the household approach were between [1]N$12,000 and N$18,000 (ca. US$950–US$1400) depending on the material, where ferrocement tanks have the lowest investment costs. Model calculations indicated that all pilot plants implemented were capable of generating positive cost–benefit ratios through the production and selling of vegetables especially when cash crops such as tomatoes and spinach are produced (Woltersdorf et al. 2014). However, monitoring has shown that due to very little previous knowledge about water management and irrigation, the actual farming benefits were initially much lower than expected. This is mainly due to the case that rainwater harvesting and irrigation were largely unknown in the region (Sturm et al. 2009) and repeating capacity development measures were necessary to form a sustainable basis for long-term farming success. Subsequently profits from irrigation farming with harvested rain- and floodwater increased from year to year with more knowledge and experience gained by the farmers.

The pilot plant for floodwater harvesting was put in operation in 2012 and could thus be monitored over a period of 3 years. In two out of three monitoring seasons, good quality floodwater was harvested and stored in the tank and ponds. The rainy season of 2012/2013 brought exceptionally low rainfall and no harvesting and irrigation was possible. Compared to the rainwater harvesting pilot plants in normal or slightly below normal rainfall seasons a larger amount of water was available but water quality was lower (with no negative effects on the crop yield). The total investment costs for the floodwater harvesting pilot plant were N$186,000 (ca. US $14,800) including N$42,000 (ca. US$3300) for the underground tank, N$23,000 (ca. US$1800) for the shade net covered pond, N$31,000 (ca. US$2500) for the corrugated iron covered pond and N$43,000 (ca. US$3400) for the greenhouse. It could be shown that the construction of ponds made of dam liner is most appropriate in rural areas as construction can completely be done without machinery and investments costs are lower as for the other options tested. Monitoring has furthermore shown that no larger cost-intensive maintenance was necessary and that the materials used can cope with the local environmental conditions such as the salt

---

[1] Currency conversion: 1 N$ = 0.08 US$ (http://www.oanda.com, 29.07.2015).

content of the soil as well as high solar radiation. In terms of marketing of the vegetables produced in both cases the users found good markets directly within the villages and saved on costs for transport and were thus able to sell their produce cheaper than competitors such as supermarkets in nearby towns and villages.

For sustainable implementation, long term and intensive capacity development was necessary and provided as part of the CuveWaters project. Capacity development was provided in all aspects from tank construction to water management, irrigation and horticulture as well as book keeping and selling of the products. All people involved in the project received training on at least two topics, with each training course lasting several weeks. The capacity development measures conducted within the project (Zimmermann et al. 2012; Jokisch et al. 2014) yielded good results and Namibian staff trained during the CuveWaters project were responsible for the construction of three further communal rainwater harvesting plants financed by different Namibian institutions between 2012 and 2014. Besides providing benefits for the local communities such as income generation and improved local availability of vegetables, these rainwater harvesting plants were also used for further capacity development measures including guided tours to the plants as well as training for governmental agricultural extension staff.

For diffusion of the technology it is recommended to construct ferrocement tanks at the household level and harvesting of water from roofs of private buildings. On the communal level both for rain- and floodwater harvesting it is recommended to construct ponds covered with shade nets. Ideally both rain- and floodwater should be harvested at the same place to increase the available amount of water. Horticulture within greenhouses and the use of drip irrigation systems is also recommended due to the high evaporation rates in the region.

Conclusion

The climate of central-northern Namibia is highly variable, where both droughts and floods place a high risk on agricultural activities. Climate change is predicted to increase rainfall variability, leading to a shorter rainy season, longer interseasonal dry spells and more frequent heavy rainfall events that can lead to floods. The project "CuveWaters—Integrated Water Resources Management in Namibia" tested different approaches for rain- and floodwater harvesting in order to provide irrigation water throughout the dry season, build water buffers and to bridge interseasonal dry spells. Different options for harvesting and storing of rain- and floodwater were tested over the course of 5 years. Nearly all materials necessary for tank construction were available in the region and construction was carried out locally. Based on long-term material performance as well as cost–benefit analyses it was possible to select the ideal locally available materials, being ferrocement tanks for the household approach and ponds covered with shade net for the communal approach. By the introduction of water storage technologies adapted to the local situation, rural farmers have been provided with a further option for income generation and the availability of vegetables throughout the year for their own consumption has been improved. The higher water availability during the rainy season due to already stored water enables the users to bridge dry spells thus venture into a more reliable farming business during the rainy season. The availability of water during the dry season enables them to produce vegetables year round, thus also creating income when rain-fed crop farming is not possible. This additional water buffer also increases the local agricultural resilience to climate change.

Acknowledgements "CuveWaters—Integrated Water Resources Management in Namibia" is a joint research project of the Institute for Social Ecological Research (ISOE) and the Technische Universität Darmstadt (TUD), Chair for Water Supply and Groundwater Protection and Chair for Wastewater Technology. The project partners Namibia are the Ministry of Agriculture, Water and Forestry (MAWF), the Desert Research Foundation of Namibia (DRFN), the Gesellschaft für Internationale Zusammenarbeit (GIZ) and the Federal Institute for Geosciences and Natural Resources (BGR). The project is funded by the German Ministry for Education and Research (BMBF). Special thanks to all community members of Iipopo and Epyeshona villages, trainers and trainees, volunteers, as well as to the consultants from One World Consultants Isaac Kariuki, John Ndongo, Beatrice Wambui, and Joseph Machura.

References

Barron J (ed) (2009) Rainwater harvesting: a lifeline for human well-being. Stockholm Environment Institute. Available online http://www.unwater.org/downloads/Rainwater_Harvesting_090310b.pdf. Accessed 22 Jan 2015

Bittner Water Consult (2006) Cuvelai-Etosha Groundwater Investigation. Desk study report. Ministry of Agriculture, Water and Forestry, Windhoek

Boelee E (ed) (2013) Managing water and agroecosystems for food security. CABI, Oxfordshire

Critchley W, Siegert K (1991) Water harvesting. A manual for the design and construction of water harvesting schemes for plant production. Food and Agricultural Organization of the United Nations, Rome. Available online http://www.samsamwater.com/library/Water_harvesting_-_Critchley.pdf. Accessed 21 May 2015

de Fraiture C, Molden D, Wichelns D (2010) Investing in water for food, ecosystems, and livelihoods: an overview of the comprehensive assessment of water management in agriculture. Agric Water Manag 97(4):495–501. doi:10.1016/j.agwat.2009.08.015

Deffner J, Mazambani C (2010) Participatory empirical research on water and sanitation demand in central northern Namibia: a method for technology development with a user perspective. CuveWaters Papers, 7. CuveWaters, Frankfurt am Main

Dirkx E, Hager C, Tadross M, Bethune S, Curtis B (2008) Climate change vulnerability and adaption assessment Namibia. Final report. Desert Research Foundation of Namibia and Climate Systems Analysis Group for the Ministry for Environment and Tourism, Windhoek. Available online http://www.environment-namibia.net/tl_files/pdf_documents/strategies_actionplans/Climate%20Change%20V%20%26%20A%20Assessment%20Namibia.pdf. Accessed 11 Jun 2015

Drießen C, Jokisch A (2011) Current status of "Stengel Dams" in central-northern Namibia. Internship report. Technische Universität Darmstadt, Darmstadt, Germany

Enfors EI, Gordon LJ (2007) Analysing resilience in dryland agro-ecosystems: a case study of the Makanya catchment in Tanzania over the past 50 years. Land Degrad Dev 18(6):680–696. doi:10.1002/ldr.807

FAO (ed) (2009) FAO/WFP crop, livestock and food security assessment mission to Namibia (special report). Available online http://www.fao.org/docrep/012/ak334e/ak334e00.HTM. Accessed 29 Jul 2015

Foley JA, Ramankutty N, Brauman KA, Cassidy ES, Gerber JS, Johnston M et al (2011) Solutions for a cultivated planet. Nature 478(7369):337–342. doi:10.1038/nature10452

Gould J, Nissen-Petersen E (2003) Rainwater catchment systems for domestic supply. ITDG Publishing, London

Grey D, Sadoff CW (2007) Sink or swim? Water security for growth and development. Water Policy 9(6):S. 545. doi:10.2166/wp.2007.021

IPCC (Hg.) (2014) Climate change 2014: impacts, adaptation, and vulnerability. Working Group II Contribution to the IPCC Fifth Assessment Report, Chapter 22 Africa. Cambridge University Press, Cambridge

Jokisch A, Brenda M, Rickert P, Urban W (2014) Water harvesting from ephemeral river streams for small scale agriculture as a climate change adaption strategy in central-northern Namibia. In: Proceedings of the IWA 6th Eastern European Young Water Professionals Conference "East meets West". IWA 6th Eastern European Young Water Professionals Conference "East meets West". International Water Association: IWA Young Water Professionals, Istanbul, 28–30 May 2014

Jokisch A, Schulz O, Kariuki I, Krug von Nidda A, Deffner J, Liehr S, Urban W (2015) Rainwater Harvesting in central-northern Namibia. Institute for Social-Ecological Research (ISOE), Frankfurt am Main

Kluge T, Liehr S, Lux A, Moser P, Niemann S, Umlauf N, Urban W (2008) IWRM concept for the Cuvelai Basin in northern Namibia. Phys Chem Earth Parts A/B/C 33(1–2):48–55. doi:10.1016/j.pce.2007.04.005

Liehr S, Brenda M, Cornel P, Deffner J, Felmeden J, Jokisch A, Kluge T, Müller K, Röhrig J, Stibitz V, Urban W (2015) From the concept to the tap: integrated water resources management in northern Namibia. In: Borchardt D, Bogardi J, Ibisch R (eds) Integrated water resources management: concept, research and implementation. UFZ, Leipzig, Germany

Mc Cartney M, Smakhtin V (2010) Water storage in an era of climate change. Addressing the challenge of increasing rainfall variability. International Water Management Institute, Colombo (Blue Paper)

Mendelsohn J, el Obeid S, Roberts C (2000) A profile of north-central Namibia. Gamsberg Macmillan Publishers, Windhoek

Mendelsohn J, Jarvis A, Roberts C, Robertson T (2002) Atlas of Namibia. A portrait of the land and its people. Sunbird Publishers, Cape Town

Morton JF (2007) The impact of climate change on smallholder and subsistence agriculture. Proc Natl Acad Sci USA 104(50):S. 19680–19685. Available online http://www.pnas.org/content/104/50/19680.full.pdf. Accessed 28 May 2015

Mwenge Kahinda J, Taigbenu AE (2011) Rainwater harvesting in South Africa: challenges and opportunities. Phys Chem Earth Parts A/B/C 36(14–15):968–976. doi:10.1016/j.pce.2011.08.011

Namibia Statistics Agency (ed) (2011) Namibia 2011. Population and housing census main report. Available online http://cms.my.na/assets/documents/p19dmn58guram30ttun89rdrp1.pdf. zuletzt aktualisiert am 2011. Accessed 18 Jun 2015

Ngigi SN (2009) Climate change adaption strategies. Water resources management options for smallholder farming systems in sub-Saharan Africa. MDG Centre for East and Southern Africa of the Earth Institute at Columbia University, New York

Niang I, Ruppel OC, Abdrabo MA, Essel A, Christopher L, Padgham J, Urquhart P (2014) Africa: 2014. In: IPCC, Barros VR, Field CB, Dokken DJ, Mastrandrea MD, Mach KJ, Bilir TE et al (eds) Climate change 2014: impacts, adaptation, and vulnerability. Working group II contribution to the IPCC fifth assessment report, Chapter 22 Africa. Cambridge University Press, Cambridge, New York, NY, pp S. 1199–1265

Reid H, Sahlen L, McGregor J, Stage J (2007) The economic impact of climate change in Namibia. How climate change will affect the contribution of Namibia's natural resources to its economy. International Institute for Environment and Development

Stengel HW (1963) Wasserwirtschaft in S.W.A. Afrika-Verlag Der Kreis, Windhoek

Sturm M, Zimmermann M, Schütz K, Urban W, Hartung H (2009) Rainwater harvesting as an alternative water resource in rural sites in central northern Namibia. Phys Chem Earth Parts A/B/C 34(13–16):776–785. doi:10.1016/j.pce.2009.07.004

Tuinhof A, van Steenbergen F, Vos P, Tolk L (2012) Profit from storage. The costs and benefits of water buffering. 3R Water Secretariat, Wageningen

Walker B, Salt D (2006) Resilience thinking. Sustaining ecosystems and people in a changing world. Island Press, Washington, DC

Woltersdorf L, Jokisch A, Kluge T (2014) Benefits of rainwater harvesting for gardening and implications for future policy in Namibia. Water Policy 16(1):S. 124. doi:10.2166/wp.2013.061

Zeidler J, Kandjinga L, David A (2010) Study on the effects of climate change in the Cuvelai Etosha Basin and possible adaption measures. Gesellschaft für Technische Zusammenarbeit, Windhoek. Available online http://www.iwrm-namibia.info.na/downloads/climate20change20in20cuvelai20etosha20basin202.pdf. zuletzt aktualisiert am 2010. Accessed 21 Jan 2015

Zimmermann M, Jokisch A, Deffner J, Brenda M, Urban W (2012) Stakeholder participation and capacity development during the implementation of rainwater harvesting pilot plants in central northern Namibia. Water Sci Technol: Water Supply 12(4):S. 540. doi:10.2166/ws.2012.024

Chapter 4
Can Adaptation to Climate Change at All Be Mainstreamed in Complex Multi-level Governance Systems? A Case Study of Forest-Relevant Policies at the EU and Swedish Levels

E. Carina H. Keskitalo and Maria Pettersson

Abstract Mainstreaming adaptation to climate change in forest-relevant policy can be as a "most difficult" case, relevant for asking the question to extent to which adaptation can at all be mainstreamed in complex multi-level governance systems. This study examines the case of to what extent EU and national (exemplified by Swedish) legal and policy frameworks are able to integrate with each other in ways that may support climate change adaptation in forests. To move as close to the real life situation of mainstreaming challenges as possible, the study focuses on not only one area of mainstreaming or integration, but on the three broad policy areas: (a) adaptation per se; (b) forest biodiversity and habitat protection with respect to invasive species; and (c) water protection in relation to forest use. The study concludes that conflicts between international legal principles such as precaution and free trade, as well as distribution of competences at EU and national level, results in a great discrepancy in terms of opportunities for a nation to effectively act independently as well as for effectively integrating adaptation aims in the connected EU-national systems.

Keywords Adaptation • Mainstreaming • Forest • European Union • Sweden

Introduction

Integrating adaptation to climate change with broader policy and planning issues—mainstreaming—has often been described as an aim to promote climate-considerate development (Ayers et al. 2014; Kok and De Coninck 2007; Rauken et al. 2014), by increasing policy comprehensiveness and consistency, for example. However, multiple studies also note the non-sequential character of mainstreaming as well

E. Carina H. Keskitalo (✉) • M. Pettersson
Department of Geography and Economic History, Umeå University, 901 87 Umeå, Sweden
e-mail: Carina.Keskitalo@umu.se

as its complex and interactive nature, where the existing legal and policy context in particular presents an integration challenge in terms of mainstreaming adaptation (e.g., Uittenbroek et al. 2013; Ayers et al. 2014; Rauken et al. 2014). Forest is here an urgent area given multiple impacts from climate change, such as seasonal shifts and increased risk of outbreaks of pest insects and the influx of invasive species with impact on biodiversity. Increased flooding due to extreme weather or changes in snowmelt may also lead to increased leaching of nutrients from fertilized or harvested land, adversely affecting water quality. More extreme weather also increases the risk of storms with severe windfall (IPCC 2012). For adaptation in relation to forest, it is thereby not sufficient to mainstream responses only in one of these areas, but mainstreaming needs to take place in all of them, and within complex governance systems.

However, coordination in the case of forest in Europe is also particularly difficult. The EU constitutes a complex multi-level governance system with varying competences on areas that relate to forest (e.g., Keskitalo 2010 ed; Knieling and Leal Filho 2012 ed). Forest is thus covered by both EU-level, national and voluntary (certification) instruments. There is no coherent legal instrument covering the forest sector at the international or EU level—as it is one of the sectors in which the national level has retained the legislative authority—however, the sector is still significantly impacted by other EU legal acts, in particular the EU Nature Legislation, the Water Framework Directive and the Floods Directive, as well as climate policy (cf. Ellison 2010). Areas intersecting with forest, such as invasive species protection, flood and water management, and biodiversity, are typically covered by both EU and national law. The international and, particularly, the EU role in the context of forest is thus to provide guidelines for national forest policies, rather than to create a legal basis for forestry issues in the Union (COM(2010)66 final). This already complex situation is added to by the importance of voluntary market-based certification systems, which are now widely adopted and include a level of environmental consideration above state law.

As a result, mainstreaming forest-relevant adaptation can be seen as a "most difficult" case, relevant for considering the question of how well adaptation can at all be main streamed in established and highly complex multi-level governance systems. To this end, this study reviews the example of to what extent EU and national (exemplified by Swedish) legal and policy frameworks are able to integrate with each other in ways that may support climate change adaptation in forests. To move as close to the real life situation of mainstreaming challenges as possible, the study focuses on not only one area of mainstreaming or integration, but on the three broad policy areas outlined above: (a) adaptation per se; (b) forest biodiversity and habitat protection with respect to invasive species; and (c) water protection in relation to forest use.

Theoretical Framework: Mainstreaming of Climate Change Adaptation as a Multi-Level Legal and Policy Issue

Adaptation can be defined as the range of actions that may be taken to deal with the societal consequences of climate change (Smit and Wandel 2006). Mainstreaming adaptation is often discussed as a way of integrating adaptation to existing frameworks of policy and practice (e.g., European Commission 2013). It has been argued that to improve climate integration, climate change should be integrated with all sector policies. Such mainstreaming, based on the concept of environmental policy integration and applied to climate, should increase comprehensiveness and consistency, "increase policy coherence, minimize duplications and contradictory policies, avoid mal-adaptation, deal with trade-offs and capture the opportunities for synergistic results in terms of increased adaptive capacity" (Rauken et al. 2014, para. 2; cf. Kok and De Coninck 2007). Thus, "[m]ainstreaming involves the integration of information, policies and measures to address climate change into ongoing development planning and decision-making" (Ayers et al. 2014: 38).

However, how this process is to take place remains poorly understood. It has been noted that processes that support mainstreaming adaptation are often hampered by the sectoral nature of policy as well as by the fact that systems are not perfectly adapted in the present (Uittenbroek et al. 2013; cf. Næss et al. 2005, 2006). Ayers et al. (2014; cf. Uittenbroek et al. 2013) also note that many approaches to assessing mainstreaming at the national level (e.g., Huq and Ayres 2008) are limited in that they propose a linear sequence of capacity building, training, information to key policy-makers, and incorporation of results into policy, whereas in fact the process is not linear, but rather "made up of a patchwork of processes, stakeholders and approaches that converge or coexist" (Ayers et al. 2014: 48; cf. Wellstead et al. 2013). While environmental policy integration literature, which problematizes the potential for integration, has gone beyond linear sequencing, environmental policy integration studies have largely been single-sector studies (seldom including multi-sectoral issues such as, for example, adaptation in relation to forest, or even bioenergy). Environmental policy integration literature also largely takes the existing system as given, and focuses its efforts on what factors can within this context be seen as supporting integration of environmental policy. As a result, environmental policy integration literature has shown that factors supporting mainstreaming relate to amongst other the use of environmental knowledge, existence of monitoring mechanisms, balance of power between environmental regulators and the target sector, access of diverse actors to the policy process, technological potential for win-win solutions, and the role of attempts at policy integration (e.g., Söderberg 2011; Brouwer et al. 2013).

Rather than ask what factors support mainstreaming, this study has chosen to proceed from the case of how complexities already in the system of governance complicate mainstreaming. This is done to fulfill the aim of the paper to question whether mainstreaming is at all possible in existing systems given their complexity (apart from whether knowledge, monitoring, access or other factors may support mainstreaming in the system). To this end, this study places the focus on the conceptualization of multi-level governance, which posits that international, EU, and national legislation, as well as private and NGO initiatives, constitute a system of steering that needs to be understood in terms of how it plays out in its entirety in relation to different questions and issue areas (Marks and Hooghe 2004; Pierre and Peters 2005; Keskitalo 2010 ed; Knieling and Leal Filho 2012 ed; see also Stephenson 2013; Mwangi and Wardell 2012). A more complex overlay of international, EU, and private initiatives (Type II governance) is thereby added to more traditional Type I governance centred on the state (Marks and Hooghe 2004). The complexities in implementing cross-cutting changes in such system can be illustrated through the path-dependent character of developed governance systems, where choices or decisions made in the past affect the future span of possible, or feasible, choices (e.g., North 1994; Pierson 2000) (see Fig. 4.1 for a summary). The reminder of this section outlines the complexities in such multi-level governance systems with particular focus on the role of legal systems and associated path-dependency.

The legal status of systems is here seen as conditioning to all other development; as recognized in some environmental policy integration and mainstreaming literature (e.g., Brouwer et al. 2013), the capacity to regulate and the existence of hard (binding) regulation are crucial. As encased in governance systems, the strongest level of formal explicit requirement and evidence of capacity to regulate in the system is binding law (at EU level primarily regulations and directives), which delimit all other activities. Rather than being easily changed through linear capacity-building processes, for example, the conservative character of the law as

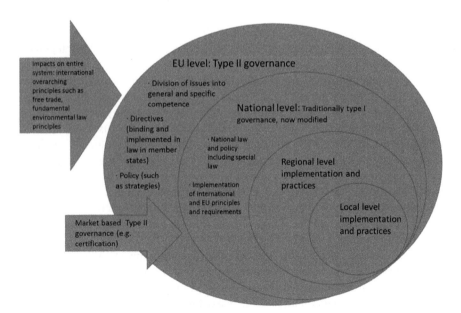

Fig. 4.1 The multi-level governance system

an institutionalized body of specified legal sources—consisting of encoded understandings of phenomena that have been possible to agree on in a social system—implies that law, as well as broader societal institutions, is characterized by path dependence. The broader social sciences assumptions in institutional literature that all institutions, whether formal law and policy or informal norms or practices, are assumed to be relatively enduring and hence resistant to change, often supporting incremental rather than transformative change (Hathaway 2003; Mahony and Thelen 2010). New inputs into these systems thus need to feed into and be considered within the existing parameters and already highly complex systems, which have largely been developed for purposes other than that of adaptation.

In addition to this path-dependent nature highlighted in institutional literature, multi-level governance literature makes clear that systems of steering seldom consist of only a few components that would need to be revised in order to take into account adaptation considerations (e.g., Ayers et al. 2014); it is rather that the patchwork of different steering approaches that characterizes the system, makes it highly complex and contextually dependent. Marks and Hooghe (2004) have described the change in governance as one from the national state system with multi-purpose, non-overlapping levels having local, regional, and national authority (what they call "Type I"), to a "Type II" governance system characterized by potentially more issue-based and cross-level organization. In practice, however, more complex stakeholder-based systems do not replace but rather supplement existing (Type I) systems, which mean that we retain a strongly path-dependent legislative system overlaid with other organizational processes, such as participatory processes. The larger EU system can be seen as a typical case of this. The EU is invested by its member states with the formal *competence* (decision-making power) to act on some issues and not others; and should under the principle of subsidiarity, only act "if and insofar as the objectives of the proposed action cannot be sufficiently achieved by the member states" (TFEU 2008). In the case of forest, the historical insistence on the resource as a commodity that belongs to the sovereign state has undermined EU competence, resulting in an almost complete absence of community legislation for the sector. However, this does not necessarily limit complexity: as legislation has not been developed in a way that could strongly support sustainable forest management, supplementary processes such as market-based forest certification have developed, introducing a third-party auditing system for assessing the inclusion of certain environmental and social requirements in forestry (cf. Cashore et al. 2004). All of these factors bring the system further from one in which decision-making is "linear" (Ayers et al. 2014), or where only improved knowledge, monitoring, or access would be sufficient to support integration of new issue areas into this system.

In addition, significant complexity often already exists in the national system—for example, in the distribution of authority between local, regional and national levels—indicating a deviation from the "capacity-building-to-policy" sequence that is built into the system setup. National legal systems typically build on *overarching principles*—such as, for instance, integrating principles required at international and EU levels, for example, that of free trade. In addition, fundamental principles of environmental law, such as the "precautionary principle" (that measures shall be taken to prevent also potential harm) or the "polluter pays principle" (that polluters are responsible for any damage caused by their activity) which have been recognized as significant for integrating environmental issues, are not necessarily recognized by the legal system as a whole. Rather, they may be trumped by other, overriding principles, such as principles for ownership (ownership rights). Moreover, the principle that special law prevails over general law (i.e., *lex specialis*), entails that e.g., general consideration rules applicable to most activities affecting the environment can be superseded by specific sectoral legislation, thus bypassing precautionary requirements. Per se, this limits overarching environmental policy integration at system level.

The complexity of the legal system is further compounded by the complexity of the policy system, which includes principles concerning the interpretation and development of the law. For the forestry sector, policy development is of prime importance since the area itself is virtually unregulated at the EU level. While forestry is still very much subject to supranational standards, such as EU nature conservation legislation, the influence of such legislation can be reduced as a result of the strong sovereignty of the forest sector and existing national law as well as policy and practice. Sectors with strong institutionalized interests that are important nationally or in terms of GNP, can be expected to have traditionally been granted more leeway (which is also encased legislatively), and this may be difficult to shift as situations (such as regulations instituting increased environmental concern) change (Keskitalo and Pettersson 2012). For instance, in Sweden's forestry sector, the protection of private ownership has traditionally been very strong and has resulted in comparatively more limited regulation than in many other sectors. As a result, Swedish forest legislation is expressed in the form of a framework law that leaves open the means by which the aims of the law are to be accomplished, and makes policy and advice development one of the principal means by which forestry can be influenced. This also means that the role of existing practices will have a large steering effect on the way in which any policy is implemented (e.g., Appelstrand 2007).

A legal and policy analysis of multi-level governance in any given case is thus necessary and will, in any defined review of cross-cutting requirements, indicate implementation difficulties, potentially larger the more sub-sectors or issue areas are involved.

Materials and Methods

This study utilizes the forest sector as an example of a major multi-use sector in which separate (and often dispersed) legal and policy frameworks need to manage multiple interests, such as biodiversity and habitat protection, including invasive species risks. Given the complexity of reviewing a number of intersecting policy areas, the study utilizes a comparison of several case studies on forest use in relation to adaptation policies, forest biodiversity and habitat protection in relation to invasive species, and water protection at both the EU and one selected national level case.[1] Further sectors could be envisioned, which means that the study even given this breadth does not capture the full complexity of mainstreaming adaptation. Also, while taking into consideration these three areas that impact adaptation in forest use at multiple levels, implementation is described for only one country, Sweden, which means that complexity would also increase with consideration of the full EU27.[2]

Sweden was chosen on account of the considerable importance of forest-based sectors to the state (Eurostat 2011), coupled with a significant level of private forest ownership and the extensive historical role of the forest industry (which makes forest difficult to regulate directly from the state level). This makes Sweden both an important multi-use land-use case and a potentially highly difficult case for adaptation with regard to this sector. The main legislation on forest management in Sweden is the Forestry Act—a framework law based on "freedom under responsibility" that requires forestry to consider multiple interests, focusing particularly on production and biodiversity protection. Most of its substantive rules are fleshed out in ordinances and prescriptions issued by the sector authority. The Swedish legal framework for forestry, however, includes more than just the Forestry Act: it also encompasses environmental legislation—for example, regarding habitat protection and protection against invasive species—making the legal map complex and difficult to navigate. In addition, fully 50 % of Swedish forest land is privately owned, making adaptation to climate change in forestry largely the responsibility of individual forest owners, which are subject to information campaigns by the Swedish Forest Agency (Swedish Forest Agency 2013; Keskitalo et al. 2012).

The complexity of studying several sub-sectors relevant to policy makes it necessary to draw upon dedicated studies in each of these domains that highlight adaptation aspects. With regard to adaptation policy, the study on the national level draws upon a policy document and interview study undertaken in 2008–2009 comprising a total of 25 semi-structured interviews on the national level and lower and coded in relation to a multi-level governance framework (reported in

---

[1] Given the complexity of forest as a field for policy-making, it should be expected that also areas other than the three treated here will have an impact on adaptation relevant to forest: most notably, general forest policy (included here only in its relation to each subfield), rural policy, policies on storm risk, and pest outbreak management, and private sectors such as insurance. Thus, despite including three areas of particular relevance to adaptation, the study will still underestimate, rather than overestimate, the complexity of mainstreaming adaptation. However, it can serve as an indication of the large requirements in each of these fields and, when taken together, in the three cases.

[2] In order to govern forest issues as a whole and mainstream adaptation in this context, this study thus indicates only one of the specific cases that need to be understood in order to mainstream adaptation. While the focus on detailed case-studies in only one country may seem limited, the multiple concerns as regards correlating and integrating adaptation within existing frameworks, which are likely to have a fundamental impact on the potential for implementing adaptation policies at the local level in different states, are illustrated.

full in Keskitalo 2010: major policy documents and legislation developed during this time remain relevant today, since current legislation was developed during this period). The study of adaptation policy at the EU level is based on a review of EU adaptation policy development, in particular the 2013 strategy. Forest biodiversity, habitats, and invasive species legislation and policies on the Swedish and EU levels are drawn from legal and policy document surveys (Pettersson and Keskitalo 2013; Pettersson and Keskitalo 2012; Pettersson 2013) including updating based on EU-level developments in 2013. All legal analysis are qualitative studies based on positive analytical jurisprudence, defined as the study of the concept or nature of law, i.e., the 'existing legislation' (e.g., Austin 1832; Kelsen 1941; Hart 1961), understood here not only as the legal texts and associated preparatory works and case law, but also voluntary instruments like certification that have a significant (and persistent) influence on the legal situation. The assessment regarding the potential use of novel instruments extends the methodology to include elements of normative and constructive jurisprudence, which allows for normative statements about how the law ought to be designed (cf. Austin 1832), whereas policy analysis extends the document study to developing areas of regulation as well as to advice and implementation of legal requirements. Finally, studies of water protection draw on the EU Water Framework Directive and its implementation in Swedish legislation and policy as well as implications for implementation drawn from a study based on 34 interviews with organizations active in forest certification focused on ongoing debates about buffer zones and coded in relation to a multi-level governance framework (see Keskitalo and Pettersson 2012). In all cases, the interviews were semi-structured and undertaken in person, recorded and fully transcribed as a basis for analysis.

The results below are organized to outline existing law or directives (if any) and policy at EU and national level, respectively, and also to outline the role of forest in these documents.

Results

Table 4.1 summarizes the results in tabular form.

The Context of EU Governance: Adaptation Policy as an Area for Ongoing Integration

Adaptation policies at both the EU and the Swedish levels largely reflect the complexity described above, the ongoing nature of adaptation issue development, and the fact that implementation will need to be developed within each specific sector. Up to January 2013, EU policy was relatively non-formalised, and the

Table 4.1 Level and degree of regulation in policy areas related to adaptation in forest

Policy area related to forest		Level and degree of regulation					
		EU			Sweden		
		Legislation	Policy	Role of forest and climate, respective	Legislation	Policy	Role of forest and climate, respective
Adaptation to climate change		–	EU Adaptation Strategy	Forest mention in strategy (climate main focus)	Climate Bill	More comprehensive Commission on Climate and Vulnerability	Forest mentioned in both, implementation largely left to sector—role of informal practices important (climate main focus)
Forest biodiversity	Habitat protection	Comprehensive, relatively well developed area protection measures	(Part of comprehensive, relatively well developed area protection measures)	Significant role of forest, role under climate change less developed	Comprehensive, well developed area protection measures	(Part of comprehensive, well developed area protection measures)	Significant role of forest, role under climate change less developed
	Invasive species	No comprehensive framework	E.g., list of major risk species	Forest not targeted per se, only selected species; role of climate not integrated	Prohibited from acting unilaterally under EU law	Existing, noting limitations of EU framework	Forest not targeted per se; role of climate not integrated
Water protection in relation to forest use		Comprehensive: EU Water Framework Directive	(Part of comprehensive framework and supporting this)	Implications for forest; climate included through later measures (e.g., Floods Directive)	Interpretation of EQS under Water Framework Directive limits impact	Existing in several sectors (e.g., Swedish Forest Agency)	Implications for forest drawn, but largely following established practice and existing decision-making structures

policies in the Green and White Papers on adaptation focused in particular on adaptation measures being developed within each member state, and thus without obligative impact on states (Ellison 2010). As the next step of development, the EU Strategy on adaptation to climate change launched in January 2013 includes an outline of the aims of all European states in adopting comprehensive adaptation strategies, supporting better informed decision-making, and promoting adaptation in key vulnerable sectors (COM(2013) 216 final). Here, forestry is mentioned as one of the key vulnerable sectors (e.g., SWD(2013) 131 final), and is discussed at length in a related impact assessment document targeting different sectors (SWD (2013) 132 final). The document notes that Regulation (EC) 2152/2003, repealed by Regulation (EC) No 614/2007, established a scheme of forest monitoring that included the effects of climate change, and refers to the broad EU Forest Strategy adopted later in 2013 (COM(2013) 659 final).

However, these documents contain, at most, general guidelines, under which member states commit to the aims of sustainable management that have been determined nationally. The associated impact assessment document further notes that the changes in rural development policy support will support adaptation in forestry, as will the adoption of a new plant health law (delayed and under development) and the development of a specific legislative instrument on invasive species (SWD(2013) 132 final). The Strategy thus constitutes the first overarching approach to mainstreaming adaptation in the EU. However, given the principle of subsidiarity and the shared responsibility of member states, both the Strategy and its implementation will depend on existing legislation and international initiatives, and thus constitute general non-binding aims rather than legal requirements. The document also notes a number of barriers to adaptation in forestry that are potentially related both to differences between states, in terms of forest owners as a group as well as other forestry interests, and to the competing and multiple jurisdictions over the forestry sector and difficulty with which policy decisions reach individual forest owners (SWD(2013) 132 final). As with this document, some previous studies have noted that adaptation policies (not necessarily expressed as formal strategies or specific formal policies) in European countries suffer from being overly general and having few clearly implemented and planned measures. A similar difficulty in reaching diversified and fragmented forest-owner structures has also been observed (e.g., Keskitalo 2011).

At the Swedish level, the issue of adaptation has been managed largely by the Commission on Climate and Vulnerability (2007) and the Climate Bill (Government Offices of Sweden 2009a). The Commission was to some extent the result of concerns over flood risk in south-western Sweden, and the water issue, more than forestry, was therefore highlighted in its work. The final report of the Commission, however, deals with a number of sectors in which adaptation will be relevant, forestry being one of these. It suggests, as later determined in the Bill, that forestry is subject to sectoral coordination by the Swedish Forest Agency (similar to the situation for other sectors and their sectoral agencies), including revision of legislation (ongoing). It is also noted that adaptation will to a large extent rely on the decisions of the many Swedish private forest owners, and to that end the Swedish

Forest Agency is tasked with developing public awareness campaigns. The planning system, according to which municipalities have a planning monopoly but no influence over forestry planning (even in cases where this may affect biodiversity or other aims relevant to the municipality), is thus not impacted (Commission on Climate and Vulnerability 2007). This could have a large impact on the choice between various adaptation measures, where some of the measures that could support adaptation, such as an increased proportion of mixed forests that better resist insects and storms, are often costly to the forestry sector. Others, such as increasing quick-growing forests and changing management schedules, are less demanding and thus potentially more attractive (cf. Keskitalo et al. 2012).

In conclusion, while the Swedish Commission on Climate and Vulnerability reports a number of potential impacts on and adaptations in forestry, the implementation is largely left to individual forest owners and industry. This is partly due to the large number of forest owners in Sweden, but may also reflect the institutional situation in which the sector itself is allowed to determine what action is to be taken within the framework of, primarily, the Forestry Act. In keeping with this, both the Commission and the Bill treat adaptations in forestry as an issue for the sector rather than the state, which in turn is in line with the overarching Swedish adaptation approach, involving basically no support for approaches on the municipal level, and that does not go beyond the principle of "freedom under responsibility" entrenched in Swedish forestry legislation. As a result, existing practices and informal norms in forestry become highly important: for instance, while forest certification has been seen as a way of increasing environmental consideration, so far such measures do not include adaptation considerations.

Protection of Forest Biodiversity: Caught Between Different Competences?

Climate change, including changed climatic zones, migratory and invasive species and extreme weather, implies major challenges for institutional preservation regimes. In Sweden as in most EU member states, the range of potential environmental protection measures includes both pre-existing instruments and measures imposed by the EU. The main instrument to protect forest biodiversity in Sweden is area protection, for instance the establishment of national parks, nature reserves, biodiversity protection areas and EU Natura 2000 areas. As a legal tool, area protection is comparatively well-developed and has—albeit to a varying extent—the ability to constrain land use in favor of biodiversity protection in the designated areas (Pettersson and Keskitalo 2013). However, while the current protection system is designed for clearly demarcated sites without clear connections between them (the EU Natura 2000 network in effect does not form a connected network), climate change may result in species distribution changes that will place new flexibility demands on area protection. In relation to the member state level, the existence of EU nature legislation means that to some extent national conservation policy and legislation are also controlled by the EU; Sweden, for example, no longer has the sole authority to decide whether an area should be afforded protection, the extent to which it should be protected, or whether the protection should cease. It has also been noted that climate change considerations have not been integrated into existing protection decisions, and that the currently protected sites may not be sufficient to effectively protect habitats and biodiversity under conditions of climate change.[3]

This may be the case particularly under conditions involving species movement and the introduction of invasive species—an issue that is currently fragmented and scattered across different sectoral legislation. Invasive species regulations are typically either very precise (i.e., take the form of exhaustive lists of the species subject to control) or target invasive species only indirectly (cf. the Birds and Habitat Directives). While the former is easy to comply with (the species is either on the list and therefore subject to some form of restriction, or not, in which case bringing it into the country, for example, is not prohibited), a lack of general applicability severely limits the scope of the regulations and increases the transaction costs.

With concern for the threat to biodiversity and related ecosystem services as well as the social and economic impacts posed by invasive species, and as a response to the fact that the existing EU policies left most invasive species unaddressed, a proposal for a Regulation on the subject matter was presented in 2013(COM(2013) 620 final). The Regulation entered into force on January 1, 2015 and sets out rules to "prevent, minimize and mitigate the adverse impact on biodiversity of the introduction and spread within the Union, both intentional and unintentional, of invasive alien species." (IAS) (Regulation (EU) 1143/2014, Art. 1). To achieve the objective, a "Union list" of IAS that pose a particular threat to the Union shall be adopted by the Commission by means of implementing acts. To qualify as being of Union concern, the damage caused by the species should be significant enough to justify the adoption of dedicated measures. This, in turn, is assessed on the basis of certain criteria, all of which are in line with the SPS agreement and include risk assessments.[4] Furthermore, since prevention is preferable to reaction it is considered necessary that the list of species is continuously revised and updated. In case of

---

[3] In addition, as requirements in production forest are limited and rest on consideration for other uses rather than specific legal limitations in harvesting, it has been argued by some that levels of formal protection are too low (see e.g., Lisberg Jensen 2002).

[4] The criteria for including IAS on the Union list are the core instrument of application of the Regulation. Besides being alien to Union territory, the conditions for inclusion on the list include that the species-based on available scientific evidence—are capable of establishing a viable population and spreading in the environment under current conditions and in foreseeable climate change conditions, and are likely to have a significant adverse impact on e.g., biodiversity. In addition, it must be demonstrated by a risk assessment that action at Union level is required to prevent IAS introduction, establishment or spread, and that the inclusion on the Union list is likely to effectively prevent, minimise or mitigate the adverse impact of the IAS (Art. 3).

a sudden and unexpected appearance of species that have not yet been defined as IAS, but for which there is scientific evidence of harmfulness, it is possible for member states to adopt certain emergency measures (Art. 10). In order to prevent the introduction, establishment and spread of IAS, it is possible for member states to maintain or lay down more stringent rules than prescribed by the Regulation, on conditions that the measures are compatible with the TFEU. Such action must be notified to the Commission (Art. 23).

The policy implications of the regulation are still difficult to assess. However, from a legal perspective it seems clear that, while some adaptation or protection measures beyond the EU regulation certainly can be taken, member states are still constrained by the overall WTO regime.

Water Protection in Relation to Forest Use

Water quality issues are important to forestry in particular because water quality is to a large extent impacted by different land uses, including forest management practices. The EU Water Framework Directive (WFD) calls for considerable implementation efforts and extensive participation on the part of member states to prevent deterioration and achieve good water status, both in terms of administrative and legal measures. It can, however, be considered (mainly by extension) as including attention to climate change (e.g., Quevauviller 2011, for instance, by its common implementation with the later Floods Directive).

In terms of distribution of competences, water resource management is an issue on which the EU and its member states have shared competences (Art. 4 TFEU). The issuance of the EU legislation as a Directive means that, although the objective of the directive is legally binding, the actual implementation—that is, how to achieve the objective, by what instruments, etc.—is essentially a matter for the member states (Art. 288 TFEU). This implies significant challenges, not least in connection with the integration of different types of land use. The right of each member state to choose how to implement Environmental Quality Standards (EQS) for water as one of the crucial requirements under the WFD has in Sweden resulted in a weak formulation, meaning that the possibility of imposing absolute requirements does not apply, as would have been the case with a stricter formulation of the EQS as limitation standards (Government Office of Sweden 2009b: 39–42). The risk of potential violation of the standard will thus not necessarily ward off environmentally hazardous activities or water operations, nor will it prevent a trade-off between environmental requirements and what is considered reasonable from the operator's point of view. Instead, the legal effect of the Swedish EQS for water is that they "should be followed" and that a program of measures should be implemented if deemed necessary (Ch. 5, s. 2, para. 1, p. 4, Swedish Environmental Code).

At the implementation level, although Swedish forestry policy stipulates that environmental considerations should be taken for all production forest, for example, through the introduction of buffer zones, the Swedish Forest Agency's evaluations suggest that there are some difficulties with regard to the extent to which such measures are carried out (Swedish Forest Agency 2008a, b). Interviews concerning the implementation of buffer zones provide evidence of the significant context- and site-dependence of buffer-zone width, as well as the importance of local norms and training in how buffer zones are developed. Interviews indicate that the implementation of the WFD has here mainly served to highlight issues with regard to water, as well as potential adaptations, that were already known, but that are now implemented given the increased focus on water. Examples include, for instance, the development of "blue plans" to include water issues, to correspond with the "green plans" including biodiversity that have been in existence for forest management.

Additionally, large-scale interests with diffuse emissions—such as the forest industry—have attempted to redefine how they participate in processes to ensure water quality over the long term, and also need to evaluate the consideration of harvesting and forest operations in new ways. Unable to participate in local Water Councils mandated under the WFD, as this would require larger local personnel resources than large forest companies typically have, large forest owners and others gathered in an informal self-organized Forest Water Council during WFD implementation to impact state development. These differences between state and EU frameworks that appear in organizational scale and consideration of consensus development further support more incremental change. Adaptation to climate change per se has so far not been considered when adapting regulations or advice, or implementing the WFD in this case.

Discussion

While many EU countries have developed policies for adaptation to climate change, forests often represent a minor part of these policies (Keskitalo 2011). The EU Adaptation Strategy here constitutes the first strategic approach to incorporating the various considerations related not only to forestry but to several other sectors with respect to adaptation. Thus far, however, adaptation is largely in the development stage and has at the Swedish level not yet resulted in mandatory requirements for the integration of adaptation into forest management decisions—for example, at the forest-owner level. The frameworks for biodiversity and habitat protection, protection against invasive species, and water quality then all constitute largely separate spheres of regulation, which to a large extent do not include climate change considerations; this holds true for the WFD as well, despite its relatively recent inception. For biodiversity and habitat protection, although systems based on current choices (e.g., regarding level of protection), are well developed, climatic changes will require (re-)consideration of both networking between sites and potential changes within relevant sites (cf. Wilson and Piper 2008).

At EU level, the strength of biodiversity, habitat, and water protection regimes is ultimately dependent on the individual member state's implementation of the directives, which in turn will depend on the existing institutional setting. Here, Sweden's fairly strict approach to the protection of forest biodiversity can be at least partly explained by the traditional/historical use of site protection as the main environmental conservation strategy, whereas the ecosystem approach and the new administrative structure presented by the WFD represents a new and thus institutionally more difficult strategy to implement. For invasive species, one of the areas now targeted under the EU Adaptation Strategy, considerable limitations exist with regard to enforcing any comprehensive legal framework, in particular if based on the precautionary principle, in light of overarching (WTO) restrictions on limiting free trade that enforce path-dependence in a system that is already path-dependent.

Thus, this study illustrates that there is a great discrepancy between the examined policy areas in terms of opportunities—in this case for Sweden—to effectively act independently as well as for effectively integrating adaptation aims in the connected EU–national systems. In the case of water, Sweden indeed has an opportunity to go even further in the requirements than what is demanded by the WFD (since it is a minimum directive), whereas other issues, like protection against invasive species, cannot be controlled legally beyond what is allowed under the EU and the WTO regime. In particular, conflicts between the maxim of free trade and the precautionary principle, as well as between issues of subsidiarity and proportionality stand out:

The conflict between the maxim of free trade and the precautionary principle: under Swedish law, the precautionary principle entails, first of all, that the burden of proof for taking the precautionary measures is on the 'operator', i.e., the legal entity that undertakes the activity. Secondly, Swedish environmental law adheres to a 'strong version' of the precautionary principle (cf. Ansari and Wartini 2014) meaning that precautionary measures shall be taken "as soon as there is cause to assume" that the activity can cause "damage or detriment to human health or the environment" (Ch. 2, s. 3, Environmental Code). In essence, the Swedish precautionary principle thus includes two very important aspects: precautionary measures must be taken already when there is risk of harm, and to avoid the requirements, the operator must show that there is no risk (Michanek 2007: 126). According to the SPS agreement, parties to the agreement are allowed to adopt precautionary measures "only to the extent necessary" for the protection of human, animal or plant life and health, and only if the measures are "based on scientific principles" and not upheld without scientific evidence (SPS Agreement, Art. 2, paragraph 2). If the relevant scientific information is insufficient, provisional measures may be adopted, pending additional information and a more objective assessment of the risks (SPS Agreement, Art. 5, paragraph 7). Since "[p]recautionary measures by definition involve situations of scientific uncertainty" (Wirth 2013: 1154), the formulation in the SPS agreement implies—at best—a weak version of the precautionary principle. In addition, the burden of proof under the Agreement is on the receiving country, which thus has to show that harm will in fact be caused if appropriate precautionary measures are not taken.

In terms of how to interpret the significance of the precautionary requirements as they are expressed in the different legal frameworks, it is also important to take account of the legal acts' overall purpose and objectives. While Swedish environmental law primarily aims to promote sustainable development, and thus has implemented a precautionary *principle* to underpin the legal application, the overall purpose of the WTO is to promote and maintain free trade.[5] In this context, the SPS agreement and the possibilities to take precautionary measures constitute restrictions of the free trade (making it a little less free). Thus without detracting the ambitions of the WTO and the SPS Agreements regarding the possibility of adopting restrictive trade measures with the aim of e.g., protecting the environment, substantial discrepancies as to the meaning of the precautionary principle can still be detected. Since free trade is also the foundation of EU cooperation, legal instruments that limit free trade can, in principle, not be adopted at EU level. Thus, at the end of the day, neither Sweden nor the EU can act independently to protect themselves from the influx of species resulting from international trade. The issue of the interpretation and application of the precautionary principle in the context of controlling invasive alien species is stressed also by Swedish authorities; SEPA calls for clarification of meaning of the principle and the different standards accepted by the SPS agreement in order to investigate the possibilities of applying the precautionary principle in the management of invasive species in Sweden (SEPA 2008: 98). Thus, while the precautionary approach forms the basis for both environmental policy and law in most member states, this development is in some respects incompatible with the foundations of EU cooperation.

The issue of subsidiarity and proportionality: in areas of shared competence between the EU and the member states, legislative measures on EU level shall only be taken to the extent necessary, and may not exceed what is necessary to achieve the objectives of the Treaty of the EU. However, the responsibility for the actual implementation of EU legislation is, in many instances, left to the sovereign member states who—in case of most EU Directives—can choose among implementation measures such as statutory law, ordinances, prescriptions and even policy as a means to achieve the aim of the legal act. The choice of measures does, however, influence both the final outcome and key features of the process; whereas statutory law, besides its democratic aspects, grants stronger power to enforce measures, ensure public participation and appeal decisions; legal sources of a lower rank, as well as policies, do not have the same capacities. The implementation of EU legislation, such as the WFD, will thus differ between different member states, as will the decision whether, and if so which, adaptation measures should be taken. As a result, EU frameworks on water, including common implementation with potentially stronger adaptation measures in the Floods Directive, will be limited in the extent to which they can mainstream issues in member states.

---

[5] It should be noted here that the WTO Agreement allows for precautions in line with sustainable development; the preamble explicitly adheres to the sustainable development concept (WTO Agreement, Ansari and Wartini 2014).

That said, while the interpretation, and subsequently the application, of the law may be either extensive or narrow (Mahony and Thelen 2010: 13), the WFD certainly provides room for more profound changes that can be implemented by other actors—if, as relevant for the national case here, national path dependencies can be overcome. Significant variations in implementation in all of these cases may exist between countries. For instance, as Brouwer et al. note with regard to the Water Framework Directive, and in a study of Catalonia, Sweden, Scotland, Italy and Poland, "we have seen wide variation ... ranging from the Polish Warta, where climate change impacts and contradictions are mostly ignored, to Catalonia where climate change impacts are extensively considered and most contradictions are revealed" (Brouwer et al. 2013: 148), with Sweden in this case ranking somewhere in the middle.

Conclusions

The possibility of mainstreaming adaptation strategies in different sectors is to a significant extent determined by the interrelationship between the EU and its member states. While EU law indeed rests on the principle of subsidiarity, EU cooperation is still primarily based on the principles of free trade, and the division of competence and the governance structure of the EU are therefore decisive factors in terms of how much leeway is left for member states to institute their own adaptation measures, especially across different sectors. With regard to land use, forest and spatial planning, the EU's *lack* of formal competence implies that the governance of these areas is solely a matter for the member states, whereas in other areas, such as trade and competition, the EU has exclusive competence to legislate. This dichotomy creates barriers with respect to the potential for implementing necessary climate change adaptation measures and entails that adaptation measures in relation to forest will largely remain under the national governance structure, thus including existing institutionalization and practices, also under the new EU Adaptation policy. In Sweden, the role of incremental or substantive change will thus largely depend on the strong Type II governance (Hooghe and Marks 2001) that characterizes the forest sector, which is particularly apparent with regard to the different actors involved in implementation, the limited legal effect of the EQS, and the role of industry in determining the scope of the implementation of, for example, the WFD. A coherent take on mainstreaming of adaptation to climate change is thus in this stage not possible.

The requirements of adaptation to climate change thus bring to the fore numerous principles, rules and constraints which affect the implementation of measures to safeguard forest systems and biodiversity against the adverse effects of climate change, such as the increased influx of invasive species, changes in habitats and changing hydrological conditions. This has several important implications for adaptation research. First, the large focus on social vulnerability and community adaptation studies, as well as on approaches to social or institutional determinants or perspectives on adaptation (e.g., Pelling 2011; see also Rodima-Taylor et al. 2012), and on bottom-up and largely local systems developing change and implementing learning may be highly circumscribed in a way that makes barriers such as path-dependences extremely important to include in all analyses: learning or a given sequence of mainstreaming cannot be assumed. Indeed, it has been suggested that understandings of the way in which adaptation can be developed in policy often suffer from a "structural–functional logic" that treats "policy-making as an undifferentiated and un-problematic output of a political system responding to input changes and/or system prerequisites" (Wellstead et al. 2013, para. 1). Limitations to transformative impacts on systems thus need to be theoretically considered not only in relation to e.g., learning, resilience, and adaptive management frameworks, but also in ways that inherently start from and consider the complexities of the multi-level governance system and its existing bodies of regulation as a baseline. This study underlines the importance of both established social science and legal conceptions in order to comprehend the "ecosystem" of regulation (as complex as existing ecosystems). In particular, the study highlights the distance between incremental and transformative change, with reference primarily to system path dependencies and thus incremental change as explanatory variables as to why adaptation policies in many countries have yet to reach the implementation stage (cf. Keskitalo 2010 ed; Swart et al. 2009). By highlighting institutional barriers to implementation of adaptation, both theoretically and on the basis of case studies, also openings for change identified; while barriers can be very far reaching, and require confronting important interests, such as existing land use, or trade, legislative measures offers opportunities for institutional change, regardless of whether the formulation of the rules is strong or weak (e.g., Mahony and Thelen 2010; Hathaway 2003). In the case of IAS, the precautionary principle will remain subordinate to the principle of free trade since the criteria for listing under the new EU regime will have to comply with the requirements of the SPS agreement, and member states will still not be able to go further in their adaptation or protection measures (at least not without scientific evidence, risk assessments, and authorization) than what is allowed under the regulation. However, with regard to the WFD, the regulatory framework does allow for further progress, provided that the institutional path dependencies are overcome (see e.g., Keskitalo and Pettersson 2012).

Acknowledgements We are grateful to the FORMAS and the Future Forest programme (funded by the MISTRA Agency for Swedish Strategic Environmental Research, the forest industries, Umeå University, and SLU) for funding our work.

References

Ansari AH, Wartini S (2014) Application of precautionary principle in international trade law and international environmental law. A comparative assessment. J Int Trade Law Policy 13 (1):19–43

Appelstrand M (2007) Miljömålet i skogsbruket—styrning och frivillighet. Lund Studies in Sociology of Law 26. Media-Tryck Sociologen, University of Lund, Lund

Austin J (1832) The province of jurisprudence determined. University of London, London, Printed by Richard Taylor

Ayers JM, Huq S, Faisal AM, Hussain ST (2014) Mainstreaming climate change adaptation into development: a case study of Bangladesh. WIREs Clim Change 5:37–51

Brouwer S, Rayner T, Huitema D (2013) Mainstreaming climate policy: the case of climate adaptation and the implementation of EU water policy. Environ Plann C 31:134–153

Cashore B, Auld G, Newsom D (2004) Governing through markets—forest certification and the emergence of non-state authority. Yale University Press, New Haven, CT

COM(2010)66—Green Paper on Forest Protection and Information in the EU: preparing forests for climate change SEC(2010)163 final

COM(2013) 216 final (2013) Communication from the Commission to the European Parliament, the Council, the European Economic and Social Committee and the Committee of the Regions. An EU strategy on adaptation to climate change. http://eur-lex.europa.eu/LexUriServ/LexUriServ.do?uri=CELEX:DKEY=725522:EN:NOT

COM(2013) 620 final (2013) Proposal for a Regulation of the European Parliament and of the Council on the prevention and management of the introduction and spread of invasive alien species

COM(2013) 659 final (2013) Communication from the Commission to the European Parliament, the Council, the European Economic and Social Committee and the Committee of the Regions. A new EU forest strategy: for forests and the forest-based sector. http://eur-lex.europa.eu/LexUriServ/LexUriServ.do?uri=COM:2013:0659:FIN:en:PDF

Commission on Climate and Vulnerability (2007) Sweden facing climate change—threats and opportunities [in Swedish]. Swedish Government Official Report SOU 2007:60, Stockholm

Council Directive 92/43/EEC of 21 May 1992 on the conservation of natural habitats and of wild fauna and flora. Off J Eur Union

Directive 2000/60/EC of the European Parliament and of the Council establishing a framework for the Community action in the field of water policy. Off J Eur Union

Directive 2007/60/EC of the European Parliament and of the Council of 23 October 2007 on the assessment and management of flood risks. Off J Eur Union

Directive 2009/147/EC of the European Parliament and of the Council of 30 November 2009 on the conservation of wild birds. Off J Eur Union

Ellison D (2010) Addressing adaptation in the EU policy framework. In: Keskitalo ECH (ed) Developing adaptation policy and practice: multi-level governance of climate change. Springer, Berlin

Environmental Code (1998) Miljöbalken, svensk författningssamling SFS 1998:808

European Commission (2013) Climate policy mainstreaming. http://ec.europa.eu/clima/policies/brief/mainstreaming/. Accessed 16 Dec 2013

Eurostat (2011) Forestry in the EU and in the World. A statistical portrait. Eurostat, 2011th edn. European Commission, Brussels

Government Offices of Sweden (2009a) En sammanhållen klimat-och energipolitik. Klimat. Regeringens proposition 2008/09:162. Government Offices of Sweden, Stockholm

Government Offices of Sweden (2009b) Prop. 2009/10:184. Åtgärdsprogram och tillämpningen av miljökvalitetsnormer

Hart HLA (1961) The concept of law. Clarendon Press, Oxford

Hathaway OA (2003) Path dependence in the law: the course and pattern of legal change in a common law system. In: Olin JM (ed) Center for Studies in Law, Economics, and Public Policy Working Paper. Yale Law School

Hooghe L, Marks G (2001) Types of multi-level governance. European Integration Online Papers (EIoP) 5(11). http://eiop.or.at/eiop/texte/2001011a.htm

Huq S, Ayres J (2008) Streamlining adaptation to climate change into development projects at the national and local level. In: European Parliament (ed) Financing climate change policies in developing countries. European Parliament, Brussels, pp 52–68

IPCC (2012) Managing the risks of extreme events and disasters to advance climate change adaptation. In: Field CB, Barros V, Stocker TF, Qin D, Dokken DJ, Ebi KL, Mastrandrea MD, Mach KJ, Plattner G-K, Allen SK, Tignor M, Midgley PM (eds) A special report of Working Groups I and II of the Intergovernmental Panel on Climate Change. Cambridge University Press, Cambridge, p 582

Kelsen H (1941) The pure theory of law and analytical jurisprudence. Harv Law Rev 55(1):44–70

Keskitalo ECH (2010) Adapting to climate change in Sweden: National Policy development and adaptation measures in Västra Götaland. In: Keskitalo ECH (ed) The development of adaptation policy and practice in Europe: multi-level governance of climate change. Springer, Dordrecht, pp 189–232

Keskitalo ECH (2010, ed) The development of adaptation policy and practice in Europe: multi-level governance of climate change. Springer, Dordrecht

Keskitalo ECH (2011) How can forest management systems adapt to climate change? Possibilities in different forestry systems. Forests 2(1):415–430

Keskitalo ECH, Pettersson M (2012) Implementing multi-level governance? The legal basis and implementation of the EU Water Framework Directive for forestry in Sweden. Environ Policy Gov 22:90–103

Keskitalo ECH, Eklöf J, Nordlund C (2012) Climate change mitigation and adaptation in Swedish forests: promoting forestry, capturing carbon and fuelling transports. In: Järvelä M, Juhola S (eds) Energy, environment and human response in northern Europe. Springer, Berlin

Knieling J, Leal Filho W (2012, eds) Climate change governance. Springer, Dordrecht

Kok MTJ, De Coninck HC (2007) Widening the scope of policies to address climate change: directions for mainstreaming". Environ Sci Policy 10:587–599

Lisberg Jensen E (2002) Som man ropar i skogen: Modernitet, makt och mångfald i kampen om Njakafjäll och i den svenska skogsbruksdebatten 1970–2000. Lund University, Lund

Mahony J, Thelen K (2010) A theory of gradual institutional change. In: Mahony J, Thelen K (eds) Explaining institutional change. Ambiguity, agency, and power. Cambridge University Press, New York, NY

Marks G, Hooghe L (2004) Contrasting visions of multi-level governance. In: Bache I, Flinders M (eds) Multi-level governance. Oxford University Press, Oxford

Michanek G (2007) Sweden. In: de Sadeleer N (ed) Implementing the precautionary principle. Approaches from the Nordic Countries, EU and USA. Earthscan, London. ISBN 1-84407-312-2

Mwangi E, Wardell A (2012) Multi-level governance of forest resources (Editorial to the special feature). Int J Commons [Online] 6(2):79–103

Næss LO, Bang G, Eriksen S, Vevatne J (2005) Institutional adaptation to climate change: flood responses at the municipal level in Norway. Glob Environ Chang 15:125–138

Næss LO, Thorsen Norland I, Lafferty WM, Aall C (2006) Data and processes linking vulnerability assessment to adaptation decision-making on climate change in Norway. Glob Environ Chang 16:221–233

Regulation (EU) No 1143/2014 of the European Parliament and of the Council of 22 October 2014 on the prevention and management of the introduction and spread of invasive alien species. Off J Eur Union

North DC (1994) Institutional change: a framework of analysis. http://ideas.repec.org/p/wpa/wuwpeh/9412001.html#provider

Pelling M (2011) Adaptation to climate change: from resilience to transformation. Routledge, New York, NY

Pettersson M (2013) Kontroll av främmande arter (i skogsmiljöer)—Behovet av rättsliga reformer. Europarättslig Tidskrift No. 3

Pettersson M, Keskitalo ECH (2012) Forest invasive species relating to climate change: the EU and Swedish regulatory framework. Environ Policy Law 42(1):63–73

Pettersson M, Keskitalo ECH (2013) Adaptive capacity of legal and policy frameworks for biodiversity protection considering climate change. Land Use Policy 34:213–222

Pierre J, Peters BG (2005) Governing complex societies—trajectories and scenarios. Palgrave MacMillan, Basingstoke

Pierson P (2000) Increasing returns, path dependence, and the study of politics. Am Polit Sci Rev 94(2):251–267

Quevauviller P (2011) Adapting to climate change: reducing water-related risks in Europe—EU policy and research considerations. Environ Sci Policy 14:722–729

Rauken T, Mydske PK, Winsvold M (2014) Mainstreaming climate change adaptation at the local level. Local Environ. doi:10.1080/13549839.2014.880412

Regulation (EC) No. 614/2007 of the European Parliament and of the Council concerning the Financial Instrument for the Environment (LIFE+). Off J Eur Union

Regulation (EC) No. 2152/2003 of the European Parliament and of the Council of 17 November 2003 concerning monitoring of forests and environmental interactions in the Community (Forest Focus). Off J Eur Union

Rodima-Taylor D, Olwig MF, Chhetri N (2012) Adaptation as innovation, innovation as adaptation: an institutional approach to climate change. Appl Geogr 33:107–111

Smit B, Wandel J (2006) Adaptation, adaptive capacity and vulnerability. Glob Environ Chang 16:282–292

Söderberg C (2011) Institutional conditions for multi-sector environmental policy integration in Swedish bioenergy policy. Environ Polit 20(4):528–546

Stephenson P (2013) Twenty years of multi-level governance: 'Where does it come from? What is it? Where is it going?'. J Eur Public Policy 20(6):817–837

Swart R, Biesbroek R, Binnerup S, Carter TR, Cowan C, Henrichs T, Loquen S, Mela H, Morecroft M, Reese M, Rey D (2009) Europe adapts to climate change: comparing national adaptation strategies. PEER Report No 1. Partnership for European Environmental Research, Helsinki

SWD (2013) 132 final (2013) Commission Staff Working Document. Impact assessment—Part 2. Accompanying the document Communication from the Commission to the European Parliament, the Council, the European Economic and Social Committee and the Committee of the Regions. An EU Strategy on adaptation to climate change. http://ec.europa.eu/clima/policies/adaptation/what/docs/swd_2013_132_2_en.pdf

SWD/2013/131 final (2013) Commission staff working document. Summary of the impact assessment accompanying the document communication an EU strategy on adaptation to climate change. http://eur-lex.europa.eu/LexUriServ/LexUriServ.do?uri=CELEX:52013SC0131:EN:NOT

Swedish Environmental Protection Agency (2008) Nationell strategi och handlingsplan för främmande arter och genotyper. Naturvårdsverket rapport 5910

Swedish Forest Agency (2008a) Generell hänsyn—en fallstudie. Skogsstyrelsen, Jönköping. http://www.skogsstyrelsen.se/episerver4/dokument/mn/Z_Ost/080404_Fallstudie.pdf

Swedish Forest Agency (2008b) Natur-och kulturmiljöhänsyn vid föryngringsavverkning—resultat från Skogsstyrelsens rikspolytaxinventering (R1), 1999–2006. Swe-dish Forest Agency, Jönköping. http://www.skogsstyrelsen.se/episerver4/dokument/sks/aktuellt/press/2008/Generell%20hänsyn/PM%20Polytaxresultat%20miljöhänsyn%202008.pdf

Swedish Forest Agency (2013) Skog och miljö. http://www.skogsstyrelsen.se/Global/myndigheten/Skog%20och%20miljo/ENGLISH/retrieve_file.pdf. Accessed 15 Nov 2013

Swedish Forestry Act (1979) Skogsvårdslag, svensk författningssamling SFS 1979:429

TFEU (2008) Consolidated version of the treaty on the functioning of the European Union. http://eur-lex.europa.eu/LexUriServ/LexUriServ.do?uri=OJ:C:2008:115:0047:0199:en:PDF

Uittenbroek CJ, Janssen-Jansen LB, Runhaar HAC (2013) Mainstreaming climate adaptation into urban planning: overcoming barriers, seizing opportunities and evaluating the results in two Dutch case studies. Reg Environ Change 13:399–411

Wellstead AM, Howlett M, Rayner J (2013) The neglect of governance in forest sector vulnerability assessments: structural-functionalism and "black box" problems in climate change adaptation planning. Ecol Soc 18(3):23ff

Wilson E, Piper J (2008) Spatial planning for biodiversity in Europe's changing climate. Eur Environ 18:135–151

Wirth D (2013) The World Trade Organization dispute concerning genetically modified organisms: precaution meets international trade law. Vermont Law Rev 37(4):1153–1188

World Trade Organisation (1998) Understanding the WTO Agreement on sanitary and phytosanitary measures. Switzerland, Geneva

World Trade Organisation, Agreement on the application of sanitary and phytosanitary measures. http://www.wto.org/english/docs_e/legal_e/15-sps.pdf

World Trade Organisation, Agreement establishing the World Trade Organization. http://www.wto.org/english/docs_e/legal_e/04-wto.pdf

Chapter 5
A Novel Impact Assessment Methodology for Evaluating Distributional Impacts in Scottish Climate Change Adaptation Policy

Rachel M. Dunk, Poshendra Satyal, and Michael Bonaventura

Abstract While it is widely recognised that the impacts of both climate change and the policy response will be distributed, there is an absence of complete information regarding the socio-economic and geographic patterning of such impacts in the intra-national context. This paper seeks to address this gap, presenting a climate justice toolkit (indicator set and guidance) that enables the consistent assessment of distributional impacts of climate policy, and thus allows cumulative impacts to be assessed across the broad suites of policies that comprise national adaptation programmes. The objective in so doing is to inform the selection of appropriate policy options and to identify situations where supplementary policy may be required to redress negative or inequitable impacts.

Drawing on a pilot impact assessment of the Scottish Climate Change Adaptation Programme, this paper discusses the rationale behind the development of a climate justice indicator set and presents a guide to the screening of large policy suites for potential cumulative impacts across communities of living (households) and working (private and public sectors). The methodological perspectives discussed can be useful in studies of climate change and social issues elsewhere, particularly in assessing the likely 'winners' and 'losers' of policy implementations.

Keywords Climate justice • Climate policy • Adaptation • Vulnerability • Impact assessment • Scotland

---

R.M. Dunk (✉)
Manchester Metropolitan University, School of Science and the Environment, Manchester M1 5GD, UK
e-mail: r.dunk@mmu.ac.uk

P. Satyal
School of International Development, University of East Anglia, Norwich NR4 7TJ, UK

M. Bonaventura
Crichton Carbon Centre, Crichton University Campus, Dumfries DG1 4ZZ, UK

© Springer International Publishing Switzerland 2016
W. Leal Filho et al. (eds.), *Implementing Climate Change Adaptation in Cities and Communities*, Climate Change Management, DOI 10.1007/978-3-319-28591-7_5

Introduction

Fairness and social justice concerns are now part of the debate on global environmental change (Walker 2012). This is particularly true of climate change, a multifaceted issue with various distributed impacts that can be considered under the frame of 'climate justice' (Lindley et al. 2011; Marino and Ribot 2012; Walker 2012; Walker and Bulkeley 2006).

When considering climate change, two broad categories of impacts can be identified. Firstly, the impacts of climate change itself, and secondly the impacts of policies intended to address climate change (Lindley et al. 2011; Marino and Ribot 2012). Impacts from climate change arise from changes such as increasing temperatures, changing precipitation patterns, rising sea levels, and increasing frequency or severity of events such as storms, floods, droughts, and heat waves. This gives rise to one dimension of climate justice: justice in the distribution of climate change impacts on different individuals and groups (Lindley et al. 2011). Policy impacts are associated with both the mitigation and adaptation responses to climate change. These give rise to two further dimensions of climate justice: justice in the distribution of the burdens and benefits of mitigation and adaptation (Lindley et al. 2011). In addition to these dimensions of outcome or distributive justice, there is also the dimension of procedural justice regarding access to and participation in the climate policy process (Walker 2012). These various dimensions of climate justice are often interlinked, both with each other and with other aspects of vulnerability (Fig. 5.1).

Outcome justice typically examines the distribution of burdens and benefits across society in order to assess whether or not pre-defined target groups are being unfairly exposed to burdens or deprived of benefits, where the target groups

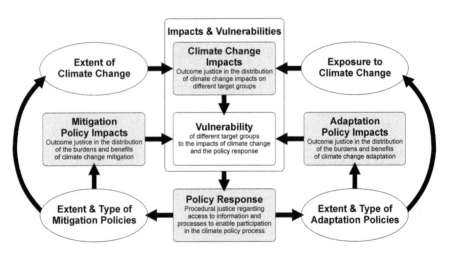

Fig. 5.1 Interlinkages between dimensions of climate justice and vulnerability in the policy response to climate change

of interest are most often those considered to be vulnerable or disadvantaged in some way (c.f. Skodvin et al. 2010). Thus, while there is considerable debate regarding climate justice in policymaking (e.g., Forsyth 2014), a central question in the development of just or fair policy remains how best to distribute burdens, as under a changing climate few may benefit, while many will be harmed and face new burdens (Duus-Otterström and Jagers 2012; Lindley et al. 2011; Marino and Ribot 2012).

In the international climate policy context, dominant approaches to distribute the response burden and determine a fair national contribution are based on responsibility for greenhouse gas (GHG) emissions, capability to respond, and vulnerability to climate change (e.g., Adger et al. 2006; Fussel 2010). While research has largely focused on international climate justice, particularly in relation to the consequences of climate change for the Global South, national and subnational responses also raise justice concerns, where the impacts of climate change and climate policies, and vulnerability to those impacts, varies not only between countries but also within countries. Climate justice in these intra-national contexts, and particularly in the Global North, is a relatively new area of policy research (Aitken et al. 2015). To date, research has considered the fair distribution of international climate finance within countries (e.g., Barrett 2014), taken sectorial perspectives (e.g., Furman et al. 2014; Popke et al. 2014; Schaffrin 2013; Walker and Day 2012), or examined climate justice at the urban level (e.g., Bulkeley et al. 2014). Thus, while it is widely recognised that the impacts of national and regional climate policy programmes will be distributed, there is an absence of complete information regarding the socio-economic and geographic patterning of such impacts and a lack of an integrated and practical set of climate justice indicators that can be applied in policy assessment.

This work seeks to address this gap, presenting a novel impact assessment methodology that enables the distributional impacts of national climate policy programmes to be evaluated across a wide range of target groups. Such information is necessary in order to (a) identify particularly vulnerable (or resilient) groups where the policy response may result in harm (or deliver benefits); (b) identify the opportunities where responding to climate change can also deliver substantial social co-benefits; and (c) develop targeted awareness campaigns (Chalmers et al. 2009; Johnson et al. 2009; Lindley et al. 2011). Furthermore, the successful implementation of climate policies rests on their political and public acceptability, where acceptability can reflect, *inter alia*, perceptions of fairness (e.g., Dreyer et al. 2015; Visschers and Siegrist 2012). We suggest that the direction of policy impact ('winner' or 'loser'), the relative magnitude of impact across different groups in society, and the change in the distributional gap between the status quo and the policy case ('better' or 'worse'), will all influence perceptions of fairness. Thus a climate policy assessment that extends the notion of 'winners' and 'losers' to a consideration of 'betters' and 'worses' (where a 'better' policy programme closes the distributional gap when compared to the status quo, while a 'worse' increases it) would also help to minimise policy implementation risk.

Drawing on a pilot impact assessment of the Scottish Climate Change Adaptation Programme (SCCAP; Scottish Government 2014a) we present a climate justice indicator set and guidance that enables the consistent assessment of distributional impacts of climate policy across Scottish society. The innovative aspects of the impact assessment methodology are twofold. Firstly, it assesses cumulative impacts across the broad suite of policies that comprise the national adaptation programme. Secondly, by considering a wide range of target groups it attempts to take a comprehensive view of those cumulative impacts across society. The purpose of such an assessment is to provide a more comprehensive understanding of distributed impacts in order to inform the selection of appropriate policy options and to identify situations where supplementary activities or policies may be required to redress negative or inequitable impacts, thereby minimising implementation risk.

We first briefly introduce the Scottish context before defining our target groups of interest and discussing the climate justice aspects and indicators that were selected for each of these groups. Following this, we present an overview of the impact assessment process, providing a rationale for our approach and outlining the three main stages of the process: identification of high priority policies, the impact assessment survey, and the assessment of cumulative impacts. Finally, we discuss key issues and present recommendations with regard to the conduct of similar impact assessments based on the experience gained in this pilot study.

The Scottish Context

Climate justice has a high profile in Scotland, where in recent years the Scottish Government has made a number of statements on climate justice, established a Climate Justice Fund to help the world's poorest communities, and hosted an international climate justice conference (BBC 2012; Scottish Government 2011, 2013).

In terms of a domestic response to climate change, The Climate Change (Scotland) Act (2009) sets a legally binding target to reduce GHG emissions by at least 80 % by 2050 against a 1990 baseline. In setting this target Scotland is taking action in line with the United Nations Framework Convention on Climate Change (UN 1992) and the conclusions of the Stern Review (Stern 2006) that (in the interests of fairness and equity) reflects both historical responsibility for emissions and capability to reduce them. Furthermore, Scotland has made a commitment to consider consumption based GHG emissions thereby addressing a key issue of responsibility in international GHG accounting (Davis and Caldeira 2010; Steininger et al. 2014). With respect to adaptation, the Act adopts a risk based approach, requiring an adaptation programme to be developed every 5 years to address the risks identified in successive UK Climate Change Risk Assessments (CCRAs). The first CCRA was published in January 2012, where the Report for Scotland highlighted 140 biophysical, ecological and socio-economic climate change risks (DEFRA 2012), and the first SCCAP, which includes 139 policies and proposals, was published in 2014 (Scottish Government 2014a).

Meeting the obligations set out in the Act has required, and will continue to require, rapid roll-out, successful implementation, and widespread uptake of a broad suite of measures. Success will depend on active participation across all parts of society, an outcome that rests in part on the acceptability of mitigation and adaptation measures, which in turn depends on whether these actions are perceived to be both fair and necessary. Conducting a climate justice impact assessment in this context can help to minimise implementation risk through assessing how burdens and benefits of climate policies are shared between different target groups, and whether such a sharing arrangement is fair or not.

Target Groups

A novel and important aspect of the process outlined here was the consideration of climate policy impacts across a wide range of target groups. Specifically, the explicit consideration of communities of living, working and place (Fig. 5.2):

Communities of Living were defined as those with shared interest, position (life-stage) or circumstance (experience), where this assessment focused on households.

Communities of Working were defined as groups of active practitioners, frequently sharing a trade or profession, where this assessment focused on the private and public sectors.

Communities of Place were defined as groups sharing a common geography.

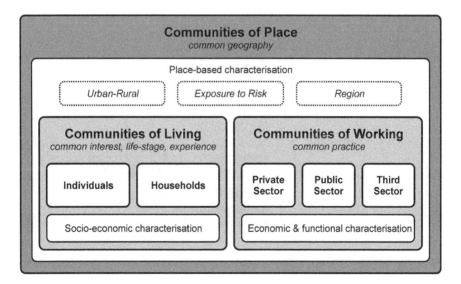

Fig. 5.2 Target groups within communities of living, working and place

Thus, the target groups of interest were households, the private sector and the public sector, where these groups were further characterised according to a range of place-based, socio-economic, and functional aspects and indicators, as reviewed in the following sections.

Communities of Place: Locational Aspects and Indicators

Location is a key factor in climate change risk, being a determinant of the likelihood of exposure to impacts and the magnitude of potential consequences. Thus, under the risk based approach to adaptation adopted by the Scottish Government, policy impacts will also be geographically distributed, where, for example, investment in infrastructure will reflect the level of climate change risk. With respect to exposure, flood risk will be greater in coastal locations, along river valleys, and in cities, while the effects of rising temperatures and heat waves will be exacerbated in cities due to the urban heat island effect (Lindley et al. 2011; SEPA undated). The magnitude of potential consequences is a function of the vulnerability and value (social, economic or environmental) of the systems at risk (SEPA undated). With respect to vulnerability, location plays a key role in defining social adaptive capacity, particularly in terms of general accessibility and access to services and other community facilities (Chalmers et al. 2009; Lindley et al. 2011; Fig. 5.3). With respect to value, the number of people at risk, the value of assets at risk, the costs of disruption, and the value of natural and cultural heritage, will also vary with location. To represent this geographical distribution of potential impacts we adopted two place-based aspects:

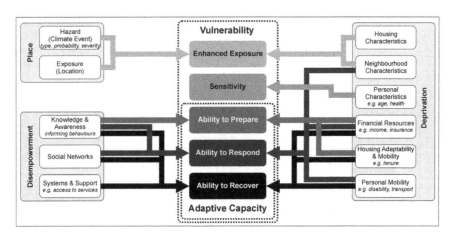

Fig. 5.3 Elements of vulnerability to climate change (after Chalmers et al. 2009; Lindley et al. 2011)

Table 5.1 Urban–rural indicators

Scottish sixfold Urban–rural classifications	Urban–rural indicators
Large urban areas (≥125,000 people)	Urban
Other urban areas (10,000–124,999 people)	
Accessible small towns (3000–9999 people, drive time ≤30 min to urban area)	Small towns
Remote small towns (3000–9999 people, drive time >30 min to urban area)	
Accessible rural (<3000 people, drive time ≤30 min to urban area)	Accessible rural
Remote rural (<3000 people, drive time >30 min to urban area)	Remote rural

Flood Risk: While recognising that exposure to flood risk is complex, we adopted four broad area type indicators to serve as a first order proxy for coastal and fluvial flood risk: *Islands*, *Coastal areas*, *Floodplains* and *Other inland areas*.

Urban–rural: Many of the locational elements of climate risk discussed above are related to position on the urban–rural transect. Some risks are exacerbated by the urban environment (pluvial flooding, heat waves), while factors influencing the magnitude of consequences vary with settlement population and degree of remoteness. For example, access to services, the number of people at risk, the economic costs of disruption, and the economic value of assets at risk, will all largely increase with settlement population. General accessibility and access to services will tend to decrease with increasing remoteness, while the social cost of disruption is likely to be higher in remote locations where access to alternative services and availability of alternative transport routes are limited (SEPA undated). To represent this variation we aggregated the Scottish sixfold Urban–rural classification (Scottish Government 2014b) into four groups: *Urban*, *Small town*, *Accessible rural*, and *Remote rural* (Table 5.1).

Communities of Living: Household Aspects and Indicators

As discussed above, a key interest with respect to the Household Impact Assessment was to enable an assessment of whether climate policies closed the distributional gap between climate winners and losers. In this context, two frameworks of climate vulnerability and disadvantage in the UK were particularly useful in identifying potential aspects and indicators. Firstly, in a report examining the differential impacts of climate change in the UK, Chalmers et al. (2009) defined socially vulnerable groups as those who were:

Living in places of risk—defined by location and the likelihood, type and severity of exposure to climate events.

Socially deprived—those who have pre-existing health problems, poor financial resources, poor quality of home or workplace, or a lack of mobility.

Disempowered—those with low awareness, or lacking access to social networks, systems and support.

Secondly, Lindley et al. (2011) defined climate disadvantage as a function of the likelihood and degree of exposure to a climate event and vulnerability to such events, where the major dimensions of vulnerability were:

Enhanced exposure—aspects of the physical environment which tend to exacerbate or mitigate the severity of climate events.

Sensitivity—personal biophysical characteristics that affect the likelihood that a climate event will have negative welfare impacts.

Social Adaptive Capacity—personal, environmental and social factors that enable individuals or communities to prepare for, respond to, and recover from climate events.

Figure 5.3 illustrates the relationships between these frameworks and the factors that influence vulnerability of individuals, households and communities to climate change impacts. Based on this model, we selected nine Household Aspects and 40 associated Indicators (Table 5.2). The locational aspects are discussed above, and the remaining household aspects are briefly discussed below.

Occupant Profile: Households can be characterised based on the composition of their inhabitants, where some household types are more vulnerable due to the presence of the young or elderly, or due to caring responsibilities or stretched resources. In a similar study examining the distributional consequences of UK

Table 5.2 Household aspects and indicators

Aspects	Indicators
Occupant Profile	Working age adults—no children, couples with children, single parent families, pensioners
Equality Groups	Disability and long term illness, gender, sexuality, race and ethnicity, religion and belief
Household Income	low income (deciles 1–3, £0–412 per week), medium income (deciles 4–7, £413–650 per week), high income (deciles 8–10, >£650 per week)
Level of Awareness	No awareness, some awareness, good awareness
Mode of Transport	Reliance on private transport, reliance on public transport, cycling, walking
Dwelling Type	Basement or ground-level flat, mid-level flat, upper-level flat, mid-terrace house, end-terrace house, detached house, non-permanent (e.g., caravans)
Tenure Type	Owned outright, mortgage owner occupied, private rented, social rented
Urban–rural	Urban, small town, accessible rural, remote rural
Flood-Risk	Coastal areas, Islands, flood plains, other inland areas

5 A Novel Impact Assessment Methodology for Evaluating Distributional Impacts... 83

Table 5.3 Occupant profile indicators

Fahmy et al. (2011) household types	Occupant profile indicators
Single pensioner	Pensioners
Pensioner couple	
Single parent families	Single parent families
Couple with children	Couple with children
Single working age adult	Working age adults
Couple no children	
Three or more adults	

mitigation policies to reduce household carbon emissions, Fahmy et al. (2011) used seven household types and we initially adopted the same classification. However, no distinction was made between pensioner households, or between childless working age adult households in the impact assessment survey, thus the indicator set was effectively reduced to a fourfold characterisation (Table 5.3).

Equality Groups are those groups for which there is a legal obligation to prevent discrimination. All groups considered within mandatory Equality Impact Assessments (EQIA) were considered here such that the assessment could also serve as a screening EQIA for the policy suite. Of these indicators, age, disability and long term illness, gender, and ethnicity all have a bearing on vulnerability (Lindley et al. 2011).

Household Income represented financial resources. We adopted a threefold banding based on household income deciles, where the values shown here are equivalised incomes before housing costs for 2012/13 (Scottish Government 2014c).

The *Level of Awareness* of climate change impacts and responses is a critical determinant of a households' adaptive capacity. In the context of this impact assessment it was considered essential to determine if the impact of a policy was sensitive to householder awareness in order to identify potential cases where targeted awareness campaigns could minimise burdens or maximise benefits.

Mode of Transport utilised to travel to work or access services was used as a proxy for personal mobility, where we note *Disability and long term illness* (included in *Equality Groups*) can be used as an indicator for a lack of personal mobility.

Dwelling Type primarily represented key housing characteristics that influence the degree to which householders are exposed to climate change impacts. For example, basement and street-level dwellings have enhanced exposure to flood risk, while upper floor flats have enhanced exposure to heat waves (Lindley et al. 2011). Secondarily, whether a dwelling is flatted, terraced or detached can also influence housing adaptability as modifications to the dwelling or environs may require changes to commonly held parts. Finally, information about climate change risks may not reach those in non-permanent dwellings, who may have recently moved to, or only temporarily be living in an area.

Tenure Type was used as the main proxy for housing adaptability and mobility, recognising also that some climate policies target specific tenure types (e.g., social housing), and noting that the leasehold system does not apply in Scotland.

Communities of Working: Aspects and Indicators

This section presents the characterisation of the private and public sector target groups and examines the different types of impact that climate policies may have on these groups.

Public and Private Sector Target Group Characterisation

For the private sector, we used the main sections of the UK Standard Industrial Classification (UKSIC 2007) to identify target groups based on area of operation or field of involvement (Table 5.4). However, using UKSIC (2007) to characterise public sector target groups would not have enabled separation of impacts at the desired level (e.g., national and local government, responder agencies, or health and social services). An alternative characterisation representing different levels and functions of government was therefore developed (Table 5.4).

Types of Impact

Both climate change and the policy response can exert a number of different types of impact across the economy. For example, the UKCIP Business Areas Climate Assessment Tool (BACLIAT) identifies six broad areas of impact that can be used

Table 5.4 Classification of private and public sector target groups

Private sector target groups	Public sector target groups
A: Agriculture, Forestry and Fishing	Agencies and Directorates of the Scottish Government
B, D and E: Mining, Quarrying and Utilities	Non-departmental Public Bodies/Regulatory Bodies
C: Manufacturing	Local Government
F: Construction	Fire and Rescue Service
G: Wholesale, Retail and Motor Trades	Police Service
H: Transport and Storage, including Postal	Education
I: Accommodation and Food Services	Health Services
J: Information and Communication	Social Services
K: Finance and Insurance	Defence
L: Property	
M: Professional, Scientific and Technical	
N: Business Administration and Support Services	
R, S, T and U: Arts, Entertainment, Recreation and Other Services	

to assess climate risks and opportunities at the level of the firm or sector (UKCIP 2011):

- Markets—changing demand for goods and services.
- Process—impacts on production processes and service delivery.
- Logistics—impacts on supply chain, utilities and transport infrastructure.
- People—implications for employees and customers.
- Premises—impacts on building and landscape design, construction, maintenance and management.
- Finance—impacts on investment, insurance, costs, and liabilities.

Similarly, the World Resources Institute Corporate Ecosystems Services Review identifies five major categories of impacts on organisations arising from ecosystem change (Hanson et al. 2012):

- Operational—relating to the day-to-day activities, expenditures and processes of a company.
- Market and Product—relating to product and service offerings, customer preferences, and other market factors.
- Regulatory and Legal—relating to laws, government policies and court actions.
- Financing—relating to the cost and availability of capital from investors.
- Reputational—relating to brand, image, or relationship with stakeholders.

However, this study was primarily concerned with the impact of the policy, as opposed to the impact of climate change (or associated ecosystem changes). Thus using these classifications as a starting point, we identified four major types of policy impact (Table 5.5). Financial, Compliance, and Employment and Skills impacts were applicable to both the private and public sectors. However, there is a notable difference between Market Dynamics and Service Demand, where an increase in demand for goods and services in the private sector would typically be viewed as positive, whilst an increase in demand for public sector services could be viewed as negative.

Other Distributed Aspects for the Private and Public Sector

We also considered the potential for distributed impacts across different locations and, for the private sector only, according to business size (Table 5.6).

The ability of a business to respond to new policy and legislative changes can be expected to relate (at least in part) to the size of the business. For example, compliance with some policies will incur a time or financial cost, or require particular expertise, where small businesses are resource constrained (Litvak 1992). Conversely, small businesses may be better positioned to exploit new opportunities as they may be more flexible, more open to taking risks, and more innovative (e.g., Bras 2006; OECD 2000; Spivack 2013). Here we adopted a fourfold categorisation of business size based on the widely used European Union (2003) definitions (Table 5.6).

Table 5.5 Types of climate policy impact on the private and public sectors

Impacts on the private sector	Impacts on the public sector
Market Dynamics: how the policy cluster might impact the market	*Service Demand*: how the policy cluster might impact service provision
Emergence of new or demise of existing markets, growth or contraction in demand for new and/or existing goods and services, changes in production processes, service delivery, supply chain, increased/decreased risk of access to raw materials	Emergence of new or demise of existing services, growth or contraction in demand and delivery of new and/or existing services, changes in service delivery
Financial: whether the policy cluster has any implications for financial risk	
Balance sheet impacts as a result of changes in the status of assets and liabilities, profit and loss adjustments as a result of increased/decreased costs and/or revenue enhancement through new goods and services, better access to capital, or, conversely, more competitive funding criteria, additional risks with corresponding increases in insurance premiums, etc	
Compliance: whether the policy cluster would create or remove any compliance obligations	
Any increase or decrease in reporting requirements and/or support staffing required to comply with the policy. [NB—For the Public Sector this refers to compliance of the public sector with the policy—the administration of policies targeting Households or Business would be a Service Demand impact.]	
Employment and Skills: how the policy cluster might impact employment prospects or skills development	
Creation of new or reduction of existing employment opportunities, shifts in employment demographics, shifts in geographic distribution of employment, requirement for new skills or re-skilling of the workforce	

Table 5.6 Other private and public sector aspects and indicators

Aspects	Indicators
Business size (private sector only)	Micro businesses (<10 employees), small (<50 employees), medium (<250 employees), large (≥250 employees)
Urban–rural	Urban, small town, accessible rural, remote rural
Flood-Risk	Coastal areas, Islands, flood plains, other inland areas

The Impact Assessment Process

Rationale for Approach

The size and complexity of a national adaptation programme makes an assessment of cumulative impacts a significant challenge where, for this pilot study, it was considered infeasible to conduct a full assessment of the 139 policies and proposals in the SCCAP. Furthermore, sponsorship from senior civil servants was essential to gain access to policy analysts and to champion positive engagement in the process, thus the sponsors needs and concerns had to be addressed. In particular, it was considered essential to make efficient use of policy analysts time. We therefore developed a three-stage process:

Stage 1: Identification of High Priority Policies.
Stage 2: The Impact Assessment Survey and Validation Review.
Stage 3: Assessment of Cumulative Impacts.

Each of these stages are described below.

Stage 1: Identification of High Priority Policies

All policies and proposals were screened against criteria developed and agreed with the sponsors (Fig. 5.4) to identify high priority policies, defined here as those specific to Scotland with a high risk of significant impact(s) on household groups, or industry sectors, or public sector bodies, or across different locations.

While the focus of this assessment was on the potential for distributed impacts, the screening process also identified policies with potential issues of procedural justice (Fig. 5.4)—that is fairness in providing information and opportunities necessary for people to participate in decisions about their environment. In the context of climate change adaptation, this could be considered to extend to the provision of information and opportunities necessary for people to prepare for, respond to, and recover from climate change impacts (Fig. 5.3). These policies were therefore flagged as requiring particular consideration of how to ensure information was readily accessible and clearly communicated to vulnerable and hard to reach groups.

Once identified, high priority policies were clustered into groups that targeted a similar area of activity and were considered likely to have broadly similar impacts (Fig. 5.4). This allowed policies to be evaluated collectively, thereby minimising the risk of repetitive questioning during the assessment process. Policy analysts then reviewed the prioritisation for each policy area, and approved the final selection of high priority policies and policy clusters taken forward to the impact assessment survey.

Stage 2: The Impact Assessment Survey and Validation Review

A survey instrument comprising a self-administered questionnaire was selected as it allowed for flexibility in the timing of response. In designing the questionnaire, the aim was for the questions to be as simple as possible to answer while allowing complexities to be identified for further exploration in the validation review. As such, all questions included a closed-ended element to evaluate the impact and an open-ended element that elicited a brief explanation of the response. The final questionnaire was reviewed by the sponsors prior to distribution to the policy analysts for each cluster.

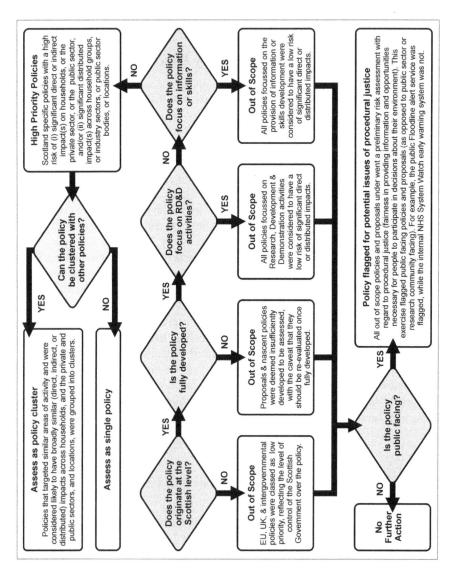

Fig. 5.4 Screening decision tree for identification of high priority policies

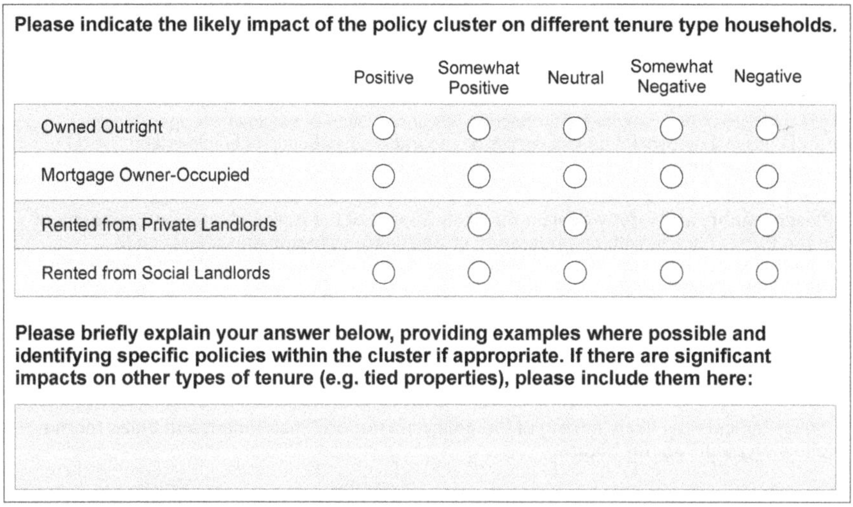

Please indicate the likely impact of the policy cluster on different tenure type households.					
	Positive	Somewhat Positive	Neutral	Somewhat Negative	Negative
Owned Outright	○	○	○	○	○
Mortgage Owner-Occupied	○	○	○	○	○
Rented from Private Landlords	○	○	○	○	○
Rented from Social Landlords	○	○	○	○	○

Please briefly explain your answer below, providing examples where possible and identifying specific policies within the cluster if appropriate. If there are significant impacts on other types of tenure (e.g. tied properties), please include them here:

Fig. 5.5 Example question layout from the household impact assessment survey

The Household Impact Assessment investigated the potential for distributed impacts across the nine household aspects listed in Table 5.2. All impacts were evaluated on a five-point scale from positive to negative. An example question layout is shown in Fig. 5.5.

The Private and Public Sector Impact Assessments investigated the impact types listed in Table 5.5 across the industry sectors and public bodies listed in Table 5.4. Financial, Compliance, and Employment and Skills impacts were evaluated on a five-point scale from positive to negative. To avoid any ambiguity, Market Dynamics and Service Demand impacts were evaluated on a five-point scale from increase to decrease. An example question layout is shown in Fig. 5.6. Respondents were also asked to evaluate impacts across the distributed aspects listed in Table 5.6, where the question style was the same as that used in the Household Impact Assessment.

Following preliminary analysis of the responses, a validation review was conducted. All respondents were provided with a copy of the results and the opportunity to modify their response. Telephone interviews were conducted to address any issues of inconsistency (between qualitative and quantitative responses provided in the survey and/or information provided in public consultation documents and other impact assessments) or uncertainty (where qualitative responses were lacking in support of a quantitative response, or where respondents highlighted uncertainties or difficulties in ascribing a particular impact type). The survey responses were then updated and re-analysed accordingly.

Fig. 5.6 Example question layout from the public sector impact assessment survey

Stage 3: Assessment of Cumulative Impacts

For the Household Impact Assessment, cumulative impacts across each Aspect were determined by assigning a numerical value to the impact of each policy cluster on each indicator (negative $= -2$, neutral $= 0$, positive $= 2$) and taking an arithmetic average of the impacts exerted by all clusters. However, this simple analysis provided only limited information about the impacts exerted on particular (vulnerable) households. To address this, impacts were also aggregated across Aspects to assess impacts on Household Profiles.

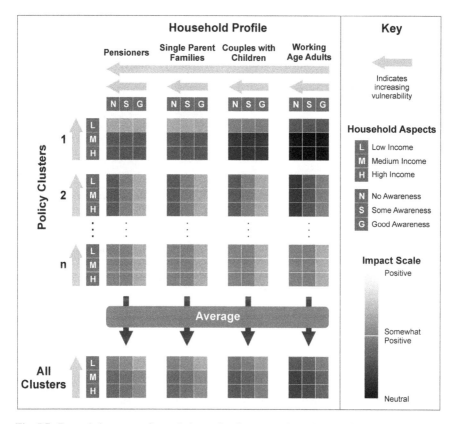

Fig. 5.7 Example heat map of cumulative policy impacts on household profiles

For example, aggregating across Occupant Profile, Level of Awareness and Household Income, created a set of 36 Household Profiles (from Pensioners with No awareness and Low income to Working age adults with Good awareness and High income). Once aggregated, cumulative impacts on Household Profiles were evaluated and the results presented as a heat-map to provide a visual summary for policymakers and other stakeholders. Figure 5.7 presents an example (theoretical) heat-map for the 36 Household Profiles described above.

Figure 5.8 presents a more detailed example of impact aggregation and its application by considering impacts on Pensioner Households (a vulnerable group) for two theoretical policy cases.

In Case I, the policy exerts a positive impact on Pensioners, and is distributed across both Level of Awareness and Household Income. The aggregated impacts are positive for all Pensioner Households, becoming increasingly positive as income decreases and awareness increases. Thus, in this case the impact distribution could be considered 'just' (positive impacts increase with increasing financial vulnerability, thus the policy acts to reduce the financial distributional gap), where a

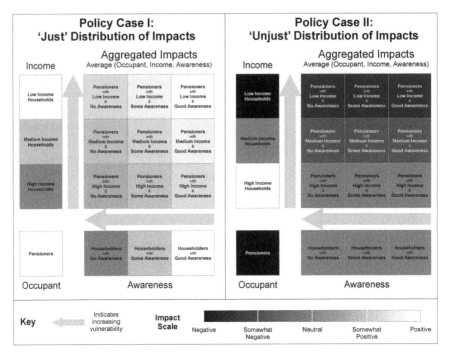

Fig. 5.8 Impacts on pensioner households for two illustrative policy cases

targeted awareness campaign would ensure all Pensioner Households maximise potential benefits.

In Case II, the policy exerts a negative impact on Pensioners and is distributed across Household income. The aggregated impacts are neutral for Pensioners with High income, but become increasingly negative as household income decreases. As the policy is neutral with respect to Level of Awareness, there is no option to minimise negative impacts through a targeted awareness campaign. Thus in this case the impact distribution could be considered 'unjust' (negative impacts increase with increasing financial vulnerability, thus the policy acts to widen the financial distributional gap) and would require redressing.

For the Private and Public Sector Impact Assessments, we evaluated both cumulative impacts of multiple policy clusters on each private sector industry or public body and aggregated impacts of policy clusters across the private or public sector as a whole. Results were presented as ranked heat-maps to provide a visual summary for policymakers and other stakeholders that enabled clear identification of those policies which exerted the greatest impact and those industry sectors or public bodies that were most impacted (Fig. 5.9, Table 5.7). For policy clusters (rows) the final column presents the aggregated impact of each cluster across All Sectors or All Bodies. For private sector industries or public bodies (columns), the

5 A Novel Impact Assessment Methodology for Evaluating Distributional Impacts... 93

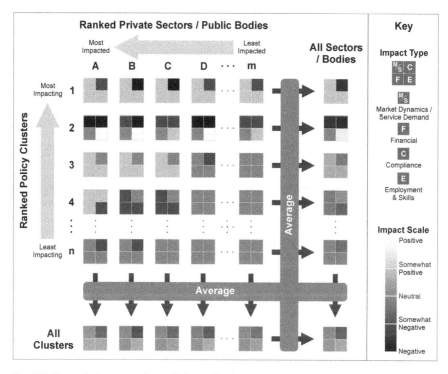

Fig. 5.9 Example heat map of cumulative policy impacts aggregated across the private or public sector

Table 5.7 Ranking policy clusters and private sectors or public bodies

Policy clusters	Private sectors or public bodies
A: The no. of non-neutral Market Dynamics/ Service Demand, Financial, Compliance and Employment and Skills impacts exerted by the Cluster divided by no. of possible impacts (4m)	A: The no. of non-neutral Market Dynamics/ Service Demand, Financial, Compliance and Employment and Skills impacts experienced by the Sector/Body divided by no. of possible impacts (4n)
B: The no. of Sectors/Bodies impacted by the Cluster divided by no. of Sectors/Bodies (m)	B: The no. of Clusters exerting an impact on the Sector/Body divided by no. of Clusters (n)
Max (A + B) = Rank 1 *to* Min (A + B) = Rank n	Max (A + B) = Rank 1 *to* Min (A + B) = Rank m

final row presents the cumulative impact experienced by the industry sector or public body from All Clusters. The impacts shown in the final row and final column presents cumulative impacts from all Clusters aggregated across all Sectors or Bodies.

Key Issues and Recommendations

The following section identifies some key issues and presents recommendations with regard to the conduct of similar impact assessments based on the experience gained in the pilot assessment of the SCCAP.

Screening Criteria

During the Prioritisation Review, a number of policies were placed out of scope as they stemmed from EU level policy. Given the importance of the issues addressed by these policies (e.g., river basin management plans, flood risk management) to national adaptation we suggest the screening criteria should be revised. In the Scottish context, high priority policies should be redefined as those that pertain to devolved powers (i.e., where transposition and/or implementation fall under the remit of the Scottish Government) in addition to those that originate at the Scottish level.

Survey Design

A number of policy clusters with a focus on land use and the natural environment were reported as having no impact on communities of living and/or working. During the validation review it became clear that this reflected the applicability of the selected aspects and indicators, as opposed to being a true 'no impact' case. As nature provides many services to society, we suggest the inclusion of an additional section in the survey addressing impacts on nature could be of benefit. Potential indicators include health indicators and ecosystem service indicators based on the Natural Capital Asset Index (SNH 2012) or the National Ecosystem Assessment (UNEP-WCMC 2011).

Reference Points

A key reference point for this type of assessment is the base-case against which impacts are evaluated, where the preferred option would be the 'no adaptation' case. However, given the nature of the SCCAP, which brought together a diverse range of pre-existing, amended and new policies, this was somewhat difficult to define. The analysts were therefore given a choice of base-case scenario, where the majority showed a preference for evaluation against the policy status quo, with some evaluating impacts against the stated aims of the policy. The implication of the former choice is that some burdens may be excluded on the basis that they are not new. The implication of the latter choice is that the policy may deliver significant benefits when compared to the no-adaptation case, yet be assessed as having negative impacts if it falls short of the stated aims. A second key reference point is the period over which impacts are evaluated, where financial impacts on the private and public sectors were particularly sensitive to this factor. This reflected the relationship between the financial burden of adaptation and the rate of adaptation, where proactive investment in adaptive measures in the short term (burden) mitigates the risk of incurring greater costs in the future (benefit). It is noted that it is essential that all impacts are evaluated against the same base-case and over the same time period prior to assessment of cumulative impacts.

Uncertainty and the Need for Validation

Multiple respondents highlighted significant uncertainty in their assessment of potential impacts due to a lack of information regarding both the impacts and the views/resilience of the target groups with respect to those impacts. Indeed, one respondent described the current position, where knowledge of impacts within the policy making process is confined to the views of the policy analysts, as 'the beginning of the journey', where extension of this type of impact assessment to a wider stakeholder community could prove beneficial in stimulating debate and informing policy development. Perhaps one of the clearest examples of this is that the assessment of locational impacts tended to be based on an assumption that remote communities have a higher resilience to climate change impacts, and therefore have a higher tolerance to (greater willingness to accept and/or ability to cope with) disruption associated with extreme weather events. The duration of any disruption was also identified as a critical issue, where the tolerance threshold of remote communities was viewed to be higher than that of urban communities. We suggest that it would be of particular benefit to explore these assumptions further and build a greater understanding of the extent (or lack) of resilience across different communities of place.

Timing of the Impact Assessment

Some policy analysts suggested this type of assessment would be of most benefit during policy development, whilst others identified significant uncertainties in potential impacts, and highlighted the need to reassess policies in the post-implementation phase. It is our view that consideration of impacts on society is of benefit at all stages in the policy lifecycle, where the approach presented here provides the basis for a consistent assessment of the potential for distributed impacts during climate policy development, appraisal and evaluation.

Summary

This paper has presented a climate justice toolkit comprising an indicator set and an impact assessment methodology for the screening and evaluation of large policy suites for cumulative distributed impacts across households and the private and public sectors. Building on the experience gained from a pilot impact assessment of the Scottish Climate Change Adaptation Programme, we have identified two key areas for future work. Firstly, the impact assessment could be usefully extended to include policy impacts on the many services nature provides to society. Secondly, the impact assessment process should be extended to solicit the views of the wider stakeholder community. This is required in order to validate (or repudiate) policymakers assumptions and assessment of impacts, and would also prove beneficial in stimulating debate and informing policy development. While this study has focused on adaptation in Scotland, the methodological perspectives discussed in this paper can be useful in studies of climate change and social issues elsewhere, where the indicators can be adapted to reflect local issues of concern or extended to include consideration of climate mitigation policies.

Acknowledgements The authors would like to acknowledge support from ClimateXChange, Scotland's Centre of Expertise on Climate Change, which is funded by the Scottish Government.

References

Adger WN, Paavola J, Huq S (2006) Toward justice in adaptation to climate change. In: Adger WN, Paavola J, Huq S, Mace M (eds) Fairness in adaptation to climate change. MIT Press, Cambridge, MA

Aitken M, Christman B, Bonaventura M, van der Horst D, Holbrook J (2015) Climate justice begins at home: conceptual, pragmatic and transformative approaches to climate justice in Scotland. Scott Aff (in press)

Barrett S (2014) Subnational climate justice? Adaptation finance distribution and climate vulnerability. World Dev 58:130–142. doi:10.1016/j.worlddev.2014.01.014

Bras B (2006) Infusing sustainability in small- and medium-sized enterprises. In: Abraham MAA (ed) Sustainability science and engineering: defining principles. Elsevier, Amsterdam, pp 443–466. doi:10.1016/S1871-2711(06)80029-2

British Broadcasting Corporation (BBC) (2012) Alex Salmond in climate justice call to world leaders. http://www.bbc.co.uk/news/uk-scotland-scotland-politics-16584424. Accessed 16 Jan 2012

Bulkeley H, Edwards GAS, Fuller S (2014) Contesting climate justice in the city: examining politics and practice in urban climate change experiments. Glob Environ Chang 25:31–40. doi:10.1016/j.gloenvcha.2014.01.009

Chalmers H, Anderson M, Houghton T, Maiden T, Parham S, Pilling A (2009) Differential social impacts of climate change in the UK. SNIFFER, Edinburgh. http://www.sniffer.org.uk/files/7513/4183/8010/UKCC22_LiteratureReview_web.pdf

Davis S, Caldeira K (2010) Consumption-based accounting of CO_2 emissions. Proc Natl Acad Sci USA 107:5687–5692. doi:10.1073/pnas.0906974107

Department for Environment, Food and Rural Affairs (DEFRA) (2012) A climate change risk assessment for Scotland. http://www.defra.gov.uk/environment/climate/government/risk-assessment/

Dreyer SJ, Teisl MF, McCoy SK (2015) Are acceptance, support, and the factors that affect them, different? Examining perceptions of U.S. fuel economy standards. Transp Res Part D: Transp Environ 39:65–75. doi:10.1016/j.trd.2015.06.002

Duus-Otterström G, Jagers SC (2012) Identifying burdens of coping with climate change: a typology of the duties of climate justice. Glob Environ Chang 22:746–753. doi:10.1016/j.gloenvcha.2012.04.005

European Union (2003) Commission recommendation of 6 May 2003 concerning the definition of micro, small and medium-sized enterprises (2003/361/EC). Off J Eur Union L124:36–41

Fahmy E, Thumim J, White V (2011) The distribution of UK household CO_2 emissions: interim report. Joseph Rowntree Foundation, York. http://www.jrf.org.uk/sites/files/jrf/carbon-reduction-policy-full.pdf

Forsyth T (2014) Climate justice is not just ice. Geoforum 54:230–232. doi:10.1016/j.geoforum.2012.12.008

Furman C, Roncoli C, Bartels W, Boudreau M, Crockett H, Gray H, Hoogenboom G (2014) Social justice in climate services: engaging African American farmers in the American South. Clim Risk Manag 2:11–25. doi:10.1016/j.crm.2014.02.002

Fussel H (2010) How inequitable is the global distribution of responsibility, capability, and vulnerability to climate change: a comprehensive indicator-based assessment. Glob Environ Chang 20:597–611. doi:10.1016/j.gloenvcha.2010.07.009

Hanson C, Ranganathan J, Iceland C, Finisdore J (2012) The corporate ecosystem services review: guidelines for identifying business risks and opportunities arising from ecosystem change. Version 2.0. World Resources Institute, Washington, DC

Her Majesty's Stationary Office (2009) Climate Change (Scotland) Act (2009). HMSO, London

Johnson V, Simms A, Cochrane C (2009) Tackling climate change: reducing poverty. New Economics Foundation, London. http://www.neweconomics.org/publications/entry/tackling-climate-change-reducing-poverty

Lindley S, O'Neill J, Kandeh J, Lawson N, Christian R, O'Neill M (2011) Climate change, justice and vulnerability. Joseph Rowntree Foundation, York. http://www.jrf.org.uk/publications/climate-change-justice-and-vulnerability

Litvak I (1992) Public policy and high technology SMES: the government embrace. Can Public Adm 35:22–38. doi:10.1111/j.1754-7121.1992.tb00677.x

Marino E, Ribot J (2012) Adding insult to injury: climate change and the inequities of climate intervention. Glob Environ Chang 22:323–328. doi:10.1016/j.gloenvcha.2012.03.001

Office for National Statistics (ONS) (2007) UK Standard Industrial Classification (UKSIC 2007). http://www.ons.gov.uk/ons/guide-method/classifications/current-standardclassifications/standard-industrial-classification/index.html

Organisation for Economic Co-operation and Development (OECD) (2000) Small and medium-sized enterprises: local strength, global reach. Policy Brief, June 2000. http://www.oecd.org/regional/leed/1918307.pdf

Popke J, Curtis S, Gamble DW (2014) A social justice framing of climate change discourse and policy: adaptation, resilience and vulnerability in a Jamaican agricultural landscape. Geoforum (in press). doi:10.1016/j.geoforum.2014.11.003

Schaffrin A (2013) Who pays for climate mitigation? An empirical investigation on the distributional effects of climate policy in the housing sector. Energy Build 59:265–272. doi:10.1016/j.enbuild.2012.12.033

Scottish Government (2011) First Minister Alex Salmond Speech to Communist Party Central School, China. Adam Smith and Climate Justice. http://www.gov.scot/News/Speeches/china-cp. Accessed 6 Dec 2011

Scottish Government (2013) Scotland driving forward climate justice. http://news.scotland.gov.uk/News/Scotland-driving-forward-Climate-Justice-502.aspx. Accessed 10 Oct 2013

Scottish Government (2014a) Climate Ready Scotland—Scottish Climate Change Adaptation Programme. http://www.gov.scot/Publications/2014/05/4669

Scottish Government (2014b) Scottish Government urban rural classification 2013–2014. http://www.gov.scot/Publications/2014/11/2763

Scottish Government (2014c) Poverty and income inequality in Scotland: 2012/13. http://www.gov.scot/Resource/0045/00454875.pdf

Scottish Environment Protection Agency (SEPA) (undated) National Flood Risk Assessment—methodology. http://www.sepa.org.uk/media/99914/nfra_method_v2.pdf

Scottish Natural Heritage (SNH) (2012) Scotland's Natural Capital Asset (NCA) Index, 2012 version. http://www.snh.gov.uk/docs/B814140.pdf

Skodvin T, Gullberg AT, Aakre S (2010) Target group influence and political feasibility. J Eur Public Policy 17:854–873. doi:10.1080/13501763.2010.486991

Spivack RN (2013) Small business participation in the advanced technology program research alliances. J Innov Entrep 2:19. doi:10.1186/2192-5372-2-19

Steininger K, Lininger C, Droege S, Roser D, Tomlinson L, Meyer L (2014) Justice and cost effectiveness of consumption-based versus production-based approaches in the case of unilateral climate policies. Glob Environ Chang 24:75–87. doi:10.1016/j.gloenvcha.2013.10.005

Stern NH (2006) Stern review: the economics of climate change. HM Treasury, London

UKCIP (2011) BACLIAT vulnerability assessment. http://www.ukcip.org.uk/wizard/future-climate-vulnerability/bacliat/

United Nations (UN) (1992) United Nations framework convention on climate change. FCCC/INFORMAL/84 GE.05-62220 (E) 200705

United Nations Environment Programme World Conservation Monitoring Centre (UNEP-WCMC) (2011) UK National Ecosystem Assessment. UNEP-WCMC, Cambridge. http://uknea.unep-wcmc.org/Resources/tabid/82/Default.aspx

Visschers VHM, Siegrist M (2012) Fair play in energy policy decisions: procedural fairness, outcome fairness and acceptance of the decision to rebuild nuclear power plants. Energy Policy 46:292–300. doi:10.1016/j.enpol.2012.03.062

Walker G (2012) Environmental justice: concept, evidence and politics. Routledge, Abingdon

Walker G, Bulkeley H (2006) Geographies of environmental justice. Geoforum 37:655–659. doi:10.1016/j.geoforum.2005.12.002

Walker G, Day R (2012) Fuel poverty as injustice: integrating distribution, recognition and procedure in the struggle for affordable warmth. Energy Policy 49:69–75. doi:10.1016/j.enpol.2012.01.044

Chapter 6
Climate Variability and Food Security in Tanzania: Evidence from Western Bagamoyos

Paschal Arsein Mugabe

Abstract Achieving food security in its totality continues to be a challenge not only for the developing nations, but also for the developed world. The difference lies in the magnitude of the problem in terms of its severity and proportion of the population affected. In developed nations the problem is alleviated by providing targeted food security interventions, including food aid in the form of direct food relief, food stamps, or indirectly through subsidized food production. Similar approaches are employed in developing countries but with less success. The discrepancy in the results may be due to insufficient resource base, shorter duration of intervention, or different systems most of which are inherently heterogeneous among other factors.

In this study; results of regression model on production entitlement suggest that climate variability has relatively minor effects on food production. It shows moderate positive relationship but not significant; where temperature has more effects than that of rainfall on production. However the difference is statistically insignificant. The relatively minor impact of weather variations on food production, combined with the analysis of other qualitative analysis such as household survey, observation, discussions, participatory techniques and thematic interviews in Western Bagamoyo district, Coast region in Tanzania reveal how extreme climatic events affect rural food security entitlements of production, labor, exchange and trade. It also reveals households adaptation strategies implemented by the communities during droughts/floods which serve as a foundation for planning responses to future climate variability and change. Results of this study suggest that food security in Tanzania where droughts and floods are expected to become more severe due to climate change could be enhanced by adoption of Climate Smart Agriculture (CSA) practises to increase productivity, adaptation and mitigation of Green House gases (GHGs). It was revealed that agriculture as their main source of income contributes 69.7 % of income to household's food security.

P.A. Mugabe (✉)
Institute for Environment and Sanitation Studies, College of Basic and Applied Sciences, University of Ghana, Legon, Accra, Ghana
e-mail: paschalmugabe@hotmail.com

Keywords Climate variability • Food security • Adaptation • Mitigation • Climate smart agriculture

Introduction

Climate change has been defined by the Intergovernmental Panel on Climate Change as "any change in climate over time, whether due to natural variability or as a result of human activity" (Watson and Albritton 2001: 22). Climate variability refers to variations in the climate on all temporal and spatial scales beyond that of individual weather events (Christensen et al. 2007). Food security is the state achieved when food systems operate such that 'all people, at all times, have physical and economic access to sufficient, safe, and nutritious food to meet their dietary needs and food preferences for an active and healthy life' (FAO 1996).

In the wake of the continuing debate on the effects of climate change on households' wellbeing, this study considers the effects of short-term weather variations (climate variability), as an indicator of climatic change on food security of the rural households in Bagamoyo district (Western Bagamoyo), Tanzania. This paper highlights the extent to which environmental change affects food security, particularly climate variability. Particular concerns are on food entitlement which is an essential guide of this study. The objective is to examine the extent to which climate variability affects food security in Tanzania. The aim is to learn adaptation measures adopted by rural communities during shocks, and their weaknesses for policy recommendations.

This paper attempts to address these challenges in a number of ways. The contribution to the existing literature is threefold. First of all, the base is estimations on high-quality household survey data on Tanzania, covering a representative sample of individuals, households and groups from the Western part of Bagamoyo (more than 600 individuals). Merging these data with a high-resolution of climatic and agricultural data of the district, which enables to test different climate measures, and use both temperature and rainfall as the main explanatory variables. Secondly, the unquestionable advantage of Bagamoyo setting to which estimations are applied is its diversity of agroclimatic conditions and ecological zones, including different types of forest and coastal environments, which proves to be essential for analysing the impact of climate on livelihoods. Finally, not only examine whether there is any impact of climate on agriculture, but also propose learn adaptation techniques and mechanism through which adverse weather conditions induce human health.

Climate Change/Variability and Food Security in Sub-Saharan Africa with Reference to Tanzania

Tanzania covers a total area of 945,000 km² with the mainland covering 939,702 km². The land area of the mainland is 881,289 km² while 58,413 km² are inland lakes. The available land for cultivation is 40 million hectares and cultivated land is about 5.2 km². Forests and woodland occupy 50 % of the total area and 25 % is wildlife reserves and national parks. The coastline extends 800 km from 40S to 100S. Except for the coastal belt most of the country is part of the Central African plateau lying between 1000 and 3000 m above sea level. It shares borders with eight countries. Its neighbours include Kenya and Uganda in the North, Rwanda, Burundi and Democratic Republic of Congo in the West, Zambia and Malawi in the South West and Mozambique in the South. Mainland Tanzania borders the main water bodies of Africa. To the east is the Indian Ocean, to the north Lake Victoria, to the west Lake Tanganyika and to the south-west Lake Nyasa. Mainland Tanzania also has the highest point in Africa. The snow caped Mount Kilimanjaro is 5950 m high (URT 2007).

Climate ranges from tropical to temperate in the highlands. Country wide, the mean annual rainfall varies from 500 to 2500 mm. The average duration of the dry season is 5–6 months. Average annual precipitation over the entire nation is 1042 mm. Average temperatures range between 24 and 34 °C, depending on location. Within the plateau, mean daily temperatures range between 21 and 24 °C. Natural hazards include both flooding and drought. Within the country, altitude plays a large role in determining rainfall pattern, with higher elevations receiving more precipitation. Generally speaking, the total amount of rainfall is not very great. Only about half the country receives more than 762 mm annually (Mwandosya et al. 1998). Tanzania's precipitation is governed by two rainfall regimes. Bimodal rainfall, comprised of the long rains of Masika between March and May and short rains *Vuli* between October and December, is the pattern for much of the northeastern, northwestern (Lake Victoria basin) and the northern parts of the coastal belt. A unimodal rainfall pattern, with most of the rainfall during December–April, is more typical of most of the southern, central, western, and southeastern parts.

Generally the climate of Tanzania including Bagamoyo District is characterized by two main rain seasons namely the long rains and the short rains which are associated with the southward and northwards movement of the (Intercontinental Tropical Continental Zone (ITCZ). The long rains (*Masika*) begin in the mid of March and end at the end May, while the short rains (*Vuli*) begin in the middle of October and continues to early December (Lyimo et al. 2013). Most of the rainfall is convective in nature and distinctly organized. The study by Mongi et al. (2010) reveals that late rainfall onset and early withdraw (cessation) are becoming common in the study area. Such situation has also been reported to be common in most parts of Tanzania (Mongi et al. 2010). Analysis of annual rainfall time series for Bagamoyo for a period of 87 years (1920–2007) indicates a normal trend ($R^2 = 9E-06$) with

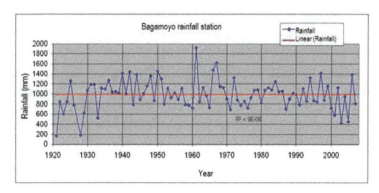

Fig. 6.1 Temporal annual variation of rainfall, Bagamoyo rainfall station. *Source*: Tanzania Meteorological Agency (TMA 2008a, b)

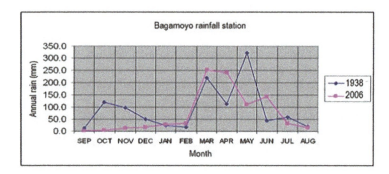

Fig. 6.2 Mean monthly rainfall variation. *Source*: TMA 2008

high inter-annual variability. Therefore, there is pronounced variability over time with relatively constant pattern. It is evident that there is no overall trend of annual rainfall in Bagamoyo. Annual rainfall appears to have fluctuated about a mean of 1000 mm. However, it is noted that as from 1999 to 2008 rainfall has been decreasing over time (Lyimo et al. 2013) (Figs. 6.1 and 6.2).

Poverty, Tanzania is one of poor countries in the world with 42 % of the total population and 50 % of the rural population live below the poverty line and with 20 % of the entire population surviving on <US$1 per day. It is the poor that are the most vulnerable to climate variability effects because they are particularly sensitive to, and have the least capacity to adapt to such effects. Within the rural zones, there are limited livelihood sources. As a result, the majority of the population relies on rain-fed agricultural activities and/or exploitation of natural resources to generate income streams and maintain livelihoods thus creating a vicious circle of increasing vulnerability to climate variability.

There is a strong link between climate and Tanzanian livelihoods because Tanzania depends heavily on rain-fed agriculture making rural livelihoods and food consumption highly vulnerable to climate variability such as shifts in growing season conditions, storage systems, and infrastructure. For example, from 1996 to 2003, there has been an observed decline in rainfall of 50–150 mm per season (March–May) and corresponding decline in long-cycle crops (e.g., slowly maturing varieties of sorghum and maize) across most of eastern Africa (Funk et al. 2005). It is from the light of the problems that this project seeks to examine the effects of climate variability on food security of the livelihood people at Bagamoyo district. Attainment of the Millennium Development Goals, particularly the first goal of eradicating extreme poverty and hunger, in the face of climate change will therefore require science that specifically considers food insecurity as an integral element of human vulnerability within the context of complex social, economic, political and biophysical systems, and that is able to offer usable findings for decision-makers at all scales (Boko et al. 2007).

Adaptation Techniques Practiced

Adaptation (CSA practices) measures reported in many parts of the country are short duration crop varieties; vegetable production; integrated soil fertility management; soil and water conservation techniques, crop associations; use of animal manures, manure and fertilizer mix, use of pesticides; composting; restitution of crop residues to the soil; restoration of degraded lands; agroforestry; association of crops with legume tree; assisted natural regeneration; use of lowlands; small scale irrigation; agricultural mechanization and cloud seeding. In the livestock sector, they include use of high yielding livestock breeds tolerant to stress, and poultry production. In the sub-humid zone, the system of rice intensification is promoted. In both the sub-humid-humid zones, and the arid lands (fed by rivers) the main adaptation/mitigation measures reportedly in use include sunken beds/earth bunds farming; short duration and drought tolerant crops; adjusting of farming calendars; dry season cropping; increased processing of crop produce; increased processing of livestock produce; intercropping; crop diversification and multi-storey tree crop farming.

Food Security Policy in Tanzania

The Tanzanian rural food security vision is of a people centered poverty-free society, based on full and equal access to food and nutrition for all, and to the resources necessary to achieve the same; control over key resources; full participation in decision-making on policy-making, implementation and monitoring; and the strengthening of sustainability and self-reliance from the grassroots to the national to the global level (Mbilinyi et al. 1999). The long term development agenda is expressed in the Tanzania Development Vision 2025. The vision expresses the resolve to eradicate poverty and attain sustainable development. Agriculture is given a prominent role in economic growth and poverty reduction since the rural sector contains the majority of the poor population. The initiative recognizes the importance of food and nutrition security, climate change adaptation and improving survival, health, nutrition and well-being, especially for children, women and vulnerable groups. Tanzania Food Security Implementation Plan (TAFSIP) (2011), a 10-year investment plan maps the investments needed to achieve the Comprehensive Africa Agriculture Development Program (CAADP) target of 6 % annual growth in agricultural sector Gross Domestic Product (GDP). The United Republic of Tanzania (URT) will pursue this target through allocating a minimum 10 % of its budget to the agricultural sector (Mbilinyi et al. 1999).

Tanzania's agriculture sector has passed through a number of stages; PEAPA's (Political Economy of Agricultural Policy in Africa) second round of comparative fieldwork examined the relationship between Comprehensive Africa Agriculture Development Programme (CAADP) and agricultural policy in selected countries. The CAADP compact led the Tanzanian government to formulate the Tanzania Agriculture and Food Security Investment Plan (TAFSIP) in November 2011, which is essentially an enhanced version of the Agricultural Sector Development Programme (ASDP) of 2006. While expanding ASDP's scope and projected cost, TAFSIP retained ASDP's overall state led agricultural development model, focusing on inputs and productivity rather than markets and value chains. In 2008, a Tanzanian private-sector-led agricultural strategy known as "Kilimo Kwanza" (Agriculture First) was launched. From mid-2009 the ruling elite embraced KK as the government's vision for agricultural transformation. Nevertheless, in claiming policy leadership, TAFSIP 'incorporated' ASDP while dismissing KK as a mere 'slogan' (Cooksey 2013). The Tanzania Agriculture and Food Security Investment Plan (TAFSIP) is an historic initiative that brings all stakeholders in the agricultural sector both in the mainland and in Zanzibar to a common agenda of comprehensively transforming the sector to achieve food and nutrition security, create wealth, and poverty reduction.

Materials and Methods

Study Area

Bagamoyo district is located in Coastal region of Tanzania north of the city, Dar es Salaam. It is one of six districts of the Coast ("Pwani") Region of Tanzania. It is bordered to the North by the Tanga Region, to the West by the Morogoro Region, to the east by the Indian Ocean and to the South by the Kibaha District. The District has an area of 985,000 ha. Bagamoyo Town is roughly 70 km from Dar Es Salaam City Centre. Bagamoyo District as a whole had a total population of approximately 311,740 according to census in 2012. The district is composed of 22 wards (URT 2013), which have a total of 82 villages. Bagamoyo was among 13 districts in

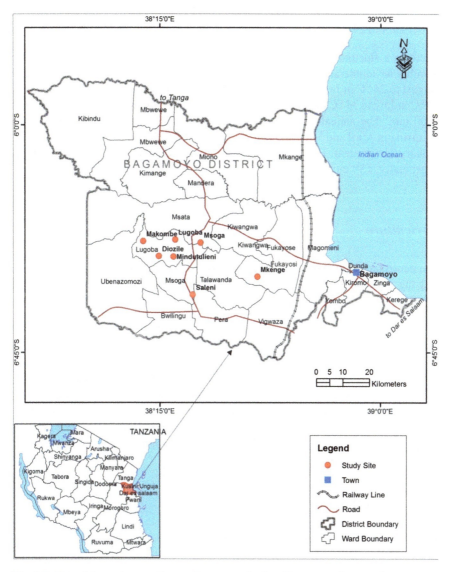

Fig. 6.3 Map of the study area. *Source*: Bagamoyo district office map and survey department

Tanzania where the stakeholders' consultation during NAPA preparation were undertaken (Mahenge 2012). The National Adaptation Plan of Action (NAPA) reveals that coastal and marine resources for generations have had profound influences on the socio-economic well-being and health status of not only the immediate communities but also those far removed from them Bagamoyo district has very few existing adaptation activities cited in NAPA hence a need to suggest other potential adaptation activities e.g., raise awareness on climate change and establishment of protected areas is highly needed (Fig. 6.3).

Climate

There is a fluctuation throughout the year in the maximum and the minimum monthly air temperatures. Minimum mean temperature varies from 18 °C in July/August to 24 °C in January/February; the maximum mean temperature ranges from 29 °C in July to 32 °C in February. Annual precipitation in the area is approximately 1000 mm. The annual rainfall is bi-modal with the first wet period (long rains) occurring in April and May, and the second wet period occurring from November to January (short rains). The driest months are June–September when monthly rainfall is generally <50 mm per month.

Agriculture

The percentage distributions of the population who engage in Agriculture are 76 % and the cultivated area per district is 75,360 (9 %). The reasons for selecting Bagamoyo district includes the average characteristics of the area in the Tanzanian context in terms of several poverty indicators and the dominant role of subsistence farming as the traditional source of livelihood (PHDR 2005). Several studies carried out in Bagamoyo district focused on tourism, settlements and mangroves. Different literature revisited shows that population increase leads to expansion of farm lands and grazing lands at the expense of forest and woodland areas. The present understanding of changes as a complex and interactive system involving population change and its impacts on agriculture is inadequate calling for a new study. Following this, Bagamoyo district provides the best sites for carrying out an assessment to link the process of policy and mainstreaming of adaptation activities.

Data Collection

Participant/Community Observation

Participant observation is a classical method of anthropology through which essential information on local circumstances can be gained. It also helps with choosing central questions for the study process (Chambers 1997: 116). Being mostly accommodated in the study villages during September 2013–March 2014 enabled observing and participating in the daily life of the villagers, and helped understand diverse issues and practices in more concrete ways than what the interviews and official data alone could have provided. Direct observation was particularly useful in regard to information on climate patterns, physical structures, cropping patterns, livestock keeping, trading, food preparation and consumption, livelihood opportunities, share of labour between sexes and age groups, the role of forest products end especially charcoal production, water collection, operations of government officials, and adaptation and mitigation techniques practised by the communities. All information was written on day to day basis, videos and voice calls were taken and recorded and lastly coded. The interpretation was based on research objective understanding basing on the objective of the study.

Questionnaire Survey

A total of 300 household survey questionnaire was conducted in 7 villages-study site (see Table 6.1) and the questionnaire included questions regarding climate variability, the sources of food consumed in the household, land holdings, cultivated crops, livestock assets, food security, and incomes, natural calamities and mitigation. The questionnaire was extended with additional questions on the livelihood sources of each household member, cultivation techniques, and usage of inputs, livestock diseases, as well as agricultural output and adaptation practices. Random sampling was applied in a manner that the assistants or the researcher walked criss-cross through different sub-villages and selected approximately every 20th household for the interview. A household was chosen as the basic unit for the interviews as it has traditionally been the primary unit of production and consumption in rural Tanzania (Bryceson 1990: 19). As a minority of the families, about 15 %, are polygamous, this was not always very straightforward. Most people were interviewed at their home yard and the rule of thumb was that the household was defined to people living in that homestead and sharing the same eating pot.

The average age of the surveyed was 41–55 years, the youngest interviewee being between 18–25 and the oldest above 65 years. The sample involved about 5.0 % of the total number of households in the studied villages according to the population figure from 2012. This means a margin of error of 6.0 % with a 95 % confidence level for the questions asked from all respondents. A sampling technique used was a combination of purposive and random sampling. Purposive was used deliberately to select seven villages in the Western Bagamoyo, factors such as remoteness and distance from the main ways was considered. Simple random

Table 6.1 List of villages in Bagamoyo district involved in the study

Community/Village	Questionnaire household	FGD	PRA	Key informants
Diozile	51	4	2	9
Saleni	52	4	2	9
Msoga	25	4	2	9
Lunga	35	4	2	9
Makombe	38	4	2	9
Mkenge	49	4	2	9
Mindu Tulieni	50	4	2	9
Total	300	28	14	56

Source: Author

sampling was used to select each household for interview where by one household was selected after counting an interval of 20 households. Household survey was important because the study wanted to get deeper understanding of the climate variability and food security issues in the community. The other reason was to get information direct from the grassroots and compare with official data. One questionnaire was composed of about 93 closed and open questions on personal background, climate variability and change, food security and other livelihood issues. Statistical Package of Social Sciences (SPSS) was used to analyze data where graphs, descriptive and frequencies were used to give the results.

Participatory Research Approach

A total of 14 participatory research approach (PRA) were organized in all 7 study sites/villages (each 2 PRA exercises). Local people from different backgrounds, regardless of their social position in their communities, were asked to participate. By doing PRA, the researcher, local people met development practitioners and government officials to articulate their livelihood conditions, their needs and their views so as to further develop their village or region. The voice of the local people is central, since they are the experts of their specific livelihood and social conditions and environment (Narayanasamy 2009). PRA included a cluster of approaches and tools to address different topics of interest. For example seasonal calendar exercises focus on the annual cycles of food production, variations in food sources and availability, agricultural practices and workload, food consumption, incomes, and non-agricultural labor opportunities (Callens and Seiffert 2003). Also Historical Matrices to identify period of extreme rainfall and temperature variability, adaptation strategies used by the household were employed. A sampling technique used was purposive sampling where by participants were selected basing on age (in order to get information from the past), gender, representation of those who were not included in the household survey. The major techniques were seasonal calendar and historical matrices. A number of exercises were conducted to indicate production circles and seasons of extreme weather. Each exercise was recorded in a note book, some photos and videos were taken in the consent of the people. The reason for using this technique was to create active participation of the communities' members who understand well their environment. Also to understand the comparison between official data, household information with people's perception on the issues at hand. The technique used to analyze information was triangulation and researcher's objective interpretation basing on the issue at hand.

Focus Group Discussion

Group discussions are said to build up collective and creative enthusiasm, which leads to sharing familiarizing new ideas and concepts with an outsider who then familiarizes with them In this regard, the population was sorted into groups, and described how each group accesses food and income. What skills and assets does each group have that helps them survive each day? The groups may be defined by occupation, relative well-being (i.e., poor, better-off), gender, or ethnicity. This includes groups that appear to be the most food insecure right now, and the groups that are at risk of becoming food insecure during a pandemic. In this study, a total of 28 focus group discussion (FGD) (4 in each village) were organized. In each group, purposive sampling was applied in order ensure equal participation of certain segment of population such as gender, age groups, occupation and marital status. The questions involved were more or less the same as PRA but different in terms of how the activity was organized. For instance in FGD the role of the researcher was to introduce a topic which was then discussed by the rest of the group members. Triangulation was used to analyze information obtained.

Key Informants/Thematic Interview

Key informant interviews likewise, enable the study to obtain climate information from people with long term experience on the area as well as expert knowledge thus counterchecking the credibility of data from other sources (see Fig. 6.4). Other people were those with strong experience on agriculture in the study area. In this activity, a total of 54 key informants were interviewed. These were people

Fig. 6.4 Household questionnaire interview in Western Bagamoyo, Tanzania in March 2014. *Source*: Author

Fig. 6.5 Key informants interview in Western Bagamoyo, Tanzania in March 2014. *Source*: Author

purposively selected by researcher to help to get sensitive information about climate variability and food security situation of the community, it also involved decision makers. These were expert people at national, district, ward and village level who have special knowledge on the field. The key informers were agricultural extension officers, elders, traditional healers, charcoal producers and teachers. For instance, director of nutrition and food security at the Ministry of Agriculture and Food Security, Engineer and head of Chalinze Water Supply Authority (CHALIWASA) under Wami river with a project to supply water, a project that includes all the studied village. Bagamoyo District Head of Agriculture Unit was also interviewed. All information was written in booklet and analysis subjected to analysis (Fig. 6.5).

Demographic and Meteorological Data

Official population data from the area, including numbers of households and age groups by sexes, is available in national censuses made in 2012 (URT 2013). The nearest weather station that has been collecting both precipitation and temperature data during the studied period is at the District Agricultural Office in Bagamoyo town, about 65 km east from the study area.

Limitations

As it was difficult to get comprehensive and up to date lists of the households from the village offices due to lack of proper recording instruments, random sampling was applied in a manner that the assistants or the researcher walked criss-cross through different sub-villages and selected approximately every 20th household for the interview. Another issue, which may have more significantly affected the validity of the findings, is the representation of a whole household by one person. For instance, it is questionable to what extent the representative can give responses that correspond with the experiences of food security by all household members (Webb et al. 2006). Approaching the households as entities is also not capable of providing much detailed information on the intra-household relationships in acquiring and accessing food, especially within the limits set by the length of the questionnaire and the focus of the study. Perhaps a more meaningful indication of the survey being representative is, however, the notion that towards the latter stage of the survey exercise the variability of the answers appeared to be low, or remained mostly within certain ranges, and new issues were seldom raised in the open-ended questions. From a more qualitative point of view, this would imply that a saturation point had been well reached (see also Guest et al. 2006).

Regression analysis of milk production on monthly basis couldn't reveal accurate information as to whether monthly production of milk could be affected by climate variability. This is because the amount of milk produced by cattle depends much on a time since a cow give birth; the longer the time; the little the milk and vice versa. A closer look at the official statistics, however, revealed serious inconsistencies. Before 2012 data collection tools and storage were very unreliable. Extension Officers were writing papers and file them; something which made a lot of data to be lost or spoiled. But from 2012 the government introduced new system with well-organized monthly and annual reports from village to district. This made official data collection for scientific analysis to establish a historical trend very challenging in this study. These data sets were similarly incomplete, but in any case indicative. The data collected indicated that, neither the veterinary officer nor the district officers had preserved older statistics, only from 2012 is consistent. By supplementing the crop and livestock figures with information from the household surveys made the most salient changes in the importance and productivity of main crops and livestock categories could nevertheless be constructed.

Results and Discussion

The Extent to Which Climate Variability Affects Food Security

Labor, transfer and trade entitlement was analysed using primary data obtained from the respondents; while, secondary data are used to run regression model for production. The regression model is used at two levels of analysis; yearly production of maize at village level with the corresponding rainfall and temperature in 6 years 2008, 2009, 2011, 2012, 2013 and 2014; and monthly milk production of 24 months in corresponding rainfall and temperature between January 2013 and December 2014. The variation of number of years is due to the availability of production data. The second part of production analysis was based on primary data obtained from the field through household survey, PRA, observation which was used to compare information of the recorded data and feedback from the respondents.

Production Entitlement

Regression Analysis on Climate Variability and Annual Production of Maize at the Village Level

The Table 6.5 show the model summary statistics from the regression analysis. The table shows the R statistics, R square, adjusted R square as well as the Durbin–Watson statistics. The R statistics indicate a moderate positive correlation and this is being accounted for by 37.7 % of the variability of climate. The Durbin statistics also shows the account of the data which support the positive relationship. Durbin Watson assumes the quality of data used (Tables 6.2 and 6.3).

The ANOVA statistics from regression analysis shows a statistically no significant difference between climate variability on food production at village level. This is indicated by ($Df = 2, 3, F = 0.909, p > 0.05$). This shows clearly that rainfall and temperature has no differences in terms of their impact on food production at the village level (Table 6.4).

From the coefficients table we establish the extent of the impact of climate variability on food production at the village level. From the unstandardized

Table 6.2 Model summary

Model	R	R square	Adjusted R square	Std. error of the estimate	Durbin–Watson
1	0.614[a]	0.377	−0.038	59,303.60637	2.408

Source: Author
In a model summary we look at R square which represent 37%. Hence production can be influenced by climate variability by 37% which means there are other 63 unexplained variables. This model can explain the influence of climate variability on food security only by 37%
[a] Predictors: (Constant) Temperature, Rainfall
[b] Dependent variable: Production

Table 6.3 ANOVA[b]

Model		Sum of squares	Df	Mean square	F	Sig.
1	Regression	6.393E9	2	3.197E9	0.909	0.491[a]
	Residual	1.055E10	3	3.517E9		
	Total	1.694E10	5			

Source: Author
In ANOVA we look at P value (F). The value of P > 0.05 hence climate has no significance effects of food security hence 0.491 > 0.05
[a] Predictors: (Constant), Climate. Temperature, Rainfall
[b] Dependent variable: Production. Village

Table 6.4 Coefficients

		Unstandardized coefficients		Standardized coefficients		
Model		B	Std. error	Beta	T	Sig.
1	(Constant)	−104,346.100	158,224.937		−0.659	0.557
	Climate. Rainfall	703.737	1457.517	0.251	0.483	0.662
	Climate. Temperature	6184.033	7079.925	0.453	0.873	0.447

Source: Author
In coefficient we look at coefficients of both rainfall and temperature which show positive effects by indicating 703.737 and 6184.033 Betas. Additionally, temperature indicates more positive effects that rainfall. However their influence were not significant as 0.662 and 0.447 > 0.05
[a] Dependent variable: Production

coefficients and its correspondence beta values it can be realized that temperature has a positive impact on food production than that of rainfall, however this difference is not statistically significant. This is indicated as rainfall (Beta = 703.737, T = 0.483, p > 0.05) and temperature (Beta = 6184.033, t = 0.873, p > 0.05). This clearly shows no differences in terms of variability of climate thus rainfall and temperature on food production.

Regression Analysis of Monthly Milk Production with Climate Variability

The above table shows the model summary statistics from the regression analysis. The table shows the R statistics, R square, adjusted R square as well as the Durbin–Watson statistics. The R statistics indicate a moderate correlation and this is being accounted for by 22.4 % of the variability of climate. The Durbin statistics also shows the account of the data which support the relationship (Tables 6.5 and 6.6).

The ANOVA statistics from regression analysis shows a statistically no significant difference between climate variability (rainfall and temperature) on milk production. This is indicated by (df = 2, 11, f = 1.303, p > 0.05). This shows clearly that rainfall and temperature has no differences in terms of their impact on milk production (Table 6.7).

Table 6.5 Model summary

Model	R	R square	Adjusted R square	Std. error of the estimate	Durbin–Watson
1	0.474	0.224	0.052	53,786.83866	1.744

Source: Author

We look at P square which is 0.224. Therefore, this model can explain the influence of climate variability on food security by only 22%. Hence there are remaining 78% unexplained variables

Table 6.6 ANOVA

Model		Sum of squares	Df	Mean square	F	Sig.
1	Regression	7.537E9	2	3.769E9	1.303	0.319
	Residual	2.604E10	9	2.893E9		
	Total	3.357E10	11			

Source: Author

P value is 0.319 which is > than 0.05. Hence climate variability has no significance influence on food security

Table 6.7 Coefficients of rainfall and temperature with milk production

		Unstandardized coefficients		Standardized coefficients		
Model		B	Std. error	Beta	T	Sig.
1	(Constant)	362,912.094	232,795.043		1.559	0.153
	Rainfall	−28.876	296.904	−0.036	−0.097	0.925
	Temperature	−11,112.039	9131.384	−0.451	−1.217	0.255

Source: Author
[a] Dependent variable: total

From the coefficients table we establish the extent of the impact of climate variability on milk production. From the standardized coefficients and its correspondence beta values it can be realized that both temperature and rainfall indicate a negative impact on milk production, however this difference is not statistically significant. This is indicated as rainfall (Beta $= -0.036$, t $= -0.097$, p > 0.05) and temperature (Beta $= -0.451$, t $= -1.217$, p > 0.05). This clearly shows no differences in terms of variability of climate thus rainfall and temperature on milk production. It can also be stated that with low negative impact by rainfall and moderate negative impact by temperature as indicated in the coefficients table this are not statistically significant.

Labor Entitlement

According to the household survey, the total share of people engaged in agriculture, full or part-time, had dropped. The main source of farm labor reported was family 83 % and hired labor 17 %. Only 2.3 % of the respondents in all study villages were employed by the government. The remaining was self-informally employed in agriculture and majority of the households had one or more non-agricultural income sources, of which the most important were charcoal production, stone quarry, carpentry and other activities. Village wise comparison shows that the involvement of households in non-agricultural activities is less common in Mkenge, Mindu Tulieni and Diozile. The reason for this was due to geographical location of the villages in relation to remoteness and accessibility which determined economic diversification. The amount of land that can be sown and harvested is, clearly, tied to available and affordable labor supply. According to the interviews and discussions, planting and harvesting are both activities that require far more labor than the rest of the agricultural cycle. In Lugoba, Msoga and Makombe the rate is higher, which obviously relates to the location of Lugoba along the main highway, which provides better opportunities for petty trade, as well as the location of Makombe nearby several stone quarries. Hired labor is a major source of food entitlement to some community members (landless) who provide hired labor services to the land owners (32.3 %). The number of labor used by some cultivators to provide their farms labor were a minimum of one person and three persons maximum.

During food crisis, hired labor is an adaptation strategy to acquire food entitlement. The households' questionnaires reveal that only 2.1 % of the respondents used tractors in the cultivation which could be an obstacle of labor entitlement in subsistence farming community especially to the landless poor. Among those households which had experienced food insecurity, the most common means to get over the period was seeking for additional labor such as charcoal production (41 %), agricultural coolie work (38 %) and other casual work. Charcoal production, farm labor and employment in the stone quarrying companies are major labor entitlements to many people in Makombe village. Worldwide, an estimated 450 million people are employed as farm workers. Of these, at least 20–30 % are women, although the proportion is higher, at around 40 %, in Latin America and the Caribbean. These figures should be treated with caution, however, since much employment in the agriculture sector is informal and undeclared, and official statistics are often unreliable (Asian Development Bank 2013).

Transfer Entitlement

Helping each other during crisis is one of the strategies to transfer entitlement where by a transfer is done from one community member to another. For instance questionnaire interview conducted in February 2014 show that 80.7 % of the interviewed were on debt after borrowing either food or money to buy food. The timing for such borrowing was reported as preparing season, a period when community members were preparing their farms for new growing season (34.7 %), growing season (13 %), dry season (49.3 %) and waiting for revenue from harvests season (3 %). When asked about family members who play transfer entitlement on food and other social welfare, about 19 % of the interviewed shows that they receive remittances from their family members who stay away from the households. But 81 % did not receive and remittance. The common strategy of transfer entitlement in all studies villages was to borrow or ask help from relatives, friends, or shopkeepers, which were mentioned by 69.4 %. Receiving governmental food aid was mentioned by 20.6 %. As reported earlier, interviews show that food insecure period repeats every year and local officials have asked for food aid from the government during several years. The biggest challenge of this is that the amounts per person delivered during the shortages have usually included only few kilograms of maize and sometimes beans. Most of the recipients addressed that the amounts of food delivered are too small to improve the situation for longer than a few days. For instance, interviews and discussions reveal that vulnerable household are given priority during food aid and each member of the family gets 2 kg twice a year. In addition to that, the selected households categorized as the most vulnerable receive funds for buying food during crisis. The amount was reported to be 5.15$ per year to each member of the household. This is too small to help people to survive during food crisis.

The thematic interview and FGDs inform that Tanzania Social Action Fund (TASAF) is the main government organization working with vulnerable people in the study area. Poor families are given food and funds for food and health. TASAF has also constructed a number of hospitals and schools in almost all seven villages. The program is designed to help vulnerable groups who are categorized into children, women, disabled and old people. Social protection plays an essential role in assuring food entitlement. Such protection can be provided on an informal basis by family and community networks, by NGOs, or formally organized by government and local collectives as stipulated above. It is vital, of course, for individuals and households that cannot produce food for themselves, or who have no income to purchase food. Even for those who have productive assets and income, social protection can be essential for improvements in productive capacity, to complement income when it drops below minimum levels, and as insurance against temporary shocks from severe and sudden weather events and price increases for essential goods such as food, fuel, and electricity (Asian Development Bank 2013).

Trade Entitlement

The study shows that subsistence farming has lost its dominant role in food provisioning due to property rights transfers, the declining productivity of land, livestock losses which accelerate increasing shift of labor to non-farm sectors. Also rapid population growth has added to the pressure on land and other natural resources. The responses of the cash crops are cassava sold by 3 %, beans by 7 %, maize by 9 %, millet by 7 % and vegetables by 7 %. More than half, 57 % report the decrease, 27 % increase and 16 % don't not experienced any significant change. However, sales naturally vary from year to another, according to the yield sizes. The most significant change in regard to cash crops is the fall of cotton production, which rapidly started to concentrate in Mkenge, Makombe and Msoga.

Sometimes crop cultivators sell their crops to be able to purchase medicine during illness. Also there are common products consumed by people in the village which they do not cultivate, for instance beans, rice, fish, meat. Many respondents believe that agriculture is no longer means of subsistence therefore one need money to subsidize his family. Petty business is observed to be the major means to food entitlement in all seven villages.

Generally, in all selected villages, rural life has changed where people need money for their subsistence. When asked about access to credit over last 2 years about 28.3 % access credits while 71.7 % said don't access. The major sources mentioned by the respondents were relatives (3 %), SACCOS (14.3 %), NGOs 2 %, other local money lenders 9.3 %, local traders 3 %, and community groups (2 %). Many respondents 87 % explain the purpose of accessing credits is to get money and buy food during absence of stock in the household, about 0.3 % said business capital, and 10 % for buying seeds and agricultural inputs. The timing of credits are, planting season (49.3 %), dry seasons (21.7 %), waiting to receive money from their produce (3 %), and 24.4 % during famine. About 80.7 % during survey accepted that there were people who might be on debts for food while 19.3 % did not accept the idea. Food crops are mostly grown by households for their own use, so the majority of cultivation is clearly subsistence production. Accumulation of surplus wealth is difficult when the produced outcomes are not sold. This effectively limits the stratification of wealth within a community (Bryceson 1989).

FGDs show that that a family response in terms of payment suggests that wealth of the family has declined. Even during harvest payment is low. In some instances, the food price in Lugoba becomes more expensive than Dar es Salaam because in Dar es Salaam products are bought in bulk and with large choice hence market competition. While product in Lugoba products are brought by individuals who incur more costs of travels. In addition to food expenses, cash is increasingly needed for school fees, clothing, transport, and housing, farming inputs, trading facilities, electricity and other necessities. This is connected to a considerable change in the livelihood structure, which has brought greater involvement in cash economy. The literature related to the surge in international food prices and its impact in Tanzania is fairly limited due to its very specific nature. Hella et al. (2009) basically conduct qualitative case studies and observe that the impact of high food prices was very diverse due to the subsistence nature of the Tanzanian economy, traditional food consumption and production behaviour. These authors also find that the high food prices between 2006 and 2008 were likely to raise poverty levels in food-deficit regions.

Dessus (2008) uses a computable general equilibrium model to assess the welfare impact of rising commodity prices. His results show that high food commodity prices in 2006–2008 may have had a negative impact on all Tanzanian households in the short run, with expected potential welfare improvement in the medium and long term. Additionally, he finds that poor households are likely to be shielded from soaring prices because they are more likely to derive their income from agricultural activities and, as he also includes rising energy prices in his considerations, consume fewer oil-intensive products. Kiratu et al. (2011) studied two Tanzanian regions; the semi-arid areas of Dodoma and Kilosa Districts. They argue that during food shortages and high food prices, households tend to reduce food consumption, indicated by a declining number of meals per day, with severe consequences for the household's nutritional and health status.

In their study of "*Climate Change and Food Security in Tanzania*", (Arndt et al. 2012) estimate the impact of climate change on food security in Tanzania. Representative climate projections were used in calibrated crop models to predict crop yield changes for 110 districts in Tanzania. They find that, relative to a no climate change baseline and considering domestic agricultural production as the principal channel of impact, food security in Tanzania appears likely to deteriorate as a consequence of climate change.

Conclusion and Recommendation

In this study, regression analysis basing on village annual climate and production reveals that temperature has more impact on production than rainfall; but this difference was not statistically significant. While regression analysis on monthly climate and milk production reveals that climate have negative impact on production of milk but this was statistically insignificant. However, both analyses show positive correlation of climate variability on production. Primary data reveals that climate variability has a big impact on agricultural production. Rainfall was revealed to be the most influencing factor than temperature.

Food security through labor entitlement reveals that agriculture remains to be the most important for rural communities. However the trend shows diminishing agriculture dependence among rural people. People have engaged on other source of labor, and this has been caused by number of factors affecting agriculture such as importation of food, poor production, population increase that has let to pressure on land and diversification of economy. However, to agricultural communities other source of labor such as charcoal production, casual works remain to be adaptation techniques.

Food security entitlement through transfer was revealed to be obtained through neighbourhood, national and international level. The major one is exchange of good and services among community members during crisis. Borrowing was also revealed as one of the adaptation technique. Rural people in Tanzania live communal life where one has to share food with someone who is facing food shortage. At national level, drought has affected food availability hence the government has to provide food assistance to the rural people in the study area. During national crisis, international donors helped to provide food aid to the people in the study area. However, the biggest challenge of the support given does not cater the demand.

Food security through trade entitlement reveals that agriculture still remains to be the dominant source. However there is a declining trend in the recent years compared to the past. Emergency of demand for money to cater for social services has made people believe that agriculture is no longer a source of subsistence in a study area. Petty business has taken a dominant role. The introduction of trade and emergency of foreign products from other parts of the country to the study area imply that people just need money to get food. Many people in the rural communities have engaged in petty business. Money has become a dominant instrument to obtain food.

This case study shows the value of examining climate data to supplement, confirm and clarify community perception and direct observations. However, it also illustrates the challenges in using climate data, especially when there are gaps in the historical record, and models of future climate change show no clear patterns in the medium and longer term. Looking at the future of Bagamoyo district, there is a great deal of uncertainty about climate hazards. No single, clear message emerges from the analysis, and the data require careful reading and interpretation. It is important to be cautious in predicting the future, and to compare different sources of data. This is a particularly sensitive issue in dealing with stakeholders' perceptions of climate and food security risks. While it is important to document and acknowledge those perceptions, they need to be checked against the available data to ensure a more comprehensive view of the issues. The comparison of historical data with local perceptions reinforces the need to use multiple sources of evidence when assessing climatic changes and food security. Models can give us a sense of how patterns may change and affect food security, for example but to know whether heavy rains will cause floods, or significant damage, we need to look at the hydrology of the region, and how land use changes, for example, might have increased flood risks.

In addition, an increase in the number of hot days has serious implications for health, potentially affecting the labour availability in the area. Shifts in the rainfall patterns place additional stress on agriculture. The combined effects of multiple climate and non-climate hazards make it challenging to prioritize adaptation options. They would also complement existing community based adaptation efforts already taking place in Bagamoyo district. A further detailed technical study on the assessment and ranking of adaptation options would complement this work and would be important in making decisions.

References

Arndt C, Farmer W, Strzepek K, Thurlow J (2012) Climate change, agriculture and food security in Tanzania. Rev Dev Econ 16(3):378–393

Asian Development Bank (2013) Gender Equality and Women's Empowerment Operational Plan, 2013–2020: moving the agenda forward. Asian Development Bank, Manila

Boko M, Niang I, Nyong A, Vogel C, Githeko A, Medany M, Osman-Elasha B, Tabo R, Yanda P (2007) Africa. In: Parry ML, Canziani OF, Palutikof JP, van der Linden PJ, Hanson CE (eds) Climate change 2007 impacts, adaptation and vulnerability. Contribution of Working Group II to the Fourth Assessment Report of the Intergovernmental Panel on Climate Change. Cambridge University Press, Cambridge, pp 433–467

Bryceson DF (1989) Nutrition and the commoditization of food in Sub-Saharan Africa. Soc Sci Med 28(5):425–440

Bryceson DF (1990) Food insecurity and the social division of labour in Tanzania, 1919-85. The MacMillan Press, Hampshire, p 275

Callens K, Seiffert B (2003) Participatory appraisal of nutrition and household food security situations and planning of interventions from a livelihoods perspective. Methodological guide. Food and Agriculture Organization of the United Nations, Rome, p 70

Chambers R (1997) Whose reality counts? Putting the first last. ITDG Publishing, London, p 283

Christensen JH, Hewitson B, Busuioc A, Chen A, Gao X, Held R et al (2007) Regional climate projections. Climate change, 2007: the physical science basis. Contribution of Working Group I to the Fourth Assessment Report of the Intergovernmental Panel On Climate Change. Cambridge University Press, Cambridge, Chap 11, pp 847–940

Cooksey B (2013) The comprehensive Africa agriculture development programme (CAADP) and agricultural policies in Tanzania: going with or against the grain? FAC Political Economy of Agricultural Policy in Africa (PEAPA) Working Paper. www.futures-agriculture.org

Dessus S (2008) The short and longer term potential welfare impact of global commodity inflation in Tanzania. Policy research working paper no. WPS 4760. World Bank, Washington, DC

FAO (1996) World food summit. Rome. Retrieved from http://gidimap.giub.uni-bonn.de:9080/geomorph/themen/intergovernmental-panel-on-climate-change-ipcc/Ipccreport_figures2_11.pdf

Funk C, Senay G, Asfaw A, Verdin J, Rowland J, Michaelson J, Eilerts G, Korecha D, Choularton R (2005) Recent drought tendencies in Ethiopia and equatorial subtropical eastern Africa. FEWS-NET, Washington, DC

Guest G, Bunce A, Johnson L (2006) How many interviews are enough? An experiment with data saturation and variability. Field Methods 18(1):59–82

Hella J, Kamile I, Schulz CE (2009) Food prices and world poor-winners and losers: a case study of two villages in Tanzania. In: Haug R et al (eds) High global food prices: crisis or opportunity for smallholder farmers in Ethiopia, Malawi and Tanzania. Noragric report no. 48. Department of International Environmental and Development Studies, Norwegian University of Life Sciences. Retrieved from http://www.umb.no/noragric/article/noragric-reports-2009

Kiratu S, Lutz M, Mwakolobo A (2011) Food security: the Tanzanian case. Series on food security. International Institute for Sustainable Development, Winnipeg, MB. http://www.iisd.org

Lyimo JG, Ngana JO, Liwenga E, Maganga F (2013) Climate change, impacts and adaptations in the coastal communities in Bagamoyo District, Tanzania. Environ Econ 4(1)

Mahenge JJ (2012) Development of policy advocacy initiatives for integrating climate change adaptation and food security for coastal fishers in Tanzania: the case of Bagamoyo district

Mbilinyi M, Koda B, Mung'ong'o C, Nyoni T (1999) Rural food security in Tanzania: the challenge for human rights, democracy and development. Excerpts from report presented at the launching workshop on rural food security policy and development, Kilimanjaro Hotel, Dar es Salaam, Institute of Development Studies, Oxfam, 23 July 1999

Mongi H, Majule A, Lyimo J (2010). Vulnerability and adaptation of rain fed agriculture to climate change and vulnerability in semi-arid, Tanzania. Afr J Environ Sci Technol 4(6): 371–381

Mwandosya MJ, Nyenzi BS, Luhanga ML (1998) The assessment of vulnerability and adaptation to climate change impacts in Tanzania. Centre for Energy, Environment, Science and Technology (CEEST), Dar es Salaam

Narayanasamy N (2009) Participatory rural appraisal: principles, methods and application. Sage, New Delhi

PHDR (2005) Tanzania poverty and human development report 2005. Research and Analysis Working Group and Mkuki na Nyota Publishers, Dar es Salaam, p 98

Tanzania Meteorological Agency (TMA) (2008a) Temporal annual variation of rainfall, Bagamoyo Rainfall Station. TMA, Dar es Salaam

Tanzania Meteorological Agency (TMA) (2008b) Mean monthly rainfall variation data. TMA, Dar es Salaam

United Republic of Tanzania (URT) (2013) 2012 population and housing census: population distribution by administrative areas. National Bureau of Statistics Ministry of Finance Dar es Salaam and Office of Chief Government Statistician President's Office, Finance, Economy and Development Planning, Zanzibar

URT (2007) National adaptation plan of action (NAPA). Vice President's Office, Dar es Salaam

Watson RT, Albritton DL (2001) Climate change 2001: synthesis report: Third Assessment Report of the Intergovernmental Panel on Climate Change. Cambridge University Press, Cambridge

Webb P, Coates J, Frongillo EA, Rogers BL, Swindale A, Bilinsky P (2006) Measuring household food insecurity: why it's so important and yet so difficult to do. J Nutr 136(5):1404S–1408S

Chapter 7
The Urban Heat Island Effect in Dutch City Centres: Identifying Relevant Indicators and First Explorations

Leyre Echevarría Icaza, F.D. van der Hoeven, and Andy van den Dobbelsteen

Abstract In the Netherlands awareness regarding the Urban Heat Island (UHI) was raised relatively recently. Because of this recent understanding, there is a lack of consistent urban micro-meteorological measurements to allow a conventional UHI assessment of Dutch cities during heat waves. This paper argues that it is possible to retrieve relevant UHI information—including adaptation guidelines—from satellite imagery.

The paper comprises three parts. The first part consists of a study of suited indicators to identify urban heat islands from which a method is presented based on ground heat flux mapping. The second part proposes heat mitigation strategies and identifies the areas where these strategies could be applied within the hotspots identified in the cities of The Hague, Delft, Leiden, Gouda, Utrecht and Den Bosch. The third part estimates the reduction of urban heat generated by the increase of roof albedo in the hotspots of the six cities. The six cities hotspots are located within the boundaries of the seventeenth century city centres. In order to avoid interference with cultural values of these historical environments most likely UHI mitigation measures regard improving the thermal behaviour of the city roofs. For instance, applying white coatings on bitumen flat roofs (or replacing them by white single-ply membranes) and replacing sloped roof clay tiles by coloured tiles with cool pigments can reduce the urban heat hotspots by approximately 1.5 °C.

Remote sensing provides high level information that provide urban planners and policy makers with overall design guidelines for the reduction of urban heat.

Keywords Climate change • Urban Heat Island • Storage heat flux • Remote sensing • Climate adaptation • NDVI • Albedo

---

L. Echevarría Icaza (✉) • F.D. van der Hoeven
Faculty of Architecture and The Built Environment, Urban design department, Delft University of Technology, Delft, Netherlands
e-mail: L.EchevarriaIcaza@tudelft.nl

A. van den Dobbelsteen
Faculty of Architecture and The Built Environment, Department of Architectural Engineering and Technology, Delft University of Technology, Delft, Netherlands

Introduction

UHI Studies Despite the Lack of Micro-measurements

According to the Royal Netherlands Meteorological Institute (KNMI 2014) 40 heat waves have struck the Netherlands since the beginning of the nineteenth century (KNMI 2014). Nevertheless, Dutch urban meteorologists only started to study the Urban Heat Island (UHI) phenomenon after the heat wave of 2003, when the amount of heat-related deaths reached more than 1400 (Garssen et al. 2005) in the Netherlands and more than 22,000 across Europe (Schar and Jendritzky 2004). In the Netherlands, this relatively recent awareness of the phenomenon explains the lack of historical air temperature records to allow a consistent analysis of the UHI patterns throughout the country (Hove et al. 2011). Future climate scenarios predict that the frequency, the intensity and the duration of heat waves will increase (Meehl and Tebaldi 2004), more specifically in the Netherlands the four climate scenarios predicted for 2050 by the KNMI forecast that the average summer temperatures in the rural environment will continue to rise, and so will the amount of 'summerly days' (maximum temperature above or equal to 25 °C) per year in the rural environment across the country. Concerned by these future predictions, Dutch scientists, climatologists and urban planners have had to develop alternative ways to fill in the shortage of historical urban air temperature records, in order to study more in depth the phenomenon in different Dutch cities. Some have used hobby meteorologist's data (Hove et al. 2011; Steeneveld et al. 2011; Koopmans 2010), others have used cargo bicycles (Heusinkveld et al. 2010; Brandsma and Wolters 2012) to retrieve temperature variations through different cities during hot summer days and finally others have chosen to use satellite imagery to map land surface temperature variations during hot days (Hoeven and Wandl 2013; Klok et al. 2010).

Bridging the Gap Between Scientific and Applied Knowledge

Even though in the Netherlands the scientific community has started investigating the phenomenon already more than 5 years ago, it seems there is still a gap between the scientific knowledge developed and the urban policies of large and medium size cities, which haven't started implementing measures to mitigate urban heat yet. Precisely one of the goals of the research programme Knowledge for Climate (Knowledge For Climate 2015) is to develop not only scientific but also applied knowledge for climate proofing the Netherlands, investigating a wide variety of topics ranging from the climate adaptation for rural areas (Climate Adaptation for Rural Areas 2015) to the climate adaptation of cities (Climate Proof Cities 2014), to which this study belongs. Many Dutch cities have taken part in the Climate Proof Cities Program either as stakeholders or as case cities and would be willing to implement measures to mitigate the urban heat problem however they often lack a basic overview of the most affected neighbourhoods (to identify the areas where to concentrate the urban heat mitigation efforts), the different mitigation options (to be able to select design mitigation proposals that match best the rest of urban planning priorities) and a high level estimation of the potential heat reduction achieved (to be able to quantify the mitigation effect).

Remote Sensing as a Tool to Identify, Mitigate and Quantify Urban Heat

For this study we have chosen to use satellite imagery because on the one hand it allows mapping and analysing many heat related parameters, such as surface heat fluxes (Parlow 2003), land surface temperatures (Dousset et al. 2011), albedo (Taha 1997; Sailor 1995) vegetation indexes (Yuan and Bauer 2007; Gallo et al. 1993) and on the other hand the analysis of satellite imagery provides consistent information of several cities at the same time. Being able to analyse several heat related parameters allows not only to identify vulnerable areas within cities, but also to assess on the different potential mitigation strategies for each city analysed and to provide a high level quantification of surface materials. The possibility of producing a simultaneous analysis of several comparable cities allows the analysed cities to join efforts, share scientific knowledge and implementation strategies. In this study we have analysed the cities of The Hague, Delft, Leiden, Gouda, Utrecht and Den Bosch (Fig. 7.1a, b). These cities have in common that they have a dense historical inner-city, dating back to medieval times but mostly consisting of buildings from the seventeenth century, the 'Golden Age' of the Netherlands, when most cities expanded rapidly with stone building that often still remain. Thanks to this comparable past and development, the chosen cities are comparable, although their urban layout and recent alterations differ.

Methodology

Research Framework

Problem Statement and Objective

Since in the Netherlands UHI awareness is relatively recent, there is a lack of consistent urban micro-meteorological measurements to allow a conventional and consistent UHI assessment of Dutch cities during heat waves (Hove et al. 2011). This lack of appropriate data hampers UHI scientific studies and hinders the development of guidelines for climate adaptation in cities. Therefore, as part of this study we aim to retrieve relevant UHI information from satellite imagery, in order to help develop UHI adaptation guidelines for Dutch cities.

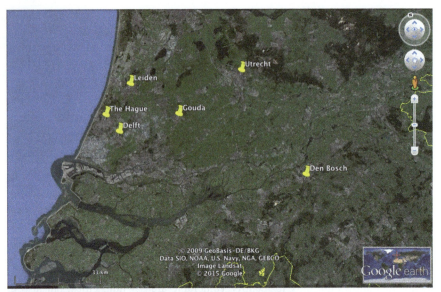

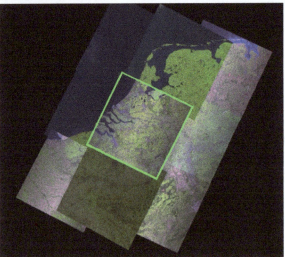

Fig. 7.1 (a) Analysed cities. Google earth image. Data SIO, NOAA, U.S. Navy, NGA, GEBCO ©2015Google. Image Landsat ©2009 GeoBasis-DE/BKG. (b) Size of the Landsat 5 TM image analysed (image extracted from the USGS Global Visualization Viewer). Courtesy of the U.S. Geological Survey. USGS/NASA Landsat

The objective of this study is twofold: to develop a method to assess the UHI phenomenon for cities with a lack of micro-meteorological datasets, and to develop a customised set of urban planning adaptation measures for the studied cities.

Research Questions

The underlying research questions for this paper are:

1. Can remote sensing help identify urban heat hotspots when there is lack of micro-measurements? If so, what are suited indicators?
2. Which are the most common heat mitigation strategies to reduce urban heat? How can we use remote sensing to identify where to implement these within the hotspots identified for the cities of The Hague, Delft, Leiden, Gouda, Utrecht and Den Bosch?
3. Could we quantify the mitigation effect of the increase of roof albedo in the identified hotspots?

Research Methodology

Research Structure

Based on the research questions this paper primarily comprised three parts.

The first part consists of a study of suited indicators to identify urban heat hotspots in areas with a lack of micro-measurements, from which a method is presented based on storage heat flux mapping. This was validated by application in two Dutch cities, The Hague and Utrecht.

The second part proposes heat mitigation strategies and identifies the areas where these heat mitigations could be applied within the hotspots identified in the cities of The Hague, Delft, Leiden, Gouda, Utrecht and Den Bosch. The third part estimates the reduction of urban heat generated by the increase of albedo in the hotspots of the six cities.

Data Collection Instruments

We have used Landsat 5 TM satellite imagery for the assessment of the tree parts of the research. Landsat is often used for UHI assessment (Bechtel 2011; Liu and Zhang 2011; Rajasekar and Weng 2009; Cao et al. 2008), for the development of mitigation strategies (Rosenzweig et al. 2006; Baudouin and Lefebvre 2014) or for the estimation of the heat mitigation effect (Odindi et al. 2015; Onishi et al. 2010). Landsat imagery has a high resolution and is open source. The raw satellite images can be downloaded from the US Geological Survey (USGS) webpage, Earth Resources Observation and Science Center (EROS). The approximate size of each retrieved scene is 170 km North-South by 183 km East-West. One Landsat image covers most of the country surface, which allows completing the simultaneous analysis of several cities at the same time (Fig. 7.1b). The sensor carried onboard Landsat 5 is Landsat Thematic Mapper (TM) which has a 16 day repeat circle. For this study we chose to analyse satellite images retrieved during the second heat wave that struck The Netherlands in 2006 (on the 16th of July at 10:33 UTC for all cities, except for Den Bosch, for which the Landsat image used was from the 25th of July, at 10:26 UTC).

Two software have been used to process the raw satellite imagery: ATCOR 2/3 and ENVI 4.7. ATCOR 2/3 was used for the atmospheric and geometric correction of the satellite imagery, as well as for the production of the albedo and surface heat flux maps (Richter and Schlapfer 2013) and ENVI 4.7 (Exelisvis 2015) was used for the analysis and enhancement of the images processed in ATCOR 2/3.

Identifying Hotspots

For the selected cities urban heat island hotspots were mapped by means of the storage heat flux. The storage heat flux was mapped using ATCOR 2/3 (Table 7.1).

The calculation of the heat fluxes is done through different models for urban and rural surfaces. In order to identify the hotspots (areas with the highest storage heat flux values) within the studies cities, we have chosen to use the model used for urban surfaces, where latent heat is usually smaller. The dominant fluxes are the storage and the sensible heat fluxes, for which Parlow's equations are applied (Parlow 1998).

$$G = 0.4 \, Rn \quad (7.1)$$
$$LE = 0.15 \, (Rn - G) \quad (7.2)$$
$$H = Rn - G - LE \quad (7.3)$$

Heat Mitigation Strategies

The analysis of remote sensing imagery can provide an overview of several urban heat related parameters: normalised difference vegetation index (NDVI), land surface temperature (LST), cool spot presence, and albedo (see Table 7.2). In this section we provide an overview of the relevance of each of these parameters. However, since in five of the six analysed cities the "storage heat flux hotspots" are within the limits of the seventeenth century city centres—and in these areas the implementation of design strategies is fairly restricted due to the historical protection of the neighbourhoods—in the following sections we have only estimated the effect on the urban heat reduction of the mitigation strategies consisting of increasing the city roofs' albedo.

7 The Urban Heat Island Effect in Dutch City Centres: Identifying Relevant... 129

Table 7.1 Hotspot identification process

Analytical phases	Scale	Parameter analysed	Methods and tools	Design guidelines
Phase 1				
Hotspot identification	City	Storage heat flux	Landsat 5 TM for July 2006. Geometrical correction and **Storage heat flux** calculation for urban environment in ATCOR 2.3.	Definition of intervention area

Table 7.2 UHI adaptation measures recapitulation chart

Analytical phases	Scale	Parameter analysed	Methods and tools	Design guidelines
Phase 2				
Adaptation measures	City	Cool corridors	Coolspot identification; Landsat 5 TM for July 2006. Geometrical correction, atmospherical correction and **Storage heat flux** for rural environment calculation in ATCOR 2.3	1/Creation of cool wind corridors connecting hotspot to natural coolspots. 2/Enhancement and preservation of the existion coolspot
	Hotspot	Albedo	Land 5 TM for July 2006. Geometrical and atmospherical correction and **Albedo** calculation in ATCOR 2.3	Proposal of surface material changes to improve albedo, depending on existing surface materials
	Hotspot	NDVI	Landsat 5 TM for July 2006 Geometrical and atmospherical correction in ATCOR 2.3. **NDVI** calculation in ENVI 4.7	Vegetation introduction in highlighted hotspots
	Hotspot	Land surface temperature	Landsat 5 TM for July 2006. **LST** calculation in ENVI 4.7	In industrial areas: deeper analysis is required to confirm the energy efficiency of the buildings with high LST on roofs

UHI Reduction Potentials

In order to estimate the UHI reduction for the six cities the implementation of the roof mitigation strategies was studied. These concern the measures that will be most likely adopted. The following methodology was adopted: the albedo maps of each hotspot were used to estimate the area of bituminous flat roofs and of clay tile sloped roofs. This area estimation was calculated using ENVI 4.7. For the estimation of the bituminous flat roofs, we considered all surfaces with albedos of 0.13–0.15, and for the clay tile surface estimation we considered all surfaces with albedos of 0.18–0.22. This way of estimating material surfaces has its limitations.

Detailed Assumptions

The following reference-based assumptions were used:

We assumed that an increase of 0.1 of the hotspot overall albedo reduces the UHI by 1 °C (Sailor 1995; Taha et al. 1988).

As a reference, we provided the maximum UHI values for the 95 percentile, calculated with hobby meteorologists data (Hove et al. 2011) for the cities of The Hague, Delft and Leiden. For the cities of Gouda, Utrecht and Den Bosch, we estimate that the max UHI will be around 5 °C as well.

For each city we have estimated the UHI reduction for several roof intervention scenarios:

- Mitigation action 1:
  Bituminous flat roof albedo (0.13–0.15) is improved by applying a white coating (albedo 0.7) or by replacing it by a white single-ply membrane (albedo 0.7). This action is likely to take place in the next 10 years. Minor repairs can be treated with white coating solutions, and major repairs will require full replacement by single ply membranes.

- Mitigation action 2:
  Clay tiles sloped roof (albedo 0.18–0.22) are replaced by coloured tiles with cool pigments (albedo 0.5). This action is likely to take place in the coming 50 years as the lifespan of clay tiles is 50 years.

- Mitigation action 1 + 2:
  Consists in improving the albedo of bituminous flat roofs by applying a white coating or a single-ply membrane, and in improving the clay sloped roof albedo by cool pigment coloured tiles.

Limitations

Landsat 5 TM is an appropriate tool to assess urban heat accumulation at city scale (sections "Identification of UHI Hotspots in Areas with a Lack of Micro-measurements" and "Heat Mitigation Strategies in Dutch Cities: Relevant Parameters") due to the resolution of its spectral bands: 30 m for bands 1–7, and 120 m for band 6 which is resampled to 30 m. However, in this study we have also used Landsat 5 TM to quantify the surface of bituminous flat roof and of clay sloped roofs (section "Quantification of Heat Reduction Through Roof Mitigation Strategies"). The resolution of Landsat 5 TM for material discrimination is a little rough. Nonetheless, the purpose of the surface estimations is to provide a high-level quantification of the mitigation effect of the proposed measures, therefore a certain degree of inaccuracy in the surface quantification is acceptable. Further, the objective of the study is not only to quantify the mitigation effect of the measures, but also to suggest a methodology that could also be replicated with finer resolution satellite imagery, allowing more accurate surface classification results.

Identification of UHI Hotspots in Areas with a Lack of Micro-measurements

Possible UHI Indicators

UHI and SUHI

The most common variable assessed through remote sensing imagery is land surface temperature (LST). However, LST assesses the surface heat island (SUHI) phenomenon, which has different characteristics than the canopy layer air temperature urban heat island (UHI).

The first important difference between SUHI and UHI is that, even though both present higher intensities on cloudless and windless days (Oke 1973; Oke 1982; Uno et al. 1988; Morris et al. 2001) SUHI has its peak during the day when the surfaces receive the maximum radiation (Carlson et al. 1981), whereas the canopy UHI has its peak at night when the surfaces start radiating the stored energy into the atmosphere (Oke 1997).

The second difference is that diurnal SUHI pattern does not match the nocturnal UHI pattern (Dousset et al. 2011; Parlow 2003; Voogt 2002; Roth et al. 1989; Price 1979). Urban planners and climatologists are typically more interested in understanding nocturnal air temperature UHI patterns because they are more strongly connected to the accumulation of heat and to human comfort; there is a high correlation between high nocturnal temperatures and excess of mortality during heat waves (Dousset et al. 2011).

Night-Time UHI

Some studies suggest that the nocturnal surface temperatures are better correlated to nocturnal air temperatures than diurnal ones, due to the stabilization of the atmosphere and to the cessation of the direct solar radiation (Nichol and Wong 2004). The reality is that most remote sensing studies on the UHI phenomenon focus on the day-time surface temperature variations due to the lack of fine resolution night-time thermal images. High-frequency thermal sensors that allow retrieving both day-time and night-time surface temperature typically have low resolutions (AVHRR: 1.1 km, Modis: 1 km) whereas finer resolution satellites such as Landsat TM (120 m) and Landsat ETM (60 m) have lower frequencies and are therefore limited to day-time observations (Nichol and Wong 2004). Thus finer resolution satellites mainly allow the retrieval of day-time LST which cannot be considered as the most relevant UHI indicator.

Heat Fluxes

Studies carried out in Basel by Parlow reveal that heat fluxes might be more relevant indicators of the UHI phenomenon than day-time surface temperature patterns (Parlow 2003). Therefore in this study remote sensing imagery is therefore used as a basis for mapping heat fluxes, more precisely storage heat fluxes.

The energy balance equation for radiant energy absorbed by heat fluxes can be written as (Asrar 1989):

$$Rn = G + H + LE \qquad (7.4)$$

where Rn is the net radiant energy absorbed by the surface; G is the storage heat flux, i.e. the energy dissipated by conduction into the ground or into the building materials; H is the sensible heat flux, that is the energy dissipated by convection into the atmosphere (its behaviour varies depending on whether the surface is warmer or colder than the surrounding air); and LE is the latent heat flux, that is the energy available of evapotranspiration.

Storage Heat Flux as Indicator of UHI

Studies on heat storage of paved surfaces in urban areas reveal that these may be the principal contributor to the nightly UHI effect (Doll et al. 1985). Net radiation, storage heat flux, latent heat and sensible heat distribution vary in the urban and rural environments. Studies on the 'Urban Energy Balance' derived from satellite data for the city of Basel (Parlow 2003) reveal that during day-time, urban pavements, industrial pavements and roofs present low latent heat fluxes, sensible heat fluxes similar to the ones obtained in their rural surroundings (among other reasons, due to the high surface temperatures of the urban surfaces) and extremely high storage heat fluxes. The heat accumulated during the day is released to the atmosphere during the night thus causing the UHI peak. Materials with higher conductivity—such as black top concrete and asphalt—present lower surface temperatures during the day, however at night—due to the different heat storage capacity of the two materials—black top concrete presented higher surface temperatures than the asphalt pavement (Asaeda et al. 1993). Therefore, storage heat flux proves to be a relevant indicator for the UHI assessment. Urban areas present higher temperatures at night due to thermal conductivity, heat capacity and thermal admittance of the built materials (Parlow 2003). Furthermore, other studies confirm that the data modelled with remote sensing imagery is in agreement with the micrometeorological in-situ measurements (Rigo and Parlow 2007).

Practical Storage Heat Flux Values

As a reference, the urban areas of Mulhouse and Basel, as well as their industrial sites, present storage heat fluxes of more than 200 W/m^2, whereas the forest areas present values of 26–50 W/m^2 (Parlow 2003). Other storage heat flux values per surface types are compiled in the ATCOR-2/3 User Guide (Version 8.0.2), which to dark asphalt areas assigns storage heat flux values of 240 W/m^2, to bright concrete values ranging from 164 to 240 W/m², to partially vegetated areas values of 185 W/m² and to fully vegetated areas values of 77 W/m² (Richter and Schlapfer 2013).

Validation Examples: Comparing Storage Heat Flux Mapping with Multi-day Mobile Observations for the Cities of The Hague and Utrecht

The Hague

Although there is a lack of ground measurements during heat waves, some research groups have attempted to map the UHI hotpots through other methods. This is the case for the maps issued by (De Groot–Reichwein et al. 2014), which for the city of The Hague forecasts the number of nights with temperatures above 20 °C for the 2050 KNMI climate scenarios W and W+ (KNMI 2006). These maps were produced as a result from the correlation between the bike measurements of the UHI for the city of Rotterdam (Heusinkveld et al. 2010), which was further extrapolated to the spatial and economic scenarios and to the climate scenarios (KNMI 2006). The number of nights with temperatures above 20 °C are higher in the area comprised between the Haagse Bos Park, Zuiderpark, Laakkwartier and Zorgvliet (see Fig. 7.2).

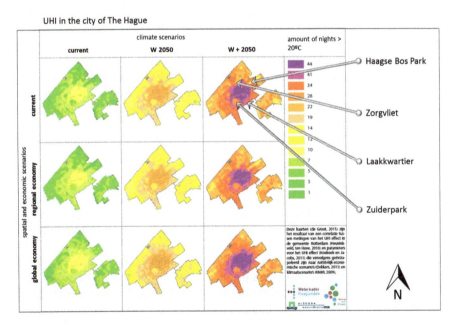

Fig. 7.2 Night-time temperature in The Hague, the Netherlands. These maps (De Groot - Reichwein et al. 2014) are the result of a correlation between measurements of the UHI effect in the municipality of Rotterdam and parameters for the UHI, which are then extrapolated to spatial economic scenarios and climate scenarios

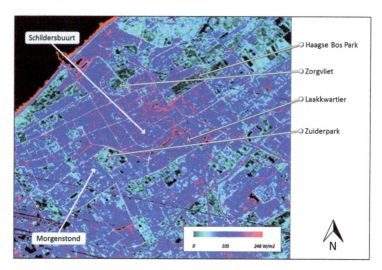

Fig. 7.3 Storage heat flux map for The Hague, 16 July 2006. Landsat image (Courtesy of the U.S. Geological Survey. USGS/NASA Landsat) further processed with ENVI 4.7 and Atcor 2.3

Figure 7.3 presents the The Hague area comprising different neighbourhoods. For the city centre an average storage heat flux of 92.82 W/m^2 was determined for the 16th of July 2006. For the Schildersbuurt area the average storage heat flux turned out to be 85.08 W/m^2. In contrast, in all scenarios the neighbourhood of Morgenstond has a lower average storage heat flux value than the area comprised between the Haagse Bos park, Zuiderpark, Laakkwartier and Zorgvliet also presents: 68 W/m^2. Morgenstond also presents less nights with temperatures above 20 °C.

These examples, together with the visual comparison of the two images show that the maps depicting the prediction of nights with temperatures above 20 °C are aligned with the results obtained when mapping the storage heat flux results during heat waves.

Utrecht

Brandsma and Wolters (2012) have attempted to map the night-time UHI intensity for the city of Utrecht and its surroundings using high-resolution multi-day mobile observations for a single transect through the city for the period of March 2006–January 2009 (see Fig. 7.4).

The areas with the highest night-time UHI—a temperature difference of around 6 °C—are those in the North-Eastern part of the city centre, which also present the highest storage heat flux values on the 16th of July 2006 (Fig. 7.5). As a reference, the average storage heat flux value in the city centre of Utrecht is 88.67 W/m^2. In contrast, the area of Nieuwegein Noord, which is mapped with an UHI of around 4.3 °C also has a considerably lower storage heat flux value: 65.8 W/m^2.

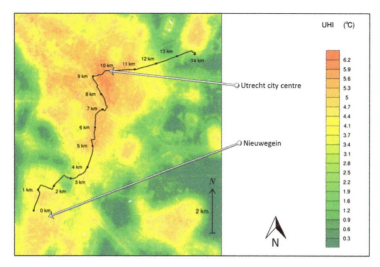

Fig. 7.4 Spatial distribution of the maximum night time UHI intensity for the city of Utrecht and its surroundings (Brandsma and Wolters 2012)

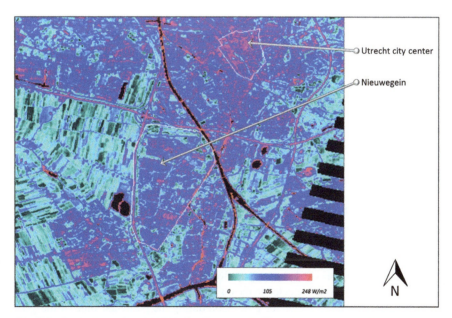

Fig. 7.5 Storage heat flux map for Utrecht, 16 July 2006. Landsat image (Courtesy of the U.S. Geological Survey. USGS/NASA Landsat) further processed with ENVI 4.7 and Atcor 2.3

These examples, together with the visual comparison of the two images show that UHI mapping based on multi-day mobile observations seems aligned with the results obtained when mapping the storage heat flux results during a heat wave.

Heat Mitigation Strategies in Dutch Cities: Relevant Parameters

Normalized Difference Vegetation Index

The NDVI is the Normalized Difference Vegetation Index which is used to quantify the vegetation density. Studies on land surface temperatures reveal that the imperviousness coefficient has a stronger linear relationship with land surface temperature values than with NDVI (Yuan and Bauer 2007), particularly in bare soil locations (Carlson et al. 1994). However, if we look at the correlation between minimum air temperatures and NDVI we observe that the difference in urban and rural NDVI is linearly related with the difference in urban and rural minimum air temperatures (Gallo et al. 1993). The NDVI variation is more strongly related with the temperature variations than with the population data used in previous studies (Gallo et al. 1993). Moreover, several studies indicate that the heat fluxes can be expressed as a function of the vegetation indexes in rural environments (Choudhury et al. 1994; Carlson et al. 1995). Therefore, NDVI can be considered as a relevant indicator for UHI studies.

NDVI Determination

The atmospheric correction of the satellite image was done in ATCOR 2/3 and the NDVI calculation was done in ENVI 4.7. using the index definition below:

$$\text{NDVI} = (\text{NIR} - \text{VIS})/(\text{NIR} + \text{VIS}) \tag{7.5}$$

where VIS is the surface reflectance in the red region (650 nm) and NIR is the surface reflectance in the near infrared region (850 nm).

Results of NDVI Analysis

The ATCOR-2/3 User Guide, version 8.2.1, serves as a reference for the different NDVI values corresponding to different surface types. For fully vegetated surfaces it establishes an NDVI of 0.78, for partially vegetated surfaces 0.33, for dark asphalt areas 0.09, and for bright concrete areas of 0.07 (Richter and Muller 2005). In Switzerland, the studies carried out by Parlow revealed that the lowest NDVI values with less than 0.2 could be detected in the city centre of Basel and Mulhouse as well as in some agricultural fields without vegetation during the particular time of the year when the satellite imagery was retrieved. The highest NDVI values reached values of up to 0.7 or more and these corresponded to forests and grassland areas (Parlow 2003). In the hotspots of The Hague, Delft, Leiden, Gouda, Utrecht and Den Bosch, the average NDVI ranges from 0.31 to 0.39. Even though the average NDVI values is pretty similar for all the hotspots, the NDVI visualisation (Fig. 7.6) suggests that there might be some consistent NDVI differences within the hotspots. These maps provide an indication of the areas with the lowest values, thus the areas where to increase the

Fig. 7.6 Normalized Difference Vegetation Index (NDVI) of the different Dutch cities. Landsat image (Courtesy of the U.S. Geological Survey. USGS/NASA Landsat) further processed with ENVI 4.7 and Atcor 2.3

vegetation. Overall, the storage heat flux hotspots of these six cities are located in the historical city centres and it seems delicate to suggest increasing NDVI at street level without analysing in detail the design implications of such a mitigation proposal. The implementation of green roofs would therefore be the most plausible option. Several studies (Kurn et al. 1994; Sailor 1995) estimate that the near-surface air temperatures over vegetated areas were 1 °C lower than background air temperatures.

Land Surface Temperature

Areas with high diurnal LST represent areas whose heat can either be released to the atmosphere and/or to the interior of buildings within the day, or during the night, depending on the roof properties of the buildings assessed.

LST Determination

The LST image has been obtained treating Landsat 5 TM imagery in ENVI 4.7, following the Yale Center for Earth Observation 2010 instructions to convert Landsat TM thermal bands into temperature. First the images are geometrically corrected and calibrated in ENVI 4.7, then the atmospherically corrected radiance is obtained applying Coll's equation (Coll et al. 2010):

$$\text{CV R2} = [(\text{CV R1} - L\uparrow)/\varepsilon\tau] - [(1-\varepsilon)*(L\downarrow)/\varepsilon] \qquad (7.6)$$

Where:
CV R2 is the atmospherically corrected cell value as radiance.
CV R1 is the cell value as radiance
$L\uparrow$ is upwelling radiance
$L\downarrow$ is downwelling radiance
T is transmittance
E is emissivity (typically 0.95)

The transmittance as well as the upwelling and downwelling radiance can be retrieved from NASA's web page (NASA 2014). Finally, the radiance can be converted into temperature (in Kelvin) as follows:

$$T = K2/[\ln((K1/CVR2) + 1)] \qquad (7.7)$$

Where:
T is degrees Kelvin
CVR2 is the atmospherically corrected cell value as radiance.

	Landsat TM	Landsat ETM
K1	607.76	666.09
K2	1260.56	1282.71

Results of LST Analysis

The land surface temperature images reveal that the storage heat flux hotpots do not necessarily present the highest diurnal LST values. As a matter of fact average land surface temperatures at 10:33 UTC in these city centers hotspots range from 36.55 to 40.8 °C (Fig. 7.7), whereas other areas of the same cities present surface temperatures of up to 50 °C. This is the case of The Brinckhorst, the Southern Transvaal, Kerketuinen en Zichtenburg in The Hague, the case of Schieweg in Delft, of the industrial area close to the Zijlkwartier in Leiden, to the Kromme Gouwe in Gouda or the industrial area between the Rietveldenweg and the Koenendelsweg in Den Bosch. These areas typically represent industrial areas that heat up very fast, but that also cool off very fast. This means that either the heat quickly penetrates into the buildings, or that it is quickly transfered back into the atmosphere. In the case of industrial buildings, which typically have bituminous sheet roofs, the heat retrieved by the roof is normally transferred to the interior of the buildings. If the industrial building does not need to preserve specific thermal conditions (storage use for example), it might not be worth it to implement any adaptation measure. If instead the building needs to preserve certain thermal conditions inside, the roof thermal behaviour could easily be improved by applying a reflective coating or surface coatings.

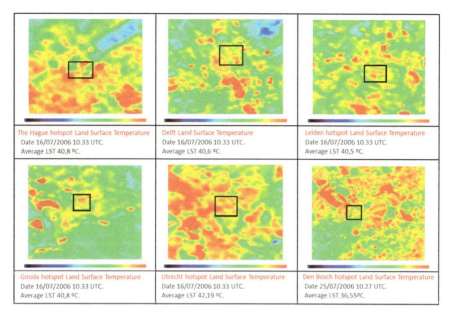

Fig. 7.7 Land Surface Temperature (LST) of the different Dutch cities. Landsat image (Courtesy of the U.S. Geological Survey. USGS/NASA Landsat) further processed with ENVI 4.7 and Atcor 2.3

Sky View Factor

The sky view factor (SVF) was defined by Oke as the ratio of the amount of the sky seen from a given point to that potentially available (Oke 1987). Its values range from 0 for full obstruction, to 1 for completely open areas. The average SVF in central parts of European cities ranges from 0.40 to 0.75, and the relationship with the nocturnal UHI was established by Oke (1981) and Park (1987), as follows:

$$UHImax = 13.20 - 10*SVF \qquad (7.8)$$

These results are aligned with the studies carried out in Gothenburg (Sweden) and in Szeged (Hungary), which find a strong relationship between the SVF and the nocturnal UHI in calm, clear nights (Svensson 2004; Unger 2009). However, other investigations carried out in Germany reveal that the nocturnal UHI is not only affected by the horizon obstructions (SVF) but also by the thermal properties of the materials, and they only find a correlation between the long-wave radiation and the UHI, but not between the UHI and the SVF (Blankenstein and Kuttler 2004). Many of these studies highlight the importance of the way the SVF is calculated, and at which height it is calculated.

Sky View Factor Determination

In the present study the calculation is done through the use of a SVF visualization tool developed by the Scientific Research Centre of the Slovenian Academy of Sciences

Fig. 7.8 Sky View Factor (SVF) of the different Dutch cities

and Arts (Zaksel et al. 2011) applied to the geographic Landsat image (Courtesy of the U.S. Geological Survey. USGS/NASA Landsat) of the concerned Dutch cities.

The Sky View Factor analysis allows to draw the city limits, to identify different urban structures within a city and to identify specific streets or urban areas with higher or lower sky view factors.

Results of SVF Analysis

The average Sky View Factor is almost the same for all hotspots (Fig. 7.8). In the hotspots of the six analyzed cities, The Sky View Factor maps do not allow to identify specific hotspots within the urban areas. The Sky View Factor analysis allows to draw the city limits, to identify different urban structures within a city and to identify specific streets or urban areas with higher or lower sky view factors.

Anthropogenic Heat Losses

Anthropogenic heat is not assessed in this paper. Average anthropogenic heat values in Europe range from 1.9 to 4.6 W/m^2 (Lindberg et al. 2013). These values increase in the urban environment reaching the 20 W/m^2 in cities as Berlin (Taha

1997). Anthropogenic heat plays a role in the formation of UHI but they are not decisive in European cities.

Coolspots and Cool Wind Corridors

Just as hotspots, coolspots can be identified through the storage heat flux mapping. Coolspots are areas with the lowest storage heat flux values; if they are situated close to a hotspot, climate adaptation measures could be focused on transporting cool air from the coolspots to the hotspot, or on getting the heat from the hotspot dissolved in the coolspots. The calculation of the heat fluxes is done through different models for urban and for rural surfaces. In order to identify the coolspots surrounding the cities, we apply the rural areas algorithm.

Coolspot Determination

Storage heat flux is mapped in this study using Landsat 5 TM imagery and ATCOR 2/3 for the storage heat flux calculation. Since the coolspots often correspond to green areas, we have applied ATCOR "rural" algorithm for the estimation of the storage heat flux which employs a parametrization with the soil adjusted vegetation index (SAVI) (Choudhury et al. 1994, Carlson et al. 1995).

$$G = 0.4 \, R_n \, (SAVI_m - SAVI)/SAVI_m \tag{7.9}$$

Where R_n represents the net radiation and $SAVI_m = 0.814$ represents full vegetation cover.

Results of Coolspot Analysis

For the hotspot of the Hague, the coolspot identified is The Haagse Bos, in Delft the Delftse Hout, in Leiden the potential cool spot is located at around 1300 m from the hotspot and corresponds to the greenfields to the South of the Kanaalweg and to the West of Zaalbergweg; in Gouda the greenfields to the South of the Gouderaksedijk (West of the Goudeseweg) although they are located 700 m away from the hotspot; in Utrecht the hinterland located to the East of the Utrecht Ring, in the areas of Fort Voordorp could also have a cooling effect on the hotspot although it is located at a distance of 2300 m from it, in Den Bosch the greenfields located to the South of the Singelgracht and to the West of the Zuiderplas could represent a natural cooling source for the hotspot (Fig. 7.9).

The identification of coolspots in the surrounding areas of the hotspots allows promoting the creation of cool wind corridors connecting the coolspots to the hotspots. In the case of The Hague, Delft, Gouda and Den Bosch the cooling

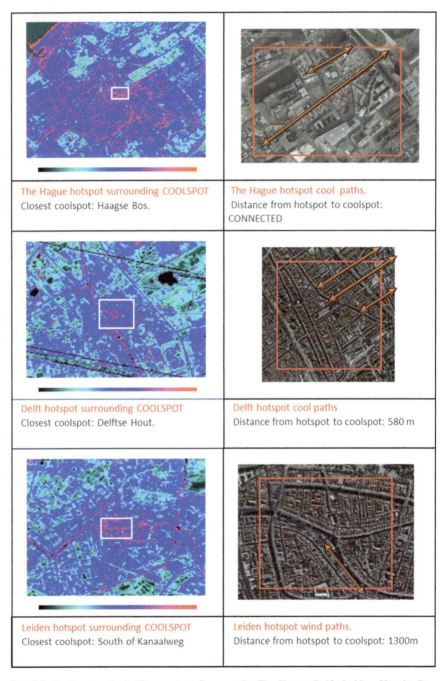

Fig. 7.9 Coolspot analysis (Storage heat flux map for The Hague, Delft, Leiden, Utrecht, Den Bosch and Gouda 16 July 2006. Landsat image courtesy of the U.S. Geological Survey. USGS/NASA Landsat. further processed with ENVI 4.7 and Atcor 2.3) and wind corridor analysis of the different Dutch cities. *Right column* google earth imagery

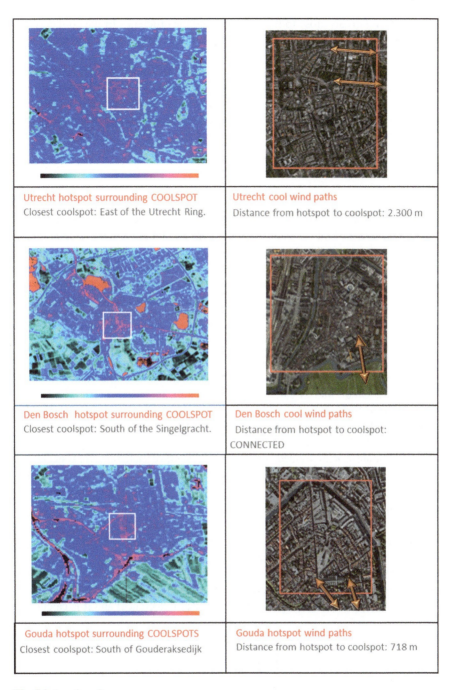

Fig. 7.9 (continued)

sources are relatively close to the hotspots, and the efficiency of the cool corridor is almost guaranteed, as they would also benefit from the urban heat island plume. The adaptation measure in this case would consist of ensuring that the selected "cool corridors" (existing streets or canals, connecting the cool and hotspots) remain cleared from obstacles to ensure the maximum wind circulation during heat waves. The cases of Leiden and Utrecht probably require deeper wind analysis studies, as in both cases the cooling source is at a distance greater than 1000 m from the hotspot.

Albedo

In the urban environment we can assume that the storage heat flux represents 40 % of the net radiation (Parlow 1998). Increasing the surface reflectance (albedo) of the urban surfaces is considered as a means to reduce the UHI since it reduces the net short-wave (solar) radiation, thus reducing the total surface net radiation. Albedo is the index representing the surface reflectance. It indicates the fraction of short-wave radiation that is reflected from land surfaces into the atmosphere. When a surface albedo is 0 it doesn't reflect any radiation, and when it is 1 all the incoming radiation is reflected to the atmosphere.

Most US and European cities have albedos of 0.15–0.20 (Taha 1997). A white surface with an albedo of 0.61 is only 5 °C warmer than ambient air whereas conventional gravel with an albedo of 0.09 is 30 °C warmer than air (Taha et al. 1992). Other studies carried out by Taha et al. reveal that increasing the surface albedo from 0.25 to 0.40 could lower the air temperature as much as 4 °C (Taha et al. 1988), or even that an increase of 0.1 of the hotspot overall albedo reduces the UHI by 1 °C (Sailor 1995).

It is important to analyse albedo surface images, since vegetated areas or water bodies might present low albedo values, yet not necessarily have a negative impact on the UHI.

Albedo Determination

The albedo of the six Dutch cities is mapped again using Landsat 5 TM imagery and importing it into ATCOR 2/3 for the albedo calculation.

In ATCOR 2/3 (Richter and Schlapfer 2013) the wavelength-integrated surface reflectance (in a strict sense the hemispherical-directional reflectance), is used as a substitute for the surface albedo (bi-hemispherical reflectance) and it is calculated as:

$$a = \left[\int_{0.3\ \mu m}^{2.5\ \mu m} \rho(\lambda) d\lambda \right] / \int_{0.3\ \mu m}^{2.5\ \mu m} d\lambda \qquad (7.10)$$

7 The Urban Heat Island Effect in Dutch City Centres: Identifying Relevant... 145

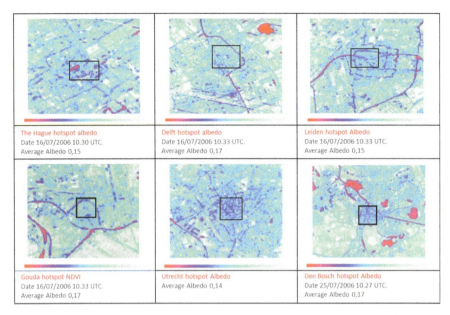

Fig. 7.10 Albedo of the different Dutch cities. Landsat image (Courtesy of the U.S. Geological Survey. USGS/NASA Landsat) further processed with ENVI 4.7 and Atcor 2.3

For Landsat 5 TM the following assumptions are made by ATCOR 2/3 for extrapolation:

– Extrapolation for the 0.30–0.40 μm region: $\rho 0.3 - 0.4\,\mu m = 0.8\,\rho 0.45 - 0.50\,\mu m$.
– Extrapolation for the 0.40–0.45 μm region: $\rho 0.4 - 0.45\,\mu m = 0.9\,\rho 0.45 - 0.50\,\mu m$.

The reflectance reduction factors in the blue part of the spectrum account for the decrease of surface reflection for most land covers (soils, vegetation). The extrapolation to longer wavelengths is computed as:

– $\rho 2.0 - 2.5\,\mu m = 0.5\,\rho 1.6\,\mu m$, if $\rho 850/\rho 650 > 3$ (vegetation)
– $\rho 2.0 - 2.5\,\mu m = \rho 1.6\,\mu m$, else

Wavelength gap regions are supplemented with interpolation. The contribution of the 2.5–3.0 μm spectral region can be neglected, since the atmosphere is almost completely opaque and absorbs all solar radiation (Fig. 7.10).

– Albedo: range 0–1000, scale factor 10, e.g., scaled albedo = 500 corresponds to albedo = 50 %.

Quantification of Heat Reduction Through Roof Mitigation Strategies

Increasing the Albedo

Delft, Leiden, Gouda, Utrecht and Den Bosch have a dense traditional seventeenth century inner-city with red ceramic roof tiles, brick street paving and canals. In order to improve the albedo in these hotspots, intervening on the brick street paving is provocative, as it is considered as part of the cultural heritage of these neighbourhoods. Instead, on the long run the existing roof tiles could be replaced by cool colour tiles, once the existing ones arrive to the end of their life cycle. Traditional tiles have albedo values that ranges from 18 to 22 %, whereas the reflectance of orange cool tiles have a reflectance of around 50 % (U.S. Environmental Protection Agency's Office of Atmospheric Programs). It is important to note that the market for these cool tiles is not consolidated yet, and that one possible municipal or regional adaptation measure would be to encourage the production or the import of these innovative products.

As far as the bitumen flat roofs are concerned the primary cool roof option for moderate repair would be to install highly reflective coatings or surface treatments that can be applied to bituminous cap sheets, gravel, metal and various single ply materials. For more extensive repairs of flat roofs the primary option is the application of thermal insulation and highly reflective single ply membranes (or - pre-fabricated sheets) generally glued to the entire roof surface (Fig. 7.11).

The results of the estimation of the UHI reduction through the implementation of albedo (sloped and flat) roof mitigation actions have an UHI reduction effect that ranges between 1.4 and 3 °C. Considering that the max UHI values for the 95 percentile, calculated with the hobby meteorologists data (Hove et al. 2011) for the cities of The Hague, Delft and Leiden ranges from 4.8 to 5.6 °C implementing roof mitigation strategies seems an efficient way of reducing the UHI in these Dutch cities.

The Hague

The UHI reduction estimation for several roof intervention scenarios can be seen in Table 7.3.

Delft

The UHI reduction estimation for several roof intervention scenarios can be seen in Table 7.4.

7 The Urban Heat Island Effect in Dutch City Centres: Identifying Relevant... 147

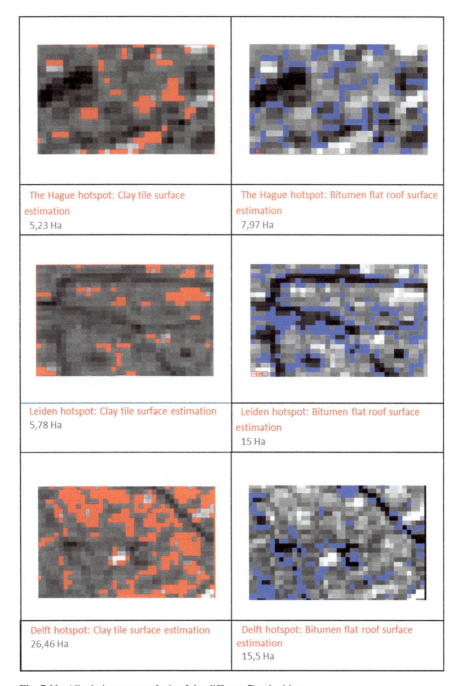

Fig. 7.11 Albedo increase analysis of the different Dutch cities

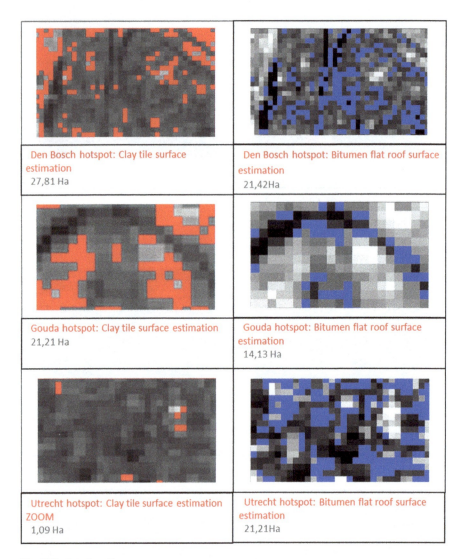

Fig. 7.11 (continued)

Leiden

The UHI reduction estimation for several roof intervention scenarios can be seen in Table 7.5.

Gouda

The UHI reduction estimation for several roof intervention scenarios can be seen in Table 7.6.

Table 7.3 Estimation of UHI reduction derived from the implementation of the roof mitigation strategies in the storage heat flux hotspot of The Hague

The Hague					
Current situation					
	Hotspot surface (Ha)	Hotspot average albedo	Estimated UHI max for 95 percentile		
	44.46	0.15	5.3		
Mitigation scenario:					
Mitigation action 1: Next 10 years	Apply over flat bitumen roofs (albedo 0.13–0.15) white coating (0.7) OR replace by white single-ply membrane (0.7)				
	Surface	Albedo of white coating or of single-ply membrane	Hotspot new average albedo	Average albedo increase	Estimated UHI max for 95 percentile
Area within hotspot with albedo ranging from 0.130 to 0.150 or highlevel estimation of **flat bitumen roof** surface	7.97	0.7	0.25	0.10	4.31
				UHI reduction:	0.99
Mitigation action 2: Next 50 years	Replace all hotspot sloped roof clay tiles (albedo 0.18–0.22) by coloured tiles with cool pigments (albedo 0.5)				
	Surface	Albedo of coloured tiles with cool pigments	Hotspot average albedo	Average albedo increase	Estimated UHI max for 95 percentile
Area within hotspot with albedo ranging from 0.185 to 0.0220 or highlevel estimation of **sloped roof clay tile** surface	5.23	0.5	0.19	0.04	4.89
				UHI reduction:	0.41
Mitigation action 1 + 2: Next 50 years					
			Hotspot average albedo	Average albedo increase	Estimated UHI max for 95 percentile
			0.29	0.14	3.90
				UHI reduction:	**1.40**

Table 7.4 Estimation of UHI reduction derived from the implementation of the roof mitigation strategies in the storage heat flux hotspot of Delft

Delft					
Current situation					
	Hotspot surface (Ha)	Hotspot average albedo	Estimated UHI max for 95 percentile		
	92.8	0.17	4.8		
Mitigation scenario					
Mitigation action 1: Next 10 years	Apply over flat bitumen roofs (albedo 0.13–0.15) white coating (0.7) OR replace by white single-ply membrane (0.7)				
	Surface	Albedo of white coating or of single-ply membrane	Hotspot new average albedo	Average albedo increase	Estimated UHI max for 95 percentile
Area within hotspot with albedo ranging from 0.130 to 0.150 or highlevel estimation of **flat bitumen roof** surface	15.5	0.7	0.26	0.09	3.91
				UHI reduction:	0.89
Mitigation action 2: Next 50 years	Replace all hotspot sloped roof clay tiles (albedo 0.18–0.22) by coloured tiles with cool pigments (albedo 0.5)				
	Surface	albedo of coloured tiles with cool pigments	Hotspot average albedo	Average albedo increase	Estimated UHI max for 95 percentile
Area within hotspot with albedo ranging from 0.185 to 0.0220 or highlevel estimation of **sloped roof clay tile** surface	26.46	0.5	0.26	0.09	3.86
				UHI reduction:	0.94
Mitigation action 1 + 2: Next 50 years					
			Hotspot average albedo	Average albedo increase	Estimated UHI max for 95 percentile
			0.35	0.18	2.97
				UHI reduction:	1.83

7 The Urban Heat Island Effect in Dutch City Centres: Identifying Relevant...

Table 7.5 Estimation of UHI reduction derived from the implementation of the roof mitigation strategies in the storage heat flux hotspot of Leiden

Leiden					
Current situation					
	Hotspot surface (Ha)	Hotspot average albedo	Estimated UHI max for 95 percentile		
	61.65	0.15	5.6		
Mitigation scenario					
Mitigation action 1: Next 10 years	Apply over flat bitumen roofs (albedo 0.13–0.15) white coating (0.7) OR replace by white single-ply membrane (0.7)				
	Surface	albedo of white coating or of single-ply membrane	Hotspot new average albedo	Average albedo increase	Estimated UHI max for 95 percentile
Area within hotspot with albedo ranging from 0.130 to 0.150 or highlevel estimation of **flat bitumen roof** surface	15	0.7	0.28	0.13	4.26
				UHI reduction:	**1.34**
Mitigation action 2: Next 50 years	Replace all hotspot sloped roof clay tiles (albedo 0.18–0.22) by coloured tiles with cool pigments (albedo 0.5)				
	Surface	albedo of coloured tiles with cool pigments	Hotspot average albedo	Average albedo increase	Estimated UHI max for 95 percentile
Area within hotspot with albedo ranging from 0.185 to 0.0220 or highlevel estimation of **sloped roof clay tile** surface	5.78	0.5	0.18	0.03	5.27
				UHI reduction:	**0.33**
Mitigation action 1 + 2: Next 50 years					
			Hotspot average albedo	Average albedo increase	Estimated UHI max for 95 percentile
			0.32	0.17	3.93
				UHI reduction:	**1.67**

Table 7.6 Estimation of UHI reduction derived from the implementation of the roof mitigation strategies in the storage heat flux hotspot of Gouda

Gouda					
Current situation					
	Hotspot surface (Ha)	Hotspot average albedo	Estimated UHI max for 95 percentile		
	30.8	0.17	5		
Mitigation scenario					
Mitigation action 1: Next 10 years	Apply over flat bitumen roofs (albedo 0.13–0.15) white coating (0.7) OR replace by white single-ply membrane (0.7)				
	Surface	albedo of white coating or of single-ply membrane	Hotspot new average albedo	Average albedo increase	Estimated UHI max for 95 percentile
Area within hotspot with albedo ranging from 0.130 to 0.150 or highlevel estimation of **flat bitumen roof** surface	4.13	0.7	0.24	0.07	4.29
				UHI reduction:	**0.71**
Mitigation action 2: Next 50 years	Replace all hotspot sloped roof clay tiles (albedo 0.18–0.22) by coloured tiles with cool pigments (albedo 0.5)				
	Surface	albedo of coloured tiles with cool pigments	Hotspot average albedo	Average albedo increase	Estimated UHI max for 95 percentile
Area within hotspot with albedo ranging from 0.185 to 0.0220 or highlevel estimation of **sloped roof clay tile** surface	21.21	0.5	0.40	0.23	2.73
				UHI reduction:	**2.27**
Mitigation action 1 + 2: Next 50 years					
			Hotspot average albedo	Average albedo increase	Estimated UHI max for 95 percentile
			0.47	0.30	2.02
				UHI reduction:	**2.98**

Table 7.7 Estimation of UHI reduction derived from the implementation of the roof mitigation strategies in the storage heat flux hotspot of Utrecht

Utrecht					
Current situation					
	Hotspot surface (Ha)	Hotspot average albedo	Estimated UHI max for 95 percentile		
	60.8	0.14	5		
Mitigation scenario:					
Mitigation action 1: Next 10 years	Apply over flat bitumen roofs (albedo 0.13–0.15) white coating (0.7) OR replace by white single-ply membrane (0.7)				
	Surface	albedo of white coating or of single-ply membrane	Hotspot new average albedo	Average albedo increase	Estimated UHI max for 95 percentile
Area within hotspot with albedo ranging from 0.130 to 0.150 or highlevel estimation of **flat bitumen roof** surface	21.21	0.7	0.34	0.20	3.05
				UHI reduction:	**1.95**
Mitigation action 2: Next 50 years	Replace all hotspot sloped roof clay tiles (albedo 0.18–0.22) by coloured tiles with cool pigments (albedo 0.5)				
	Surface	albedo of coloured tiles with cool pigments	Hotspot average albedo	Average albedo increase	Estimated UHI max for 95 percentile
Area within hotspot with albedo ranging from 0.185 to 0.0220 or highlevel estimation of **sloped roof clay tile** surface	1.09	0.5	0.15	0.01	4.94
				UHI reduction:	**0.06**
Mitigation action 1 + 2: Next 50 years					
			Hotspot average albedo	Average albedo increase	Estimated UHI max for 95 percentile
			0.34	0.20	2.98
				UHI reduction:	**2.02**

Utrecht

The UHI reduction estimation for several roof intervention scenarios can be seen in Table 7.7.

Den Bosch

The UHI reduction estimation for several roof intervention scenarios can be seen in Table 7.8.

Table 7.8 Estimation of UHI reduction derived from the implementation of the roof mitigation strategies in the storage heat flux hotspot of Den Bosch

Den Bosch					
Current situation					
	Hotspot surface (Ha)	Hotspot average albedo	Estimated UHI max for 95 percentile		
	133	0.17	5.5		
Mitigation scenario					
Mitigation action 1: Next 10 years	Apply over flat bitumen roofs (albedo 0.13–0.15) white coating (0.7) OR replace by white single-ply membrane (0.7)				
	Surface	Albedo of white coating or of single-ply membrane	Hotspot new average albedo	Average albedo increase	Estimated UHI max for 95 percentile
Area within hotspot with albedo ranging from 0.130 to 0.150 or highlevel estimation of **flat bitumen roof** surface	21.42	0.7	0.26	0.09	4.65
				UHI reduction:	**0.85**
Mitigation action 2: Next 50 years	Replace all hotspot sloped roof clay tiles (albedo 0.18–0.22) by coloured tiles with cool pigments (albedo 0.5)				
	Surface	Albedo of coloured tiles with cool pigments	Hotspot average albedo	Average albedo increase	Estimated UHI max for 95 percentile
Area within hotspot with albedo ranging from 0.185 to 0.0220 or highlevel estimation of **sloped roof clay tile** surface	27.81	0.5	0.24	0.07	4.81
				UHI reduction:	**0.06**
Mitigation action 1 + 2: Next 50 year					
			Hotspot average albedo	Average albedo increase	Estimated UHI max for 95 percentile
			0.32	0.15	3.96
				UHI reduction:	**1.54**

Conclusions of UHI Analysis of Dutch Cities

General Findings

This study has two main objectives: the first one is to develop a method for the urban heat assessment based on the analysis of satellite imagery, and the second one is to develop some customised UHI adaptation guidelines for the cities of The Hague, Delft, Leiden, Gouda, Utrecht and Den Bosch.

Satellite Imagery Analysis for UHI Assessment:
Remote sensing can effectively be used to identify urban heat hotspots in areas where there is a lack of micro-measurements. Storage heat flux seems a relevant indicator for the identification of urban areas with a high tendency to accumulate heat. Storage heat flux can be mapped using Landsat 5 TM imagery and processing it in ATCOR 2/3 and ENVI 4.7.

The use of Landsat 5 TM processed in ATCOR 2/3 and ENVI 4.7. also allows defining mitigation strategies to reduce urban heat in the identified hotspots. Mapping vegetation indexes, land surface temperature, coolspots and albedo, allows identifying areas where to implement more vegetation, areas where wind corridors (connecting hotspots to coolspots) could be created and areas where to increase the reflectance of the materials (to improve the albedo).

The same satellite imagery can be used to quantify the surface of bituminous flat roofs and of clay sloped roofs, in order to calculate a high level estimation of the mitigation effect of the increase of albedo of those surfaces.

Customised UHI Assessment for the Cities of The Hague, Delft, Leiden, Gouda, Utrecht and Den Bosch:
The storage heat flux analysis for the six cities reveals that the hotspots have average storage heat flux values that range from 90 to 105 W/m^2. See Table 7.9. Hotspot areas (areas with highest storage heat flux concentration) range from 30.8 ha in Gouda to 133 ha in Den Bosch.

The hotspots in all cities correspond with the old city centres of each city, except for the case of The Hague, where it is not related to a homogeneous urban structure. The hotspots of Delft, Leiden, Gouda, Utrecht and Den Bosch, correspond to the dense traditional seventeenth century Dutch neighbourhoods with red ceramic roof tiles, brick street paving, and canals. These are typical dwelling neighbourhoods with commercial premises in the ground floor, characterised by a high quality of life. These inner-city areas belong to representative neighbourhoods with very intense street activity (commercial, leisure, and touristic). In contrast, the hotspot of the city of The Hague corresponds with an area with bituminous flat roofs and asphalt paving.

The analysis of the vegetation index maps allows to identify areas where the average value is below 0.2, and that could eventually benefit from an increase of vegetation, whether that vegetation is implemented at street level or at roof level. The coolspot analysis reveals that the in the cities of The Hague, Delft, Gouda and Den Bosch the distance between hotspots and coolspots is below 1.000 m, which suggests that the creation of wind corridors could effficitinely contribute to the

Table 7.9 Analysis of storage heat flux hotspots in the cities of The Hague, Delft, Leiden, Gouda, Utrecht and Den Bosch

Dutch cities	Surface of the hotspot (Ha)	Storage heat flux (W/m^2)	NDVI	LST (°C)	Albedo	Average SVF	Sensible heat (W/m^2)	Coolspot stge flux (distance to hotspot m)
The Hague	44.5	101	0.31	40.8	0.15	0.5	156	26 (connected)
Delft	92.8	91.2	0.37	40.6	0.17	0.5	76.3	47 (580)
Leiden	61.7	103	0.33	40.5	0.15	0.5	173.6	40 (1300)
Gouda	30.8	96.7	0.34	40.8	0.17	0.5	168.4	41 (718)
Utrecht	60.8	94.6	0.39	42.2	0.14	0.5	184.4	47.7 (2300)
Den Bosch	133	104.1	0.34	36.6	0.17	0.5	138.8	58.7 (connected)

It is important to highlight that the storage heat flux images and values might be distorted in areas covered by water, since the algorithm used by ATCOR for the storage heat flux calculation is based on Parlow's equation $G = 0.4\ Rn$, which is a valid assumption for urban areas but not for water surfaces mitigation of the urban heat, and the analysis of the albedo maps allows to identify the areas that could benefit from an increase of the surface reflectance.

The quantification of the bituminous flat roofs and clay sloped roofs, reveals that increasing the albedo of both type of surfaces could help reduce the UHI from 1.4 to 3 °C in the analysed cities.

Implications of the Work

Using remote sensing for UHI assessment allows cities to identify the areas where to concentrate their mitigation efforts. Further, it provides them with an overview of different adaptation alternatives (vegetation, albedo, wind corridors,...) to help combine the climate mitigation efforts with other urban planning priorities, and finally it facilitates the quantification of the measures mitigation effect. Cities often only need this kind of high level overview to start taking urban heat into consideration in their urban plans. Deeper climatological studies can always be carried out to provide a more detailed assessment where needed.

The simultaneous analysis of several cities can help increase their awareness, and develop parallel mitigation plans, which can benefit from one another, sharing not only scientific and applied knowledge, but also implementation and management strategies. Remote sensing is specifically suited to carry out the analysis of several cities at the same time due to the large size of its scenes.

Replicability of the Study

The study can be replicated with a basic remote sensing and climatological knowledge and requires a certain command of the two main software utilised (ENVI 4.7 and ATCOR 2.3). One critical item is the selection of the satellite imagery, which should preferably be retrieved during a heat wave, and on a cloudless and windless day.

The satellite imagery used for this study is Landsat 5 TM however the same exercise can be performed with finer resolution satellite imagery, thus obtaining more accurate results.

Acknowledgements This research is funded by the Climate Proof Cities Consortium of the Knowledge for Climate research project (CPC 2014).

References

Asaeda T, Ca TV, Wake A (1993) Heating of paved grounds and its effect on the near surface atmosphere: exchange processes at the land surface for a range of space and time scales. In: Proceedings Yokohama Symposium, July 1993. IAHS Publ. 212, International Association of Hydrological Sciences, pp 181–187

Asrar G (1989) Theory and applications of optical remote sensing. Wiley Series in Remote Sensing and Image Processing. Wiley, New York, NY

Baudouin Y, Lefebvre S (2014) Urban heat island mitigation measures and regulations in Montréal and Toronto. Canada Mortgage and Housing Corporation

Bechtel B (2011) Multitemporal Landsat data for urban heat island assessment and classification of local climate zones. Urban Remote Sensing Event (JURSE)

Blankenstein S, Kuttler W (2004) Impact of street geometry on downward longwave radiation and air temperature in an urban environment. Meteorol Z 15:373–379

Brandsma T, Wolters D (2012) Measurement and statistical modeling of the urban heat island of the city of Utrecht, the Netherlands. J Clim Appl Meteorol 51:1046–1060

Cao L, Li P, Zhang L, Chen T (2008) Remote sensing image-based analysis of the relationship between urban heat island and vegetation fraction. The international Archives of Photogrammetry, remote sensing and spatial information sciences, vol XXXVII, part B7

Carlson TN, Dodd JK, Benjamin SG, Cooper JN (1981) Satellite estimation of the surface energy balance, moisture availability and thermal inertia. J Appl Meteorol 20:67–87

Carlson TN, Gillies RR, Perry EM (1994) A method to make use of thermal infrared temperature and NDVI measurements to infer surface soil water content and fractional vegetation cover. Rem Sens Rev 9:161–173

Carlson TN, Capehart WJ, Gillies RR (1995) A new look at the simplified method for remote sensing of daily evapotranspiration. Remote Sens Environ 54:161–167

Choudhury BJ, Ahmed NU, Idso SB, Reginato RJ, Daughtry CST (1994) Relations between evaporation coefficients and vegetation indices studied by model simulations. Remote Sens Environ 50:1–17

Climate adaptation for rural areas. http://www.knowledgeforclimate.nl/ruralareas/researchthemeruralareas/consortiumclimateadaptationforruralareas. Accessed Jun 2015

Climate Proof Cities (CPC). http://knowledgeforclimate.climateresearchnetherlands.nl/climateproofcities/background-information. Accessed Jan 2014

Coll C, Galve JM, Sánchez JM, Caselles V (2010) Validation of Landsat-7/ETM+ thermal-band calibration and atmospheric correction with ground-based measurements. IEEE Trans Geosci Rem Sens 48(1):547–555

De Groot–Reichwein MAM, Van Lammeren RJA, Goosen H, Koekoek A, Bregt AK, Vellinga P (2014) Urban heat indicator map for climate adaptation planning. Mitig Adapt Strat Glob Chang. doi:10.1007/s11027-015-9669-5

Doll D, Ching JKS, Kaneshiro J (1985) Parametrization of subsurface heating for soil and concrete using net radiation data. Bound-Lay Meteorol 10:351–372

Dousset B, Gourmelon F, Laaidi K, Zeghnoun A, Giraudet E, Bretin P, Mauri E, Vandentorren S (2011) Satellite monitoring of summer heat waves in the Paris metropolitan area. Int J Climatol 31:313–323

Exelisvis (2015) https://www.exelisvis.com/

Gallo KP, McNab AL, Karl TR, Brown JF, Hood JJ, Tarpley JD (1993) The use of NOAA AVHRR data for assessment of the urban heat island effect. J Appl Meteorol 32(5):899–908

Garssen J, Harmsen C, de Beer J (2005) The effect of the summer 2003 heat wave on mortality in the Netherlands. Euro Surveill 10(7):165–167

Heusinkveld BG, van Hove LWA, Jacobs CMJ, Steeneveld GJ, Elbers JA, Moors EJ, Holtslag AAM (2010) Use of a mobile platform for assessing urban heat stress in Rotterdam. Wageningen UR, Wageningen

Hoeven FD, Wandl A (2013) Amsterwarm: Gebiedstypologie warmte-eiland Amsterdam. TU Delft, Faculty of Architecture, Delft

Hove LWA, Steeneveld GJ, Jacobs CMJ, Heusinkveld BG, Elbers JA, Moors EJ, Holtslag AAM (2011) Exploring the urban heat island intensity of Dutch cities. Alterra report 2170. Wageningen, Netherlands

Klok L, Broeke HT, Harmelen TV, Verhagen H, Kok H, Zwart S (2010) Ruimtelijke verdeling en mogelijke oorzaken van het hitte-eiland effect. TNO Bouw en Ondergrond, Utrecht

KNMI (Koninklijk Nederlands Meteorologisch Instituut) (2006) 2050 climate scenarios. KNMI, De Bilt. http://www.knmi.nl/climatescenarios/#Inhoud_0. Accessed December 2014

KNMI (Koninklijk Nederlands Meteorologisch Instituut) (2014) http://www.knmi.nl/nederland-nu/klimatologie/lijsten/hittegolven

Knowledge for climate. http://www.knowledgeforclimate.nl/programme/background. Accessed Jun 2015

Koopmans S (2010) First assessment of the urban heat island in the Netherlands. Exploring urban heat and heat stress in the Netherlands, using observations from hobby meteorologists. BSc Thesis, Wageningen University, p 35

Kurn D, Bretz S, Huang B, Akbari H (1994) The potential for reducing urban air temperatures and energy consumption through vegetative cooling (31 pp). ACEEE Summer Study on Energy Efficiency in Buildings. American Council for an Energy Efficient Economy, Pacific Grove, CA

Lindberg E, Hollaus M, Mücke W, Fransson JES, Pfeifer N (2013) Detection of lying tree stems from airborne laser scanning data using a line template matching algorithm. In: Proceedings of ISPRS Annals II-5/W2, Antalya, Turkey, 11–13 November 2013

Liu L, Zhang Y (2011) Urban heat island analysis using the Landsat TM Data and ASTER Data: a case study in Hong Kong. Remote Sens 3:1535–1552

Meehl GA, Tebaldi C (2004) More intense, more frequent, and longer lasting heat waves in the 21st century. Science 305(5686):994–997

Morris CJG, Simmonds I, Plummer N (2001) Quantification of the influences of wind and cloud on the nocturnal urban heat islands of a large city. J Appl Meteorol 40:169–182

NASA. http://atmcorr.gsfc.nasa.gov/. Accessed 2014

Nichol J, Wong M (2004) Modeling urban environmental quality in a tropical city. Landsc Urban Plann 73:49–58

Odindi JO, Bangamwabo V, Mutanga O (2015) Assessing the value of urban green spaces in mitigating multi-seasonal urban heat using Modis land surface temperature (LST) and Landsat 8 data. Int J Environ Res 9(1):9–18

Oke TR (1973) City size and urban heat island. Atmos Environ 7(8):769–779

Oke TR (1981) Canyon geometry and the nocturnal urban heat island: comparison of scale model and field observations. J Climatol 1:237–254

Oke TR (1982) The energetic basis of the urban heat island. Q J Roy Meteorol Soc 108(455):1–24

Oke TR (1987) Boundary layer climates, 2nd edn. Routledge, London, pp 262–303, 8

Oke TR (1997) Urban environments. In: Bailey WG, Oke TR, Rouse WR (eds) Surface climates of Canada. Mc Gill-Queen's University Press, Montréal, pp 303–327

Onishi A, Cao X, Ito T, Shi F, Imura H (2010) Evaluating the potential for urban heat-island mitigation by greening parking lots. Urban For Urban Green 9:323–332

Park HS (1987) Variations in the urban heat island intensity affected by geographical environments, Environmental Research Center Papers 11. University of Tsukuba, Ibaraki

Parlow E (1998) Net radiation of urban areas. In: Proceedings of the 17th EARSeL symposium on future trends in remote sensing, Lyngby, Denmark, 17–19 June 1997. Balkema, Rotterdam, pp 221–226

Parlow E (2003) The urban heat budget derived from satellite data. Geograph Helv 58(2):99–111

Price JC (1979) Assessment of the urban heat island effect through the use of satellite data. Mon Weather Rev 107(11):1554–1557

Rajasekar U, Weng (2009) Spatio-temporal modelling and analysis or urban heat islands by using Landsat TM and ETM+ imagery. Int J Rem Sens 30(13):3531–3548

Richter R, Muller A (2005) De-shadowing of satellite/airborne imagery. Int J Rem Sens 26:3137–3148

Richter R, Schlapfer D (2013) Atmospheric/topographic correction for satellite imagery. ATCOR-2/3 User Guide, Version 8.2.1. http://www.rese.ch/products/atcor/atcor3/

Rigo G, Parlow E (2007) Modelling the ground heat flux of an urban area using remote sensing data. Theor Appl Climatol 90:185–199

Rosenzweig C, Solecki WD, Slosberg RB (2006) Mitigating New Yorks city's heat island with urban forestry, living roofs, and light surfaces. New York city regional heat island initiative final report. New York State Energy Research and Development Authority (NYSERDA), Albany, NY

Roth M, Oke T, Emery WJ (1989) Satellite-derived urban heat islands from three coastal cities and the utilization of such data in urban climatology. Int J Rem Sens 10:1699–1720

Sailor DJ (1995) Simulated urban climate response to modification in surface albedo and vegetative cover. J Appl Meteorol 34(7):1694–1704

Schar C, Jendritzky G (2004) Climate change: hot news from summer 2003. Nature 432:559–560

Steeneveld GJ, Koopmans S, Heusinkveld BG, Van Hove LWA, Holtslag AAM (2011) Quantifying urban heat island effects and human comfort for cities of variable size and urban morphology in the Netherlands. J Geophys Res 116:D20129

Svensson MK (2004) Sky view factor analysis: implications for urban air temperature differences. Meteorol Appl 11–3:201–211

Taha H (1997) Urban climates and heat islands: albedo, evapotranspiration, and anthropogenic heat. Energ Build 25:99–103

Taha H, Akbari H, Rosenfeld AH, Huand YJ (1988) Residential cooling loads and the urban heat island: the effects of albedo. Build Environ 23:271–283

Taha H, Akbari H, Sailor D, Ritschard R (1992) Causes and effects of heat islands: sensitivity to surface parameters and anthropogenic heating. Lawrence Berkeley Lab. Rep. 29864, Berkeley, CA

Unger J (2009) Some aspects of the urban heat island phenomenon. Thesis for the MTA doctor's degree

Uno I, Wakamatsu I, Ueda H, Nakamura A (1988) An observational study of the structure of the nocturnal urban boundary layer. Bound-Lay Meteorol 45:59–82

U.S. Environmental Protection Agency's Office of Atmospheric Programs, Climate Protection Partnership Division. Reducing urban heat islands: compendium of strategies cool pavements. http://www.epa.gov/hiri/mitigation/pavements.htm

U.S. Environmental Protection Agency's Office of Atmospheric Programs, Climate Protection Partnership Division. Reducing urban heat islands: compendium of strategies cool roofs. http://www.epa.gov/hiri/resources/pdf/CoolRoofsCompendium.pdf

USGS (US Geological Survey), Earth Resources Observation and Science Center (EROS) (consulted 2013). http://glovis.usgs.gov/

Voogt J (2002) Urban heat island. In: Munn T (ed) Encyclopedia of global environmental change, vol 3. Wiley, Chichester, pp 660–666

Yuan F, Bauer ME (2007) Comparison of impervious surface area and normalized difference vegetation as indicators of surface urban heat island effects in Landsat imagery. Remote Sens Environ 106:375–386

Zaksel K, Oštir K, Kokalj Ž (2011) Sky-view factor as a relief visualization technique. Remote Sens 3:398–415

Chapter 8
Planning and Climate Change: A Case Study on the Spatial Plan of the Danube Corridor Through Serbia

Tijana Crncevic, Omiljena Dzelebdzic, and Sasa Milijic

Abstract Spatial planning has an important role to play in addressing climate change issues, taking into account that contemporary global and other international and regional frameworks support an integrated approach in adapting to climate change. The aim of this paper is to emphasize the role of planning in the context of climate change with special reference to current practice and the legal and other planning frameworks in the Republic of Serbia. The example of the Spatial Plan for the Specific Purposes Area of the International Waterway E-80 Danube (Pan-European Corridor VII) is introduced, and issues related to the influence of climate change on water regimes are analysed by reviewing the presence of planning measures that cover natural and cultural heritage. The results indicate that despite the lack of an adequate legal base, current practice offers indirect support to adaptation measures through promoting ecological networks, increasing the extent of protected areas, and building systems for flood protection among others.

Keywords Climate change • Spatial planning • Adaptation measures • Natural and cultural heritage • Danube river basin

Introduction

The results of recent scientific studies indicate that the impacts of climate change are global but with regional differences. In Central and Eastern Europe, a decrease in summer precipitation is anticipated, with increased risk to health due to heat waves, a reduction in forest production, and more frequent forest fires (IPCC 2007). In Northern Europe, a reduction of heating costs is expected, with increased crop yields and forest growth, and in Southern Europe, impacts include reduced water

---

T. Crncevic (✉) • O. Dzelebdzic • S. Milijic
Institute of Architecture and Urban & Spatial Planning of Serbia, Bulevar Kralja Aleksandra 73, 11000 Belgrade, Serbia
e-mail: tijana@iaus.ac.rs

© Springer International Publishing Switzerland 2016
W. Leal Filho et al. (eds.), *Implementing Climate Change Adaptation in Cities and Communities*, Climate Change Management, DOI 10.1007/978-3-319-28591-7_8

availability, reduced hydropower potential, reduced summer tourism, and reduced crop yields (IPCC 2007). According to the latest IPCC Assessment Report, climate change will increase the likelihood of systemic failures across European countries caused by extreme climate events affecting multiple sectors (IPCC 2014). Future projections indicate increased extreme precipitation in Continental Europe. As such, the hydrology of river basins will also be affected, where the occurrence of current 100-year return period discharges is projected to increase in the Danube basin. Across most of Northern and Continental Europe, an increase in flood hazards could increase damages to crops and plant growth, and increase yield variability. The provision of ecosystem services in Southern Europe is projected to decline across all service categories in response to climate change, where other European sub-regions are projected to experience both losses and gains. There are fewer studies for cultural services, although these indicate a balance in service provision for the Alpine and Atlantic regions, with decreases in service provision for the Continental, Northern, and Southern sub-regions (IPCC 2014).

The Republic of Serbia belongs to the region of Continental (or Southeast) Europe, which is recognized as an area sensitive to climate change, with recorded total average economic losses from climate related disasters of 200 million dollars a year (RS MEMSP 2010; SEEFCCA 2012). Although predictions of the impacts on climate change are uncertain, it has been pointed out that there is already enough data to consider risks regarding the increase in air temperature, availability of drinking water, loss of biodiversity, food quality, and overall living conditions (Dzelebdzic et al. 2013). To achieve the transition to a climate-smart world it is necessary to act now, to act jointly, and to act on several fronts (World Bank 2010).

At the global level, the Intergovernmental Panel on Climate Change (IPPC) is a benchmark organisation that not only collects scientific knowledge on climate change and monitors the work of other organisations (the Global Climate Observation System, the World Climate Program and others), but also provides appropriate guidance for mitigation and adaptation (http://www.ipcc.ch). As a reaction to increasingly visible climate change impacts, the *United Nations International Strategy for Disaster Reduction* (UNISDR 2000) and the *Hyogo Framework for Action 2005–2015* can be particularly emphasised for promoting a joint effort to establish the conditions required for mitigating disaster risk and reducing losses. The UN Conference on climate change in Montreal (2005) is particularly important because of the acceptance of the pan-European emissions trading scheme and the Clean Development Mechanism, a tool to promote sustainable development and combat climate change. Within Europe, in addition to programs and activities that encourage mitigation of climate change, the *European Climate Change Programme* (2003) (http://ec.europa.eu/clima/policies/eccp/index_en.htm) also places adaptation within the legal framework and the other regulative frameworks of the European Union. The *White Paper on Adapting to Climate Change* (2009) refers to the significant role that spatial planning plays in promoting efforts to reduce greenhouse gas emissions (GHG), improve energy efficiency, and increase the uptake of cleaner fuels and renewable energy sources. Today, the premise of contemporary sustainable development requires full comprehension of climate change issues. It is therefore necessary to establish adapting to climate change as a key objective of spatial planning. To achieve this will require an assessment of vulnerability to climate change impacts in all sectors of spatial planning, while spatial plans and adaptation measures will have to be revised in relation to the latest scientific knowledge in order to ensure they are effective over the long term (ESPACE 2008).

The Republic of Serbia (RS) has signed and ratified the Kyoto Protocol (RS 1997) that, along with United Nations Framework Convention on Climate Change (UNFCCC), represents the basic regulatory framework. Serbia have no obligations to reduce emissions but do have an obligation to fulfil general obligations as well as the ability to use the *Clean Development Mechanism* (CDM). However, although Serbia, has a high potential for production of energy from renewable sources, its "characteristic is lack of developed institutions and appropriate procedures that are considered as one of the key factors for adaptation to climate change" (Pucar 2013: 59).

Taking into account the above mentioned, the aim of the paper is to present the current legal basis and practice in spatial planning in Serbia in the context of climate change. In addition to an analysis of the representation of climate change issues in legal and other regulatory frameworks for spatial planning, the Spatial Plan for the Specific Purposes Area of the International Waterway E-80 Danube (Pan European Corridor VII) (Official gazette of RS No. 14/15) is presented with an aim to give an insight into current practice.

Planning and Climate Change in Serbia

Serbia is comprised of three main landforms: in the north, the Pannonia Plain covers about 25 % of the territory, while the central and southern parts are covered by hilly areas with lower and lowland expansions. The Danube, which flows through the north of Serbia, connects Eastern and Western Europe both geographically and strategically. It serves as the pan-European water corridor (the most important European waterways) that (together with the Rhine and Main) connects the Black Sea and the North Sea, and is the backbone for the inland waterway networks. The Danube is navigable throughout its course in Serbia and has the importance of an international waterway. The total length of the Danube in Serbia is 558 km, which represents one quarter of the total waterway of 2145 km. The greatest part of its course through Serbia is through the plains, while the second part, which runs through the mountainous regions, is characterized by a high percentage of forests which represent the primary factor in protecting soils from erosion (The National Park "Djerdap", municipalities and cities of Majdanpek, Negotin, Kladovo) (Fig. 8.1).

In the period 1990–2000, land use in Serbia changed on 1.1 % of the territory, where agricultural areas were reduced by 8473 ha while the land under forest increased by 36,419 ha (First Report of RS to the UNFCCC 2011). As indicated in the First Report of RS to the UNFCCC (2011), forests are notably vulnerable

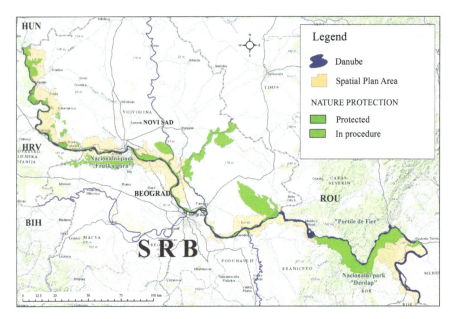

Fig. 8.1 The area of the spatial plan for specific purposes area of the International Waterway E-80 Danube (Pan-European Corridor VII)

ecosystems, while other extremely endangered ecosystems such as wetland and marshland habitats, steppes and forest steppes, sands, continental slats and high mountains habitats are also indicated, where these very sensitive habitats include habitats of relict and endemic species.

The climate in Serbia is moderate continental with more or less expressed local characteristics that mainly relate to the mountain areas (over 1000 m above sea level) where there are characteristics for continental climate. The average annual temperature is 10.1 °C and annual rainfall is 734 mm. January is the coldest month with an average temperature of −1.3 °C and the warmest is July with an average of 19.9 °C. Over the last 50 years, meteorological data indicate a continuous rise in temperature. In the autumn period, the mean annual temperature has a positive trend of 0.4 °C over the whole territory of Serbia, while in the southeast part of the country there is a negative trend of 0.05 °C per year (Popovic et al. 2005, 2009; First Report of RS to the UNFCCC 2011). A further increase in mean annual temperature of 0.8–1.1 °C by 2030 is projected, while precipitation is expected to increase slightly with values up to +5 % across most of the territory (First Report of the RS to the UNFCCC 2011).

In recent years, extreme climatic events have been evident in Serbia—this has included floods (Obrenovac, Tekija), which set off landslides and erosion, more frequent and longer periods of heat waves, droughts, and severe windstorms. In relation to the total number of natural disasters, floods are recorded most frequently (55 %) (WWF 2012). However, although the effects of extreme events are increasingly emerging, the systematic collection of impact data and development of a database for all strategic areas has not yet been established, and a full assessment of affected sectors and systems has not been carried out (First Report of RS to the UNFCCC 2011). Furthermore, the lack of financial, technical and technological capacity to manage and respond to these extreme events has also been emphasised (First Report of RS to the UNFCCC 2011).

Overview of Planning Legislation in Serbia

As discussed above, the observed impacts of climate change have already highlighted the sensitivity of natural and human systems to current and future climate change (IPCC 2014). The role of spatial and urban planning in this context requires recognition and mitigation of anthropogenic causes of climate change (GHG emissions and deforestation), applying strategic goals and interventions to promote measures such as increasing renewable energy supply, reducing the need for transportation, the use of specific construction techniques and enhanced energy efficiency, reforestation, and others. In addition, spatial planning deals with climate change impacts (droughts, intense rainfalls, sea level rise and the heat waves) by supporting adaptation measures. This includes locating development in areas that are away from the risk zones (floods, coastal erosion), water resource management, torrent control, considering the condition of river embankments, providing shelter for population at risk, and providing appropriate infrastructure and services (APA 2011; Davoudi 2009; IPPC 2014)

In contemporary spatial planning practice, areas of natural and cultural heritage are already identified as "sensitive" and specifically designated. In order to provide 'climate-proof' areas, the need to review generally accepted planning principles[1] of these areas has been highlighted. Specifically, by promoting appropriate measures in relation to: (1) *Designation of critical areas for conservation of natural resources;* (2) *Water management;* (3) *Increasing the resilience of natural areas;* (4) *Cultural heritage monitoring and the prevention of damage* (see Table 8.1).

During the last decade, the legislation in Serbia has been significantly innovated in accordance with the European Union *Acquis*. The main law that regulates spatial and urban planning is the Law on Planning and Construction (2014) and it determines the preparation of:

1. Planning documents [Spatial plans—Spatial Plan of the Republic of Serbia (SPRS), Regional Spatial Plan (RSP), Spatial Plan for the Local Government Units (SPLGU) and Spatial Plan for the Specific Purpose Area (SPASP); Urban

---

[1] Planning in these areas is based on the principles of ensuring the preservation of authenticity, integral protection of natural and cultural heritage, controlled use of resources, the minimum intervention and maintaining on an permanent basis (UNESCO 2007).

Table 8.1 Providing climate-proof areas within spatial planning—overview of the potential measures (Mittermeier et al. 2005; Bomhard 2005; Morecroft et al. 2012; Dzelebdzic et al. 2013)

Area of action	Potential measures
Designation of critical areas for conservation of natural resources	Mapping vulnerable areas; establishing rules for land use within affected areas; integration of climate change scenarios in risk evaluation and risk management processes
Water management	Establishment and implementation of appropriate measures for flood protection; water conservation of protected areas, as well as the infrastructure network for irrigation and drainage; maintenance and restoration of wetlands and water courses
Increasing the resilience of natural areas	Increasing the number of protected areas; planning ecological corridors and habitat networking and providing functional connectivity between protected areas within the landscape; maintaining the diversity of species in the community; establishment of potential refugia
Cultural heritage monitoring and the prevention of damage	Taking planning measures that treat the integral cultural and natural heritage; every cultural heritage site must be considered vulnerable to climate change; implementation of regular repairs and restoration of cultural heritage sites; promoting research on how to avoid damages

plans—General Urban Plan (GUP), General Regulation Plan (GRP) and Detailed Regulation Plan (DRP)];
2. Documents for implementation of regional plans (program for implementation of SPRS, program for implementation of RSP, program for implementation of SPASP) and
3. Urban—technical documents for implementation of planning documents (urban project, project for land parcellation and subdivision).

The law on SPRS (2010) in relation to the concept of spatial development supports two approaches. Firstly, the general concept for development based on knowledge about observed and expected climate changes and associated impacts on the availability of natural resources (Phase I). Secondly, sectorial concept development which precisely scrutinizes climate change impacts. As a priority, the SPRS, among other items, determines the preparation of multidisciplinary studies covering local climate change and the impacts of climate change on agriculture, renewable energies, water management, biodiversity, ecosystems, etc.

Table 8.2 provides an overview of the presence of climate change issues in the current legal framework covering spatial planning in Serbia. The analysis covered the overview of selected regulations and analyses of incorporation of climate change issues and the presence of general strategic trends referring to energy provision and energy needs and promotion of adaptation measures.

Table 8.2 Climate change issues in planning framework legislation

Law on Spatial Plan of the Republic of Serbia 2010–2014–2021 (Official gazette RS no. 88/10)—gives the bases for the inclusion of climate change issues in the spatial and urban planning
Law on Planning and Construction (Official gazette of RS No. 79/09, 81/09-correction, 64/10-US, 24/11, 121/12, 42/13-US, 50/13-US, 98/13-US, 132/14 and 145/14)—stands for rational use of non-renewable natural resources and renewable energy sources and promotes energy efficiency in the sense of producing "Certificate of the energy performance of the building"
Law on Strategic Environmental Assessment (Official gazette of RS no. 88/2010)—contains an obligation to consider the ways for the assessment to take into account the effects of environmental factors, which includes climate
Law on Environmental Impact Assessment (Official gazette of RS no. 135/2004.36/2009)—provides the obligation for data collection and for forecasting harmful impacts of projects on, inter alia, the climate
Law on Environmental Protection (Official gazette of RS no. 135/2004)—does not consider climate change issues specifically
Law on Emergency Situations (Official gazette RS no. 111/09, 92/11, 93/12)—include measures aiming to strengthen the resilience of the community and includes the obligation to obtain protection conditions for prevention of natural disasters (floods, droughts, torrents, storms, heavy rain, atmospheric discharges, hail or landslides, avalanches and snow layers, extreme air temperatures, ice accumulation on the watercourse) and their inclusion in planning documents
Law on Nature Protection (Official gazette. RS no. 36/2009, 88/2010)—promotes development of the areas within the program NATURA 2000 and development of ecological networks in accordance with the ***Regulation on the ecological network*** (Official gazette RS no. 102/2010)
Rule books on criteria for separation of habitat types, about habitat types: sensitive, vulnerable, rare, for the protection of priority habitat types and about measures for their preservation (Official gazette of RS no. 35/2010)—establishes the measures for the protection and preservation, and within the criteria includes "endangered habitat types—ones that are under threat of extinction due to the effects of anthropogenic and/or natural factors, in the territory of the Republic of Serbia" (Article 4)
Regulations on compensatory measures (Official gazette of RS no. 20/2010)—establishes measures for the areas that are ecologically important or protected only in cases of impacts induced by the projects, works and activities in nature (Article 1)
Law on Ratification of the European Landscape Convention (Official gazette RS, International Agreements, no. 4/2011)—there are no special provisions regarding climate change issues
Law on Cultural Property (Official gazette of RS no. 71/94)—there are no specific provisions to address climate change
Law on forests (Official gazette RS, No. 30/10 i 93/12)—Article 6, paragraph 10 states, the favourable impact of forest on the climate; according to Article 45, in the case of disruption of the biological balance and serious damage within forest ecosystems caused by natural disasters, the Ministry in charge defines protection measures, provides rehabilitation and financial support, while forest recovery is done by forest users

Considering the information presented in Table 8.2, it can be concluded that the current legal framework for spatial planning foresees the inclusion of climate change issues, primarily by representing the use of renewable energy, energy efficiency, prevention and protection against natural disasters, elimination of climate change causes, conservation and sustainable use of natural resources and in particular promotion of the ecological network NATURA 2000. In addition to these regulatory frameworks, other important strategic documents are formulated to promote sustainable development in line with contemporary frameworks and they also, in accordance with their strategic and sectorial frameworks of action, promote climate change issues.[2]

Based on the above, it can be concluded that adaptation activities are partially represented in the legal and planning framework. However, there is no corresponding national strategic document that addresses adapting to climate change. Development of a system for adaptation is still not recognised as a priority, and therefore it is not adequately embedded in the process of spatial planning. The current legal and planning basis is, it could be said, missing an adequate response to the potential risks of climate change, especially in the fields of impact research and implementation of planning instruments. As a methodological framework for planning in the context of climate change is missing, spatial planning in Serbia indirectly represents the planning framework in the context of climate change, primarily by increasing protected areas, defining the protection regime, planning of ecological corridors and networking habitats, measures for protecting ecosystems and others (Crncevic 2013). Additionally, current practice shows that, although there is legal basis, the issue of climate change is not systematically addressed in strategic environmental assessment (SEA) (Crncevic et al. 2011). In particular, a key prerequisite for adequate consideration of climate change impacts is availability of data regarding risk zones and other custom data and indicators that already integrate data on climate change (Bazik and Dzelebdzic 2011).

Spatial Plan of the Danube Corridor Through Serbia: Towards Planning Climate Resilient Areas

The Danube Corridor: The European Context

The waterways of the Danube corridor are multifunctional, serving not only as a major European transport corridor, but also as a touristic and cultural corridor, and an ecological corridor of international importance that forms an integral part of the Pan-European ecological network. There are 390 protected areas of special national and international significance within the Danube basin in Serbia. These include several national parks in addition to archaeological sites from the Mesolithic

---

[2] National Sustainable Development Strategy (Off. gazette RS no. 57/2008), National Strategy for inclusion of RS in the CDM mechanisms—waste management, agriculture and forestry (Off. gazette RS no. 8/2010), National Strategy for the sustainable use of natural resources and goods (Off. gazette RS no. 33/2012), National Environmental Protection Programme (Off, Gazette RS no. 12/10), Biodiversity Strategy of the Republic of Serbia for the period 2011–2018. (2011) and others.

(Lepenski Vir) and Neolithic (Starcevo and Vinca) periods, the Roman Road in Djerdap and several Roman forts, the Medieval sites at Smederevo, Golubac, and Vratna, and the Turkish Fetislam fortress near Kladovo (RS 2010).

Taking into the account the policy of the Republic of Serbia regarding EU integration and EU interest in the Danube corridor, the liability for protection and development of this waterway has been agreed for the RS territory. From a number of international frames, the comprehensive EU Strategy for the Danube Region (The Danube Strategy 2010) can be singled out. The Danube Strategy was adopted by the European Commission in December 2010 and its priorities are better connectivity among the Danube region, environmental protection, improving prosperity in the region, developing institutional capacities, and cooperation and safety. One of the key tasks of the Danube Strategy is the improvement of the Danube as a transport corridor within the Rhine–Main–Danube system, thereby providing better connectivity between the Baltic and Black Seas and resulting in less costly overseas trade between Europe and Asia.

Although water transport is ecologically the most favourable form of transportation, transport activities must be based on the Danube Commission Recommendations for the Prevention of Water Pollution of the Danube caused by Water Navigation (Danube Commission 1998) and negative impacts on the Danube corridor environment must be minimized in accordance with the Espoo Convention on Environmental Impact Assessment in a Transboundary Context (UNECE 1991), the Danube River Protection Convention (ICPDR 1994), and the EU Water Framework Directive (WFD; EC 2000) and Floods Directive (EC 2007) among others.

The Danube River Protection Convention is the main legal instrument for cooperation and transboundary management in the Danube basin. This Convention is implemented and coordinated by the International Commission for Protection of the Danube River (ICPDR) and provides mutual priorities and strategies for improving the quality of the Danube and its tributaries. The Republic of Serbia ratified the Danube River Protection Convention and became a member of the ICPDR in 2003 (RS 2003).

The cooperative activities of the Danube Commission, the ICPDR, and the International Commission for the Sava River Basin (which has the largest discharge of water to the Danube of any tributary), play a significant role in implementing both the Danube River Protection Convention and the WFD.

In terms of sustainable use and protection of water, a key factor is implementation of the WFD, which places an obligation on all countries of the Danube basin to prepare River Basin Management Plans (RBMPs). The WFD has a series of deadlines for the achievement of good ecological status for all surface and underground water bodies by 2015 (or good ecological potential for significantly modified and artificial water bodies). The aims, conditions and obligations defined by this Directive were developed in the first Danube RBMP (DRBMP 2009) which is a platform for coordination of activities and implementation of the WFD at the level of all 14 countries within the Danube basin. With respect to the Floods Directive, the main aim is to reduce and to manage floods risks for inland and coastal waters within EU territory, where the Directive requires identifying areas that are at risk of flooding, producing flood risk maps, and establishing flood risk management plans.

Introduction to the Spatial Plan of the Danube Corridor Through Serbia

The basic natural axis of the Spatial Plan for the Specific Purposes Area of the International Waterway E-80 Danube (Pan-European Corridor VII) (2015) is the Danube River. The Danube corridor includes significant known and potential natural resources including good quality agricultural land (Pannonian lowland), coal (Kostolac coal basin), oil (Stig and Banat), copper ore (Majdanpek and Bor), hydropower potential of the Danube (Iron Gates hydropower plants) and diverse flora and fauna (Fig. 8.1). Taking into account the remarkable natural potential and importance of the area as the most important Pan-European water transport corridor, the "immediate goal of the Spatial plan is to contribute to planning, development and protection of the Danube primarily as [a] transport corridor, but also as water, ecological, cultural and tourist corridor" (IAUS 2012). Therefore, it should be noted that the main task of the Spatial Plan is to set the conceptual basis for the integrated and safe development of the Danube as a Pan-European transport corridor, together with solutions that will enable the preservation of values and evaluation of the potential of this corridor as part of the Danube region in Serbia. This includes consideration of local community development in the coastal zone, adequate customized management system development, environmental protection, applying river information services, and others. The main goals of long-term development, use and regulation of the area, among other things, include:

- Reduction of spatial conflicts associated with the waterway, the Danube (corridor) and water management, with spatial development and the protection of resources and heritage sites in the area of direct influence of the waterway corridor;
- Environmental protection and sustainable use of natural and cultural heritage for the development of tourist destinations and nautical tourism in the Danube corridor;
- Integrated development of the waterway, other infrastructure systems and the environments, in two aspects: integration and greater role of the Danube waterway in improving traffic in the region and likewise greater impact on economic and social development of the region;
- Altering waterways to improve connectivity and increasing the involvement of other waterway networks in domestic traffic, with the aim of functional and efficient connection with the Danube;
- Strengthening functional integration and cross-border cooperation of urban centres and economic activities (particularly waterpower engineering, energy, agriculture, industry and tourism) in the Danube Corridor (IAUS 2012).

The priority of international cooperation intends to establish cooperation with the countries within the Danube river basin with the aim to achieve the objectives of the WFD and the Floods Directive for that part of the Danube river basin within the RS for the period until 2015. The plan promotes cooperation between the Danube river basin countries, particularly in relation to:

- The development and harmonization of assessment of water condition and effects of the measures taken on the water;
- The estimation of biotope elements for river ecosystem and landscape and biological quality elements for the assessment of ecological status and potential, information systems; and
- The inventory of trans-border emissions and transnational monitoring network for the Danube river basin.

Thus the Plan particularly emphasises cross-border cooperation based on joint activities and measures for the sustainable development of the area to protect and enhance natural and cultural heritage (see Fig. 8.1), especially in relation to cross-border biosphere reserves (Danube–Drava–Mura, Carpathian areas, network of Danube parks), the European green belt (green corridor along the border with Romania and Croatia) and the European cultural route (Danube *limes,* Roman Emperors Route, medieval castles and fortresses on the Danube).

Analysis of the Spatial Plan of the Danube Corridor Through Serbia

Taking into consideration the information provided in Table 8.1 regarding the selected areas of action and related measures for providing climate proof areas, the contents of the corresponding Plan sections (water management, cultural heritage and nature protection) were analysed for the presence of the given measures and then classified. The results of this analysis are presented in Table 8.3.

It can be seen that climate change issues are represented within the Plan in terms of adaptation measures for water management and increasing the resilience of natural areas. However, the use of climate change scenarios for risk assessment, the establishment of refugia, and research on the vulnerability of cultural heritage sites to climate change, are absent. In addition, the results of our analyses indicate a significant impact of the international framework in promoting climate change issues, particularly through the development of a network of protected areas NATURA 2000, and the integral protection of natural and cultural heritage. Therefore, it can be concluded that the Plan contains some elements for establishing a climate-proof area and indirectly represents the planning framework in the context of climate change.

Table 8.3 Climate change issues in the spatial plan of the Danube corridor through Serbia

Climate change issues	Spatial plan responses
Classification of critical areas for natural resources preservation	
The mapping of vulnerable areas and the establishment of rules for land use in endangered areas	– Re-cultivation of the barren land (providing functional status); protection of soil from water erosion; rehabilitation of degraded forest ecosystems
Integrating climate change scenarios in risk analysis	– Not mentioned
Water management	
Development and implementation of appropriate measures for flood protection	– Maintenance of the system for flood protection in accordance with the standards and their amendments where there is a breakthrough according to the indicators defined by criteria for protection, with an urgent remedy with all non-reliable parts and defensive line embankment (Golubac, Veliko Gradiste, etc.), reconstruction and preparation of the embankment II defensive line and preparing cassettes for localization of floods in a case of breakthrough; defining the belts of floodplain along both banks of the Danube, particularly in the area of Hydro Power Plant "Djerdap 1" and "Djerdap 2" and removal of buildings and facilities from these zones which endanger the security of protection systems (embankments and systems for coastal protection), prevent possible elevation/expansion and reconstruction of the embankment and the use of machinery during the flood defence period
Conservation of protected water areas and infrastructural irrigation and drainage	– Creating conditions for irrigation of the highest quality classes of soils in defensible coastal areas (especially on the network of channels of the system Danube Tisa Danube (DTD) as a vital resource for both transport and agriculture and water management); defining "floodplain in the spatial plans of local self-management units exiting to the Danube, according to the established principle of the Spatial Plan, that floodplain includes a belt width of 100 m from the dike base on the protected area, and this protection should encompass all areas of drainage system for the protection of coastal agricultural land"
Maintenance and restoration of wetlands and river beds (they have the role of natural barrier against floods)	– Increase the area of land which is permanently covered by water in nature reserves—the achievement of good ecological status (especially in localities Karapandza, Siroki rit (Wide Swamp), Monostorski rit (swamp) northern alternative severna varijanta, Bestrement, Strbac i Kosare) ensuring

(continued)

Table 8.3 (continued)

Climate change issues	Spatial plan responses
	constant water level in the protected part of the reserve and the revitalization of the water ecosystems—backwaters, ponds and wetlands; construction works (digging canals, culverts and setting a constitution for the water flow control) on the part of Danube (dunavci) where the water flow is difficult or impossible (Staklarski dunavac, Misvald, Lasufok, Dondo, etc.), and regulation of flow and water level in the protected area; preservation and restoration and activation of the network backwaters and other local waters as a necessary compensation measure for violating the integrity of aquatic habitats by development of river transport (air pollutants, noise, waves); conservation of natural resources, rehabilitation of degraded areas resulting from the arrangement of the waterway
Increasing the resilience of natural areas	
Increasing the number of protected areas and increasing the extent of individual protected areas	– Total extent of the protected areas and proposed protected nature reserves covered by the Regional Plan is approximately 1173 km^2 (about 1072 km^2 of existing and approximately 101 km^2 is in the process or planned for protection). Of this, about 81 km^2 comprises the immediate protected zone of the waterway (about 77.4 km^2 of existing and approximately 4 km^2 in the process or planned for protection). Protected areas that exit to the Danube banks have a total length of about 265 km (of which about 230 km of existing and around 35 km in the process or planned for protection); improvement of the actual status by increasing the area under forests and enhancement of forest infrastructure
Maintaining the diversity of species within the community	– Maintaining species diversity of wild flora and fauna and its increase by reintroduction of missing indigenous species; regeneration of indigenous forest species throughout the coastal area with breeding programmes of indigenous species of large and small deer, birds and fish; sustaining ecosystem diversity and protection of natural and agro ecosystems from invasive species of plants and animals, and other species, varieties and breeds that bring undesirable changes in natural and agro-biodiversity

(continued)

Table 8.3 (continued)

Climate change issues	Spatial plan responses
Planning ecological corridors and habitat networking	– Identification and mapping of habitats as substrates for functional establishment of a national ecological network and preserving, strengthening, numerically strengthening and the expanding populations of protected or rare, endangered and critically endangered plant and animal species; preservation or artificial establishment of ecological corridors; identification of internationally important ecological areas or habitats of European importance for the protection of wild fauna and flora by the NATURA 2000 programme
Establishing potential refugia	– Not mentioned
Monitoring of cultural heritage and the prevention of damage	
Integrated treatment of cultural and natural heritage and conducting regular repairs and restoration of cultural heritage sites	– Integral protection of material and non-material cultural values and natural values of the Spatial Plan area; undertaking technical measures for the protection of immovable cultural heritage and defining broader protection zones around cultural heritage
Consider cultural heritage vulnerable to climate change	– There is none
Promoting research on how damage can be at least sometimes avoided	– Involvement in international programs for the protection of cultural heritage and European roads of culture ("Cultural Heritage—a Bridge towards a shared future", "The Roman Emperors" and others, priority is given to research and protection of folk architecture, which should be preserved in its original purpose)

Concluding Remarks

Although climate change is a global issue, responding to climate change requires active engagement and action at regional and local levels. At the regional and local level, recent years have seen the development of legal frameworks that specify measures for climate change adaptation and for acting in cases of disasters, where the same tendency is evident in Serbian spatial planning policy.

An analysis of the Spatial Plan of the Danube corridor through Serbia demonstrated that in practice climate change issues are indirectly represented within spatial planning in Serbia. In particular, the Plan promotes measures such as the establishment of rules for land use in vulnerable areas, the establishment and implementation of appropriate measures for flood protection, preservation of protected water areas, maintenance and restoration of wetlands and river beds, increasing the number and area of protected areas, planning ecological corridors, and habitat networking.

The implementation of the Spatial Plan for the Specific Purposes Area of the International Waterway E-90 Danube represents a methodological step forward with respect to the analysis of flood zones and defining floodplains in the Danube corridor. The floodplains in the Danube corridor, in addition to water management infrastructure, has segments with high natural, cultural, and tourism capital, together with the coastal belts of the cities and settlements. The priorities related to the solution of the Danube navigability limits are defined, development of the efficient system of ports and planning of the nautical route, preservation of the natural resources and environmental protection, with emphasis on ecologically vulnerable habitats and securing the conditions for implementation. In addition, it can be considered that the Plan contains elements for establishment of a climate resilient area. This is presented through analysis of the critical places for preserving natural capital, the rules for water management and floodplains, ways for developing the resilience of natural areas, monitoring cultural heritage sites, and measures for prevention of damage.

A key pre-requisite for spatial planning to play a proactive role in responding to climate change is the availability of data in order to be able to identify and map areas at risk, both now and in the future. The evident lack of measures such as mapping of vulnerable areas and integration of climate change scenarios in the process of evaluating risks and risk management is also the basis for further adaptation to climate change. Therefore, future efforts should be directed towards the development of an appropriate methodological framework and the methods and techniques that would enable monitoring current and predicting future climate change, and associated impacts, with more certainty. This would then inform the planning regime, ensuring adaptive measures also considered the extent of future potential changes.

Acknowledgments This work is a result of research conducted within the research project "Sustainable Spatial Development of Danube Area in Serbia" (TR36036) which is financed from the program Technological development by the Ministry of Education, Science and Technological Development of the Republic of Serbia from 2011 to 2015.

References

APA (2011) American Planning Association, policy guide on planning and climate change. www.planning.org/policy/guides/pdf/climatechange.pdf. Accessed 20 Nov 2014

Bazik D, Dzelebdzic O (2011) Adapting to climate change—the new role of spatial planning (The case of the Danube region in Serbia)/Prilagodjavanje klimatskim promenama—nova uloga prostornog planiranja (primer Podunavlja u Srbiji). In: Uticaj klimatskih promena na planiranje i projektovanje. Arhitektonski fakultet Univerziteta u Beogradu, pp 66–84 (in Serbian)

Bomhard M (2005) Securing protected areas in the face of global change: lessons learned from the South African Cape Floristic Region. A report by the ecosystems, protected areas, and people project. IUCN and SANBI, Bangkok and Cape Town

Crncevic T (2013) Planiranje i zastita prirode, prirodnih vrednosti i predela u kontekstu klimatskih pormena u Republici Srbiji—Prilog razvoju metodoloskog okvira. Posebna izdanja br.72. Institut za arhitekturu i urbanizam Srbije, Beograd (in Serbian)

Crncevic T, Maric I, Josimovic B (2011) Strategic environmental assessment and climate change in the Republic of Serbia—support to development and adjustment process. SPATIUM 26. IAUS, Belgrade, pp 14–19

Danube River Basin Management Plan (DRBMP) (2009) http://www.icpdr.org/main/activities-projects/danube-river-basin-management-plan-2009. Accessed 30 Jun 2015

Davoudi S (2009) Framing the role of spatial planning in climate change. Electronic working paper no. 43. www.nlc.ac.uk/publications/working/documents/EWP43.pdf. Accessed 15 Jan 2013

Directive 2007/60/EC on the assessment and management of flood risks (2007) http://ec.europa.eu/environment/water/flood_risk/. Accessed 30 Jun 2015

Dzelebdzic O, Bazik D, Crncevic T (2013) Vulnerability of natural and cultural heritage in relation to climate change—new challenge for spatial and urban planning. In: Conference Proceedings, 2nd International Scientific Conference "Regional development, spatial planning and strategic governance". IAUS, Belgrade, pp 808–822

Environmental Agency, Halcrow (2008) ESPACE project: climate change impacts and spatial planning decision support guidance. www.espace-project.org. Accessed 20 Jan 2013

European Union Strategy for the Danube Region (2010) http://ec.europa.eu/regional_policy/en/policy/cooperation/macro-regional-strategies/danube/. Accessed 30 Jun 2015

First Report of Republic of Serbia (RS) for UNFCCC (2011) www.ekoplan.gov.rs. Accessed 20 Aug 2014

Institute for Architecture and Urban and Spatial Planning of Serbia (IAUS) (2012) Spatial plan for the specific purposes area of the International Waterway E-80 Danube (Pan-European Corridor VII), documentation base (in Serbian)

Intergovernmental Panel on Climate Change (IPPC) (2007) Climate change 2007: synthesis report, summary for policymakers. Cambridge University Press, Cambridge

Intergovernmental Panel on Climate Change (IPPC) (2014) Climate change 2014: synthesis report. In: Core Writing Team, Pachauri RK, Meyer LA (eds) Contribution of working groups I, II and III to the fifth assessment report of the Intergovernmental Panel on Climate Change. IPCC, Geneva, p 151

International Commission for the Protection of Danube River (ICPDR) (1994) Danube River Protection Convention. http://www.icpdr.org/main/icpdr/danube-river-protection-convention. Accessed 30 Jun 2015

Law on Planning and Construction (Official gazette of RS No. 79/09, 81/09-correction, 64/10, 24/11-US, 121/12, 42/13-US, 50/13-US, 98/13-US, 132/14 and 145/14)

Law on Spatial Plan of Republic of Serbia (SPRS), Official gazette of RS No. 88/10

Mittermeier RA, Robles Gil P, Hoffman M, Pilgrim J, Brookes T, Goettsch Mittermeier C, Lamoreux J, Fonseca GAB (2005) Hotspots revisited: earth's biologically richest and most endangered terrestrial ecoregions. Conservation International, Siera Madre, University of Virginia, Mexico

Montreal UN Climate Change Conference (2005) http://unfccc.int/meetings/montreal_nov_2005/meeting/6329.php. Accessed 20 Oct 2014

Morecroft MD, Crick HQP, Duffield SJ, Macgregor NA (2012) Resilience to climate change: translating principles into practice. J Appl Ecol 49:547–551

Popovic T, Radulovic E, Jovanovic M (2005) Koliko nam se menja klima, kakva ce biti nasa buduca klima? In: EnE05—Zivotna sredina ka Evropi. Beograd, pp 210–218 (in Serbian)

Popovic T, Djurdjevic V, Zivkovic M, Jovic B, Jovanovic M (2009) Promena klime u Srbiji i očekivani uticaji, RS Ministarstvo zastite zivotne sredine, Agencija za zastitu zivotne sredine,

Peta regionalna konferencija "EnE09—Zivotna sredina ka Evropi". www.sepa.gov.rs. Accessed 20 Feb 2014

Pucar M (2013) Energy aspects of development settlements and climate change—status, opportunities, strategies and legislation in Serbia. In: Pucar M, Dinitrijevic B, Maric I (eds) Climate change and the built environment, policies and practice in Scotland and Serbia. Special issues 70. Institut za arhitekturu i urbanizam Srbije, Glasgow Caledonian University, pp 57–108

Republic Serbia (2010) Pozicija Republike Srbije za ucesce u izradi sveobuhvatne strattegije Evropske Unije za region Dunava, prezentacija postubovima. http://www.srbija.gov.rs/vesti/dokumenti_sekcija.php?id=126300. Accessed 15 Jul 2015 (in Serbian)

Republic of Serbia (1997) Law on ratification of the Kyoto protocol to the United Nations framework convention on climate change (Official gazette of RS No. 2/97)

Republic of Serbia (2003) Law on ratification of the convention on cooperation for the protection and sustainable use of the River Danube (Official Gazette of the FRY—International Treaties, No. 2/03)

RS MEMSP (2010) Initial National Communication under the United Nations framework convention on climate. Republic of Serbia, Ministry of Environment, Mining and Spatial Planning. unfccc.int/resource/docs/natc/srbnc1.pdf

South East European Forum on Climate Change Adaptation (SEEFCCA) (2012) Regional climate vulnerability assessment synthesis report. SEEFCCA, Croatia, FRY Macedonia, Montenegro, Serbia

Spatial Plan for the Specific Purposes Area of the International Waterway E-80 Danube (Pan-European Corridor VII) (Official gazette of RS No. 14/15)

The Danube Commission Recommendations for the Prevention of Water Pollution of the Danube caused by Water Navigation (1998) http://www.dunavskastrategija.rs/en/?p=191. Accessed 30 Jun 2015

The European Climate Change Programme (2003) http://ec.europa.eu/clima/policies/eccp/index_en.htm. Accessed 20 Sept 2014

UNESCO World Heritage Centre (2007) Case studies on climate change and World Heritage

United Nations Economic Commission for Europe (1991) Espoo convention on environmental impact assessment in a transboundary context. http://www.unece.org/env/eia/about/eia_text.html. Accessed 30 Jun 2015

United Nations International Strategy for Disaster Reduction (UNISDR) (2000) www.unisdr.org/. Accessed 24 Oct 2014

Water Framework Directive (WFD) (2000) http://ec.europa.eu/environment/water/water-framework/index_en.html. Accessed 30 Jun 2015

White Paper on Adapting to Climate Change (2009) http://eur-lex.europa.eu/LexUriServ/LexUriServ.do?uri=COM:2009:0147:FIN:EN:PDF. Accessed 20 Nov 2014

World Bank (2010) World Development Report 2010, The International Bank for reconstruction and development. The World Bank, Washington, DC

World Wildlife Fund (WWF), Centar za unapredjenje zivotne sredine (2012) Procena ranjivosti na klimatske promene—Srbija. Beograd (in Serbian)

Chapter 9
Programmes of the Republic of Belarus on Climate Change Adaptation: Goals and Results

Siarhei Zenchanka

Abstract Having joined to the UNFCCC and the Kyoto Protocol the Republic of Belarus has developed its own programs of Climate Change Adaptation.

The first one, that is, "National program of measures to mitigate the effects of climate change", was adopted for the period of 2008–2012 years. As a result some environmental laws were adopted, different environmental projects were realized. During this period the Republic of Belarus fulfilled its obligations under the Kyoto Protocol, that is, GHG emission was reduced by a third with the increase in GDP by a factor of 2 as compared to 1990. At the same time in the process of discussion on prolongation of the Kyoto Protocol the Republic of Belarus together with Russia refused to take part in the second commitment period.

A new "State program of measures to mitigate the effects of climate change (2013–2020)"adopted in 2013 is also analyzed in this article. The goals of this program are the fulfillment of international obligations of the Republic of Belarus on the UNFCCC and the Kyoto Protocol, the implementation of measures aimed at mitigating the effects of climate change for ensuring the sustainable development of the economy, reducing greenhouse gases (GHG) emissions in order to decrease the rate and magnitude of climate change.

As a result GHG emission should be decreased by 8 % to the level of 1990, energy intensity of GDP should be decreased by 50 % in 2015 and by 60 % in 2020 to the level of 2005.

Keywords Climate change • Adaptation • State program • Strategy

---

S. Zenchanka (✉)
Minsk Branch of Moscow State University of Economics, Statistics and Informatics, Mayakovskogo Str., 127, build.2, 220028 Minsk, Belarus
e-mail: szenchenko@mesi.ru

Introduction

Climate Change is one of the main issues of twenty-first century which has significant influence on Belarus too. 41 % of Gross Domestic Product (GDP) in Belarus is produced in the branches of economy which depend on the weather; hence the climate change adaptation is very important for the country (Communication 2015).

The climate is characteristic of a specific region and doesn't depend on the state border. During the last 30 years the average annual temperature in the region has increased by 1 °C. Thus it is clear that the borders of climatic areas have moved along a meridian for approximately 150 km to the North and the plants specific to the steppe areas are present now in the South region of Belarus (Ecoproject 2005; Report 2009). The average temperature of the last winter in Belarus was about 4° warmer than in the previous years (NashaNiva 2015).

The Republic of Belarus is a small country and its input in GHG emission is rather small. Climate change in Belarus region strongly depends on the world situation. Hence, together with the decrease of GHG emissions the main strategy of the Republic of Belarus is to adapt to climate change. Climate change will have the main impact on three sectors of the Belarusian economy—agriculture, forestry and water management (Communication 2015).

The purpose of the article is to analyze the achievements of the Republic of Belarus in the implementation of the "The National Program of Measures to Mitigate the Effects of Climate Change for the Period of 2008–2012 years" in the field of adaptation to climate change and to estimate the current state of Climate Change Adaptation (CCA) in accordance with the new State Program for the period of 2013–2020.

The research methodology is based on analysis of policy documents, reports, and published materials.

Some Historical Information

The UN Conference on Environment and Development was held in Rio de Janeiro 20 years after the Stockholm conference. As a result the Declaration on environment and development (Declaration 1992) and Agenda for the twenty-first Century (Agenda 1992) were adopted. The Declaration is a set of 27 principles that largely correspond to the provisions of the Stockholm conference. At the same time the Declaration reflects the dramatic changes that took place in the world at the end of the century. The Declaration includes the idea of sustainable development which was considered in the famous Report (1987) of the World Commission on Environment and Development for the first time.

At the Conference in Rio the UN Frame Convention on Climate Change (UNFCCC 1992) was open for signing. This convention is aimed at stabilization of GHG content in the atmosphere. The discussions of the problems on climate change are held at regular Conferences of the Parties (COP) of the UNFCCC.

Significant expansion of the Convention determining the legal obligations to reduce GHG emissions took place in December 1997 at the third Conference of the Parties (COP-3) in Kyoto, Japan (Kyoto Protocol 1998). The Kyoto Protocol outlined the basic rules, but did not provide the details for their application. It also demanded a separate formal process of signing and ratification before coming into force.

Under the Kyoto Protocol the developed countries and countries with economics in transition took commitments to reduce or stabilize the emissions of greenhouse gases. Developing countries including India and China didn't take any commitments.

In accordance with the Kyoto Protocol (1998) each country must realize such policy and measures as:

- Improving energy efficiency in relevant sectors of the national economy;
- Protection and enhancement of sinks and reservoirs of greenhouse gases;
- Promotion of sustainable forms of agriculture;
- Research on and promotion of new and renewable energy, carbon dioxide sequestration technologies and innovative environmentally sound technologies;
- Progressive reduction or phasing out of market imperfections in all sectors i.e., the sources of greenhouse gas emissions and the use of market-based instruments, and others.

The Kyoto Protocol (1998) includes flexibility mechanisms of implementation:

- International emissions trading,
- Clean development mechanism (CDM), and
- Joint implementation (JI).

The Kyoto Protocol came into force in 2005 after Russia had signed it in 2004. The Republic of Belarus joined the UNFCCC in 1992 in Rio and signed the Kyoto Protocol in 2005. The Republic of Belarus took obligations on decreasing the emissions but Belarus was not included in the Annex B to the Kyoto Protocol and the country is not eligible to use flexibility mechanisms of the Kyoto Protocol.

The Kyoto Protocol was adopted until the year of 2012 and then a new document should be signed. At a conference in Montreal (Stone 2006) a working group was established to discuss the items of the protocol for the next period (2013–2017). As a result of long negotiations the action of the Kyoto Protocol was extended until 2020 and a deadline was set to adopt a universal climate agreement in 2015, which will come into effect in 2020 (COP-18 2012). The participants of UN Climate Summit 2014 in New York supported the signing of a new climate convention in Paris in 2015.

The global carbon dioxide emissions during the extension of the Kyoto Protocol for 8 years until 2020 should be decreased by 15 % only due to the lack of participation of Canada, Japan, Russia, Belarus, Ukraine, New Zealand and the United States and due to the fact that the developing countries including China (the world's largest emitter), India and Brazil are not subject to any emissions reductions under the Kyoto Protocol. It was expected that Belarus would be a party of the second commitment period of the Protocol. It could have helped Belarus to fix its obligations and present the possibility to use the flexibility mechanisms of the Kyoto Protocol. But at the consultations held in Minsk in January, 21–22, 2013 with the participation of the representatives of public authorities and experts from Belarus, Russia, Ukraine, and Kazakhstan the sides postponed the ratification of the Doha amendment to the Kyoto Protocol regulating the second commitment period of the Kyoto Protocol.

The National Program of Measures to Mitigate the Effects of Climate Change for the Period of 2008–2012 Years

"The National Program of Measures to Mitigate the Effects of Climate Change for the Period of 2008–2012 years" (Program 2008) together with "The Strategy for Reducing Emissions and Enhancing Removals of Greenhouse Gases by Sinks in the Republic of Belarus for the period of 2007–2012" (Strategy 2006) and "The National Strategy for Introduction of Integrated Environmental Permits for the period of 2009–2020" (Strategy 2009) are aimed at GHG emissions reduction and GHG absorption increase by sinks by means of development the program of synergic measures for various sectors of the national economy.

"The National program of Measures to Mitigate the Effects of Climate Change for the Period 2008–2012 years" was developed in accordance with "The National Action Plan on Rational Using of Natural Resources and Environmental Protection" (Plan 2006).

In accordance with the National Action Plan the main areas of environmental activity in Climate Change protection are:

- Preparation and submission the National Communications on action to limit greenhouse gas emissions and national inventories of emission and sinks of greenhouse gases to the Secretariat of the UNFCCC;
- Development of a legislative basis, preparation and implementation of activities within the economic mechanisms of the Kyoto Protocol.

The realization of the National program for the period of 2008–2012 was closely connected with the fulfillment of Belarus obligations on the UNFCCC and the Kyoto Protocol:

- Observation of climate and its change;
- The decrease of pollution sources and the increase of GHG absorption;
- Adaptation of economy to climate change;
- International cooperation in the field of mitigation of the effects of climate change.

In industry the environmental protection activities in the period of years 2006–2010 were directed to the reduction of air pollution, the increase of the share of secondary energy resources, the use of the alternative and renewable energy.

The implementation of the National Program (2008) was aimed at:

- Preservation of the environment for present and future generations;
- Improving the environmental situation in the cities and regions of the country;
- Reducing emissions and increasing the absorption of GHG;
- Implementation of the commitments under the UNFCCC and the Kyoto Protocol and other international agreements;
- Obtaining the economic benefit from the improved management of natural resources;
- Improving the quality of the environment;
- Reducing the cost of the production;
- Improving governance on climate change;
- Integrating issues related to climate change mitigation and adaptation to climate change in decision-making at all levels of the governance;
- Improving the provision of climate information to public authorities;
- Improving the functioning of climate observing networks;
- Establishment of a reliable data storage system of climate and climate change projections and ensuring consumer access to this information;
- Attracting foreign investments into projects aimed at reducing GHG emissions and increasing removals by sinks.

The following results were supposed to be obtained at the end of program:

- Reducing the greenhouse gas emissions for the period of 5 years (2008–2012) was to be not less than by 12 million tons of CO_2 equivalent;
- Increasing the removals by sinks;
- Reducing the emissions of the following pollutants:
  - Carbon monoxide was to be at least 600 t,
  - Nitrogenoxides—1200 t,
  - Ammonia—600 t,
  - Substances containing sulfur—not <1600 t;
- Development and implementation of measures for adapting agriculture to climate change;
- Development of recommendations and a set of measures to adapt forest economy to climate change;
- Development of water management, protection of surface and groundwater including water supply;
- Development and implementation of measures for adaptation of economic sectors to climate change;
- Development of specialized programs (subroutines) to adapt to climate change.

Results of "The National Program of Measures to Mitigate the Effects of Climate Change for the Period of 2008–2012 Years"

The results of "The National Program of measures to mitigate the effects of climate change for the period of 2008–2012 years" are presented in the National report "Sustainable development of the Republic of Belarus based on "green" economy principles" (Report 2012) and in the Sixth National communication of the Republic of Belarus under the UNFCCC (Communication 2015).

The main problems in climate change in Belarus are associated with the pollution of atmospheric air by all energy and industry entities and mobile sources. As a result of competent state policy to protect air quality conducted in the period 1990–2010 the total emissions from stationary and mobile sources decreased by a factor of 2.6 and from stationary sources by a factor of 3.1 (Report 2012).

Reduction in emissions from stationary sources has been achieved by conducting a targeted environmental policy including increased rates of environmental tax on emissions of pollutants into the air and the introduction of an automated emission accounting system for businesses, as well as increased fines for violation of environmental laws.

Positive results were facilitated by the introduction of progressive technologies and implementation of energy saving measures. While increasing the production output the major polluters were equipped both dust and gas-traps bringing the proportion of trapped and neutralized emissions from stationary sources from 77 % in 1990 to 88 % in 2010. There was a steady decline in emissions from mobile sources across the country despite the fact that over the past 20 years the Republic's car fleet has grown considerably.

Belarusian economy is characterized by a high level of energy intensity of GDP, so the country has adopted a number of programs aimed at its declining and, through the implementation of energy efficiency measures, energy intensity of GDP in 2013 decreased by 69 % compared with 2005. The share of rural and renewable energy in the total energy production is so far negligible (approximately 6 % total and 0.6 % of renewable energy sources). One of the main priorities of the energy policy in Belarus is the development of renewable energy and positive results have been obtained mainly due to energy savings, the restructuring and transfer of fuel burning boilers to natural gas, as well as the introduction of payments for emissions of the main greenhouse gases.

In accordance with "The National program on the development of local and renewable energy sources for 2011–2015" (Program 2011) the share of the country own energy resources in the balance of boiler and furnace fuels must reach at least 30 % by 2015.

One of the priorities of Belarusian policy in energy efficiency and renewable energy is the development of technical standards and regulations. The Republic has developed more than 120 technical regulations to ensure a comprehensive approach to establish requirements for fuel and energy resources, energy consuming products, insulation of buildings and structures, control and measuring equipment, use of production waste, secondary and renewable energy sources. More than 80 of these documents are harmonized with international and European standards.

Belarus seeks to prevent the entry of CO_2 into the atmosphere, and thus much attention is paid to the development of forestry, preservation of wetland ecosystems, re-wetting of exhausted peatlands. Climate change directly or indirectly (through a change in the groundwater level, fires, propagation of forest pests and worsening of disease) affects the condition of the forest vegetation, leading to changes in the composition and structure of tree plantations.

Communication (2015) considers the following adaptation measures realized in Belarus in the forest management:

- Development and implementation of sectoral strategies and the targeted program of forest adaptation to new climatic conditions;
- Afforestation, taking into account the displacement of climatic zones and the change of moisture regime;
- Transition to higher cutting age;
- Protection of forests from pests, the development of preventive measures for counteracting of the proliferation of pests unusual for Belarus;
- Rehabilitation of degraded peat lands and restoration of other natural sinks of greenhouse gases;
- Improving the fire warming system in forests and peat bogs;
- Development and implementation of research activities aimed at assessment of the impact of climate change on forest vegetation and forestry and elaboration of measures aimed at forest adaptation to such change.

The State program "The development of forestry of the Republic of Belarus for the period of 2011–2015 years" (Program 2010) supposes improving the forestry effectiveness, increasing of its productivity and improving its age and grade structure.

In agriculture, Belarus is conducting a well established public policy. Significant results were achieved in the course of implementation of "The State Program of Rural Development for the period of 2005–2010 years" (Program 2005), as well as a number of sectoral programs. A sustainable growth in agricultural production was ensured, the processing industry worked with the increased capacity, export opportunities were substantially extended, and the issue of food security was mainly solved. This growth occurred due to the increasing productivity in agriculture and animal breeding.

The measures on adaptation of agriculture to climate change involved:

- The introduction of highly productive crops and vegetables;
- The expansion of areas under the new highly productive crops (maize, millet, soybean, sugar beet, spring rape and others);
- The orchards planting with a new variety composition;
- The earlier time shift of the spring crops sowing.

The potential for production increase and costs reduction in agriculture was realized by the use of information and high technologies, improved farming culture and introduction of new forms of management. The State Program for Sustainable Rural Development for 2011–2015 stipulated further growth of crop and livestock production which was aimed at forming an effective competitive, sustainable and environmentally friendly agro-industrial complex corresponding to the international standards.

Climate change has a significant impact on the water management. First of all it is connected with the change of rivers drain, flooding and the change of the underground water level.

The measures on adaptation of water management to climate change included:

- Efficient use of all declining water quality resources;
- Widespread adoption of water-saving technologies in various sectors of the economy;
- Conversion of irrigation systems in a more technologically advanced systems with optimal water consumption for production;
- Zero wastewater management systems;
- The possibility of artificial groundwater recharge.

The State Program of Measures to Mitigate the Effects of Climate Change for the Period of 2013–2020 Years

After the completion of the National program in 2012 a new State program of Measures to Mitigate the Effects of Climate Change for the period of 2013–2020 (Program 2013) was adopted. The period of this program is equal to the second commitment period of the Kyoto Protocol.

The objectives of the State program are the fulfillment of international obligations of the Republic of Belarus on the Framework Convention and the Kyoto Protocol, taking measures to mitigate the effects of climate change, ensuring sustainable development of the economy, decreasing GHG emissions in order to reduce the rate and magnitude of climate change.

The main areas of realization of these objectives are:

- Implementation of measures to save fuel and energy resources in the energy sector;
- Stabilization of GHG emissions through the use of resource-saving technologies in energy-intensive sectors of the economy;
- Optimization of waste management;
- Improvement the quality and increase of GHG sinks.

The Republic of Belarus has assumed a number of voluntary commitments aimed at reducing energy intensity of GDP and greenhouse emission for the period of 2015–2020 years, that is

- To reduce greenhouse gas emissions in 2020 by 8 % below the levels of 1990 year;
- Greenhouse gas emissions are expected to be no more than 110 million tons by the year of 2020;
- Reduction of energy intensity of GDP over the period 2011–2015 by 29–32 % compared to 2010.

Agriculture, forestry and water management, energy, construction and social services are the most vulnerable to climate change. Key actions to adapt to climate change will include:

- Development and improvement of the criteria and conditions of the climate security;
- Assessment of vulnerability of some regions to climate change;
- Development of the sectoral strategies for adaptation to climate change and the implementation of these strategies;
- Promotion of activities related to the implementation of agricultural activities on adaptation to climate change;
- Minimization of the impact of several weather events caused by climate change including the development of methods for risk and damage assessment, as well as scenarios of adaptation to such events;
- Minimization the risk of reducing agricultural production including the decrease of the number of agricultural animals, productivity and total yield of crops;
- Introducing into the production of heat-loving species and varieties of crops with the expansion of cultivated areas;
- Assessment of the impact of climate change and potential threats to the biodiversity of the natural ecosystems of the Republic of Belarus and the development of measures for their conservation.

Conclusion

After joining the UNFCCC and the Kyoto Protocol the Republic of Belarus has made remarkable steps in climate change mitigation and adaptation. There is a sufficient institutional framework in Belarus for an annual inventory of greenhouse gases, it is updated in accordance with the requirements of the UNFCCC. Ministry of Natural Resources is the coordinating body, it maintains a national greenhouse gas inventory system, the timely collection of activity data, as well as national reporting on greenhouse gas inventory (NIR) to the UNFCCC Secretariat.

Report (2012) and Communication (2015) underline achievements of the Republic of Belarus in the measures in climate change mitigation and adaptation. As a result of the state policy to protect air quality in the period 1990–2010 the total emissions from stationary and mobile sources decreased by a factor of 2.6 and from stationary sources by a factor of 3.1

It should be noted that despite the activity of the Republic of Belarus to register carbon units, Belarus has not been included in Annex B to the Kyoto Protocol to the UNFCCC in the period of its action from 2008 to 2012. Thus the work on keeping the national registry of carbon units was discontinued prior to the signing of a new climate agreement.

In the course of implementation of the Program (2008–2012) a number of problems that hindered the development of the country were not solved. These problems were associated with inefficient economy especially in materials consumption of and energy consumption. In accordance with the Climate Change Performance Index (CCPI 2015) Belarus lost its position compared to the Climate Change Performance Index (CCPI 2014). The trends to worse indices in greenhouse gas emissions in industry, transport and construction are characteristic for Belarus.

The fact that the Belarusian government is going to introduce quotas for projects in renewable energy (Minenergo 2015) might lead to a further deterioration of Belarus position in the Climate Change Performance Index.

The objectives of the State Program (2013) are the fulfillment of international obligations of the Republic of Belarus on the Framework Convention and the Kyoto Protocol, taking measures to mitigate the effects of climate change, ensuring the sustainable development of the economy, reducing GHG emissions in order to decrease the rate and magnitude of climate change. As a result GHG emission should be decreased by 8 % to the level of 1990, energy intensity of GDP should be decreased by 50 % in 2015 and by 60 % in 2020 to the level of 2005.

References

Agenda (1992) for the 21st Century. http://www.un-documents.net/agenda21.htm. Accessed 20 Jan 2015
Climate Change Performance Index (CCPI) (2014) Results 2014. Germanwatch, Climate Action Network Europe
Climate Change Performance Index (CCPI) (2015) Results 2015. Germanwatch, Climate Action Network Europe
Communication (2015) Sixth National Communication of the Republic of Belarus. Under the United Nations Framework Convention on Climate Change, Minsk (in Russian)
COP-18 (2012) The Doha climate gateway. http://unfccc.int/key_steps/doha_climate_gateway/items/7389.php. Accessed 20 Jan 2015
Declaration on Environment and Development (Declaration) (1992) http://www.un.org/documents/ga/conf151/aconf15126-1annex1.htm. Accessed 20 Jan 2015
Ecoproject (2005) Climate change and its consequences for Belarus—sustainable development at local level. Ecoproject, Minsk
Kyoto Protocol (1998) to the United Nation framework convention on climate. http://unfccc.int/456 resource/docs/convkp/kpeng.pdf. Accessed 20 Jan 2015
Minenergo (2015) Belarusian Energy Ministry would cut state support of green energy. http://news.tut.by/economics/437452.html. Accessed 27 Feb 2015 (in Russian)
NashaNiva (2015) http://nn.by/?c=ar&i=145374&lang=ru. Accessed 27 Apr 2015 (in Russian)
Plan (2006) The National action plan on rational using of natural resources and environmental protection. BelNIC "Ecology", Minsk (in Russian)

Program (2005) The state program of rural development for the period of 2005–2010 years. http://a-h.by/s153/archives/150_ot_25_marta_2005_g._Ukaz_Prezidenta_Respubliki_Belarus.html. Accessed 20 Jan 2015 (in Russian)

Program (2008) The National Program of measures to mitigate the effects of climate change for the period of 2008–2012 years, approved by the resolution of the Council of Ministers of the Republic of Belarus, 04 August 2008, г. № 1117. http://www.government.by/ru/content/1237. Accessed 20 Apr 2015 (in Russian)

Program (2010) State program "The development of forestry of the Republic of Belarus for the Period of 2011–2015 years". www.mlh.by/docs/official/1626pr.doc. Accessed 20 Jan 2015 (in Russian)

Program (2011) National program on development of local and renewable energy sources for 2011–2015. http://www.government.by/upload/docs/file663fb27db70962e8.PDF. Accessed 20 Jan 2015 (in Russian)

Program (2013) The State Program of measures to mitigate the effects of climate change for the period of 2013–2020 years. http://pravo.newsby.org/belarus/postanovsm0/sovm378.htm. Accessed 20 Jan 2015 (in Russian)

Report (1987) Our common future: report of the world commission on environment and development. http://www.un-documents.net/ocf-cf.htm. Accessed 20 Jan 2015

Report (2009) On strategic estimations of subsequences of climate change in nearest 10–20 years for environment and economics of Joint State. Moscow. http://pogoda.by/download/report-climat-10-12.doc. Accessed 20 Jan 2015 (in Russian)

Report (2012) Sustainable Development of the Republic of Belarus based on "green" economy principles. National Report. Scientific-Research Economic Institute of the Ministry of Economy of Belarus, Minsk

Stone SJ (2006) Comment on COP 11 to the UNFCCC. Sustain Dev Law Policy Winter 2006, 45–46, 67

Strategy (2006) The strategy for reducing emissions and enhancing removals of greenhouse gases by sinks in the Republic of Belarus for the period of 2007–2012, approved by the Resolution of the Council of Ministers of the Republic of Belarus, 7 September 2006, No 1155 (in Russian)

Strategy (2009) The national strategy for introduction of integrated environmental permits for the period of 2009–2020, approved by the Resolution of the Council of Ministers of the Republic of Belarus, 25 July 2009, г. № 980. http://pravo.levonevsky.org/bazaby11/republic09/text763.htm. Accessed 20 Jan 2015 (in Russian)

UNFCCC (1992) The United Nations Framework Convention on Climate Change. http://unfccc.int/472 resource/docs/convkp/conveng.pdf/. Accessed 20 Jan 2015

Chapter 10
A Global Indicator of Climate Change Adaptation in Catalonia

Ester Agell, Fina Ambatlle, Gabriel Borràs, Gemma Cantos, and Salvador Samitier

Abstract The Catalan Office for Climate Change has been working on a pioneering and innovative document, "A Global Indicator of Climate Change Adaptation in Catalonia". The study took 83 indicators as its starting point, identifying and classifying 29 key indicators as the methodology was developed. These are grouped under the following headings: water management, agriculture and livestock farming, forestry, health, the energy sector, industry, services and commerce, tourism, town planning and housing, mobility and transport infrastructures, research, development and innovation.

The subsequent statistical analysis of these 29 indicators, with the assistance of the Catalan Institute for the Evaluation of Public Policies (Ivàlua), produced a global adaptation indicator quantifying Catalonia's capacity to adapt to climate change impacts. This global indicator is based on two factors: use of resources and environmental quality. Effectively, capacity to adapt to the impacts of climate change depends on how we use resources (primarily water and energy) and on the quality of the environment (primarily air quality).

The global indicator allows a country's capacity to adapt to climate change to be monitored over time. The absolute value of the indicator (from 0 to 10) dropped slightly between 2005 and 2011. Catalonia is working on the issue, but must continue to make every effort to improve the extent to which we are adapting to ensure that our land, natural systems and society are progressively less vulnerable to the impacts of climate change.

In developing this indicator, Catalonia has a tool which provides guidance on where to focus efforts towards developing a green, circular, low-carbon economy which is adaptable to the new conditions brought about by climate change. This is

---

E. Agell • F. Ambatlle • G. Borràs (✉) • G. Cantos • S. Samitier
The Catalan Office for Climate Change, Directorate-General for Environmental Policy, Government of Catalonia, Avinguda Diagonal, 525, 08029 Barcelona, Catalonia, Spain
e-mail: gborras@gencat.cat

opening up a range of economic and social opportunities which are currently being developed in fields such as energy efficiency, saving water, renewable energies, smart mobility, forestry management and healthcare.

The indicators included in the global indicator should be reviewed every 5 or 10 years based on new information available (in order to include more aspects in the global indicator).

Keywords Indicators • Adaptation measures • Catalonia • Use of resources • Environmental quality

Introduction

The Catalan Strategy for Adapting to Climate Change 2013–2020 (ESCACC), which was drafted by the Catalan Office for Climate Change (OCCC) and approved in November 2012 by the Catalan government, marks a significant step towards reducing Catalonia's vulnerability to the impacts of climate change.

There are two key operational objectives of the ESCACC. Firstly to generate and transfer knowledge of climate change adaptation(CONADAPT).Secondly to increase the adaptative capacity of the most vulnerable areas (the Pyrenees, the Ebro Delta and the coast), the socio-economic sectors and the natural systems in Catalonia(CAPADAPT): agriculture and livestock, biodiversity, water management, forest management, industry, services and trade, mobility and transport infrastructure, fisheries and marine ecosystems, health, energy, tourism, and urban planning and housing.

To achieve these objectives, the ESCACC proposes a total of 182 adaptation measures: 30 of these are generic and the remaining 152 are specific to sectors and systems. The generic measures include the drafting and approval of a Catalan climate change law (in progress), and the establishment of a monitoring system and indicators for the adaptation measures set out in the ESCACC, in order to evaluate how well climate change adaptation is progressing. In other words, a system to determine the effectiveness of the measures to adapt to the impacts of climate change.

An initiative aimed at establishing adaptation indicators is also envisaged within the framework of the LIFE12 project ENV/ES/000536(also known as LIFE MEDACC): *"Demonstration and validation of innovative methodology for regional climate change adaptation in the Mediterranean area"*, for which the Ministry of Land and Sustainability, through the Catalan Office for Climate Change, is the coordinating beneficiary. LIFE MEDACC proposes the definition of new adaptation measures based on the evaluation of climate change impacts and vulnerability and the evaluation of existing adaptation measures. For this purpose, LIFE

MEDACC establishes that it is necessary to compile and review methodologies by means of statistical analysis or an analysis of the existing literature for climate change in order to develop a set of indicators of adaptation to the impacts of climate change.

To summarize, there is a strategic framework for planning climate change policies (ESCACC) and a demonstration project at Mediterranean Europe level (LIFE MEDACC) that call for the establishment of a tool to assess the effectiveness of the measures to adapt to climate change impacts. The preliminary work carried out within the framework of both the ESCACC and MEDACC projects has made it possible to reach a sufficiently advanced stage such that the creation of a global indicator of adaptation to climate change impacts in Catalonia is now feasible.

Preparatory Work for the Creation of the Indicator

The adaptation evaluation, i.e. the analysis of whether or not Catalonia is making progress in adapting to climate change impacts, requires the development of an indicator set with three different levels of integration: (1) for the specific adaptation measure, whenever possible; (2) for each sector and system; (3) and lastly, for the whole of Catalonia.

Four basic criteria must be taken into account when the indicators are developed: (1) they must be easy to measure and monitor, i.e. the requisite data should be readily available; (2) there must be historical data on what is measured; (3) the indicator must be easy to interpret; and (4) the information and data must be specific to the Catalan country.

The task of evaluating the effectiveness of adaptation measures is not straightforward. This was acknowledged in a communication from the European Commission concerning the EU Strategy on Adaptation to Climate Change (COM(2013) 216): "*Monitoring and evaluating climate change adaptation policies are crucial. The emphasis is still on monitoring impacts rather than adaptation action and its effectiveness. The Commission will develop indicators to help evaluate adaptation efforts and vulnerabilities across the EU, using LIFE funding and other sources.*" In March 2014, the Commission proposed an Adaptation preparedness scoreboard based on the evaluation of five different areas as a tool for measuring the degree of progress in climate change adaptation policies in the member states of the European Union. One of these areas concerns the monitoring and evaluation of adaptation measures through indicators, but the calculation mechanisms were not specified.

The unfamiliarity of the task at hand and the lack of references therefore meant that the search for indicators was neither simple nor easy. This partly explains why some of the initial indicators were more mature than others. Thus, a preliminary task to search and select data resulted in a proposal that grouped together a total of 83 potential indicators to evaluate the effectiveness of the adaptation measures. The main sources of search of indicators were Governmental websites of each sector, as well as contrast of the indicators with several experts. The 83 selected indicators were related to 37 different adaptation measures from 12 sectors: 8 are classified as CONADAPT and 29 as CAPADAPT.

The information included in each indicator was organized in a data-sheet format[1] with the following sections:

1. Sector indicator—name of the indicator.
2. Operational objective—CONADAPT (to generate and transfer knowledge of climate change adaptation) or CAPADAPT (to increase the adaptive capacity of sectors and/or systems).
3. Measure(s) to which it responds—adaptation measure(s) to which the indicator corresponds.
4. Source—information source (direct and indirect sources are differentiated).
5. Methodology—explanation of the methodology used by the information source to obtain the numerical data.
6. Data—numerical values of the indicator by year (table).
7. Graphic representation—of the data.
8. Desired trend of the adaptation—description of the direction the indicator must take in order to achieve a more effective adaptation, i.e. whether it should increase or decrease. This is a way of explaining how to interpret the indicator and its evolution.
9. Relevance of the indicator—justification of why this indicator is useful for evaluating the specific measure. It also explains how the data should be interpreted.

The diversity of the indicators and, at the same time, the differences between qualitative and quantitative information for some of these indicators or the lack of time-based consistency of the data meant that it was impossible to respond to the key question: Is Catalonia adapting well to the impacts of climate change? Therefore, following a meeting with the Catalan Institute of Public Policy Evaluation (IVÀLUA), a preselection process was conducted. This second selection process was based primarily on the potential capacity of the indicator to quantify the outcome of adaptation actions implemented or in progress (and, therefore, on the effectiveness of the indicator to evaluate the measures). In other words, only

---

[1] It is important to note that, at this stage, some of the data sheets on the indicators could summarize more than one indicator. For example, the indicator "Household consumption" (Water management) includes data broken down by the Metropolitan Area of Barcelona and by Catalonia as a whole.

indicators that directly measured the outcome of the application of the measure were included (e.g. the increase in mortality rate due to heat waves following implementation of the Action Plan to Prevent the Effects of Heat Waves on Health—POCS), while indicators that measured a sector or system's sensitivity or degree of exposure were rejected (e.g. the population ageing index). Indicators that were more qualitative in nature, such as planning tools that incorporate climate change impacts and adaptation (forestry plan, tourism plan, etc.), were also retained in the preselection process. During this process, the initial 83 indicators were reduced to a set of 50.

Synthetic Adaptation Indicator

Following the collaboration with IVÀLUA, in order to obtain a synthetic indicator of climate change adaptation, entailed a third selection process: only those indicators with a series of historical data based on at least 10 consecutive years were chosen. This process reduced the number of indicators to a total of 29. Four indicators were discarded because of being qualitative and the rest because they did not have 10 years consecutive data.

In order to achieve the objective mentioned above, the most appropriate statistical technique was found to be principal component analysis (PCA), a procedure related to factor analysis. The purpose of factor analysis is to analyse the structure of interrelations between a number of variables (indicators, in our case) and define common dimensions, thus producing a lower dimensional space. Principal component analysis, in particular, aims to reduce the dimensionality of the data matrix in order to obtain a lower number of new variables (Z_j) or principal components with the following characteristics:

(a) The principal components are linear combinations of the original variables.
(b) The principal components are not correlated with each other.
(c) The number of principal components must be simultaneously small (so that the analysis is effective) and sufficient (to absorb most of the information on the original variables). There are several criteria to determine the number of factors to incorporate. One of the most widely used criteria is to keep factors that have a characteristic value greater than one, or factors that explain more than 20 % of the total variance.

Thus, the calculation of the first component (or factor) is performed as a linear combination of the original variables that retains the maximum amount of total variance. In the calculation of the second component (or factor), the same procedure is performed (linear combination of the original variables to retain the maximum amount of total variance of the part not included in the first), and so on.

Each variable (indicator) has a relative contribution to each factor. This contribution expresses the correlation between this variable (indicator) and the factor. A high relative contribution of the variable indicates a strong correlation between the variable and the factor. In other words, the variable is important for the interpretation of the factor. This contribution can be positive or negative, depending on whether that variable increases or reduces the value of the factor.

Results

This methodology was used to perform the principal component analysis of the annual values of 29 indicators[2] categorized into 10 groups (see Table 10.1). Of the 50 indicators originally provided, those that did not present sufficient variability were discarded in advance, either because the information was not annual, because the indicator was only qualitative, or because there were insufficient observations. With these special cases in mind, all biodiversity indicators and the majority of agriculture and livestock indicators had to be discarded for the analysis. Of the 29 indicators, 20 were more related to climate change adaptation and 9 were more related to mitigation. However, the inclusion of some indicators more related to mitigation was accepted because of the necessity of integrating adaptation and mitigation policies.

In order to standardize the information, the values of all variables were converted to values of 0–1. Using the statistical program Stata, two factors that explained 100 % of the variability of the original information were obtained. The first factor explained 61 % of the variability and the second factor 39 %.

The significance of the two factors was interpreted as follows: the first factor evaluates the use of resources (primarily water and energy), while the second factor evaluates environmental quality (primarily atmospheric emissions). Tables 10.2 and 10.3 show the contribution of each variable to each of the factors.

Indicators with a strong contribution to each factor are marked in green colour. Values below −0.8 or above 0.8 were considered to be strong contributions. Thus, examples of indicators that strongly affect factor 1 (the use of resources) are "Number of special regime facilities in Catalonia" (−0.9900) and "Household consumption (l/inhab./day): Metropolitan Area of Barcelona" (0.9929). In the case of factor 2 (environmental quality), those with most influence are "Catalan

---

[2] To carry out the principal component analysis, the presentation of the data for some indicators had to be adapted. For example, the data for the indicator "Total overnight stays in hotels in Catalonia" (Tourism) were broken down by the four quarters of the year. When entering the variables in the model, the data were entered as a percentage of overnight stays in the third quarter/year total, which was actually the most relevant information for the indicator. This is why the names of some indicators were adapted.

Table 10.1 Name of the indicators included in the analysis

Abbreviations[a]	Definition and units
pa1	Total agricultural output of dryland crops with added value (t): olives + grapes
pe1	% Consumption of electricity from renewable sources
pe2	Number of special regime facilities in Catalonia
pe6	Primary energy intensity (energy content of GDP) (toe/million € in the year 2000)
pg1	Household consumption (l/inhab./day): Catalonia
pg2	Household consumption (l/inhab./day): Metropolitan Area of Barcelona
pgf2	Timber harvesting for firewood and biomass in Catalonia (t)
pgf3	Production of forest products (other than timber and firewood) (t): cork + truffles and other fungi + pine nuts
pgf5	Hectares burned per fire (%)
pi1	Water consumption: amount billed to industry and services (m^3)
pi2	Final energy consumption of the industrial and service sectors (ktoe)
pi3	GHG emissions from the industrial sector (thousands of t of CO_2-eq)
pi4	Imports of oil extraction and refining, coal (millions of euros)
pm1	Passengers on Renfe and FGC trains (thousands)
pm2	Goods on Renfe and FGC trains (thousands of tonnes)
pm3	Passengers on buses (thousands)
pm4	Energy consumed by transport (ktoe)
pm5	GHG emissions from transport (kt CO_2-eq)
prd1	Domestic expenditure on R&D/GDP (%)
ps6	At-risk-of-poverty rate: after social transfers in Catalonia
ps8	Green area per inhabitant in the city of Barcelona (m^2/inhabitant)
ps9	Catalan air quality index ICQA (% (satisfactory + excellent))
ps10	Maximum value of ozone emissions ($\mu g/m^3$)
pt2	Total overnight stays in hotels in Catalonia (% 3rd quarter/year total)
pt5	Foreign tourists' reasons for travelling to Catalonia (% professional tourism)
pt14	Snow cannons on Catalan ski resort (km skislope/cannon)
pu2	Volume of water billed in the household sector in Catalonia (m^3)
pu3	Final energy consumption of the household sector in Catalonia (ktoe)
pu4	GHG emissions from the residential sector (t CO_2-eq)

[a]The letter that follows the "p" indicates the sector or system to which the indicator refers (a: agriculture and livestock; e: energy; g: water management; gf: forest management; i: industry, services and trade; m: mobility and transport infrastructure; rd: research, development and innovation; s: health; t: tourism; u: urban planning and housing). The number indicates the indicator's assigned position in the list of 50 indicators selected prior to the factor analysis

air quality index ICQA (% (satisfactory + excellent))" (−0.9972) and "GHG emissions from the residential sector (t CO_2-eq)" (0.9822).

Finally, to avoid overweighting groups with a greater number of indicators, the influence of each of the 10 groups (systems and sectors) was evaluated. Thus, the weighting of natural systems and socio-economic sectors based on their vulnerability to the impacts of climate change (and in accordance with the ESCACC

Table 10.2 Relative contribution of each indicator to factor 1, which evaluates the use of resources (primarily water and energy)

Code	Indicator	Factor 1
prd1	Domestic expenditure on R&D/GDP (%)	−0.9973
pe2	Number of special regime facilities in Catalonia	−0.9900
pa1	Total agricultural output of dryland crops with added value (t): olives+grapes	−0.9596
pm3	Passengers on buses (thousands)	−0.9286
ps6	At-risk-of-poverty rate: after social transfers in Catalonia	−0.9039
pe1	% consumption of electricity from renewable sources	−0.8903
ps8	Green area per inhabitant in the city of Barcelona (m^2/inhabitant)	−0.7287
pu3	Final energy consumption of the household sector in Catalonia (ktoe)	−0.4418
pi4	Imports of oil extraction and refining, coal (millions of euros)	−0.2640
pt2	Total overnight stays in hotels in Catalonia (% 3rd quarter/year total)	−0.0745
ps9	Catalan air quality index ICQA (% (satisfactory + excellent))	0.0745
pu4	GHG emissions from the residential sector (t CO_2-eq)	0.1877
pgf5	Hectares burned per fire (%)	0.3279
pm5	GHG emissions from transport (kt CO_2-eq)	0.3516
ps10	Maximum value of ozone immissions ($\mu g/m^3$)	0.4148
pm4	Energy consumed by transport (ktoe)	0.4518
pt14	Snow cannons on Catalan ski resort (km skislope/cannon)	0.7051
pgf3	Production of forest products (other than timber and firewood) (t): cork+truffles and other fungi+pine nuts	0.7516
pi2	Final energy consumption of the industrial and service sectors (ktoe)	0.8110
pgf2	Timber harvesting for firewood and biomass in Catalonia (t)	0.8613
pi1	Water consumption: amount billed to industry and services (m^3)	0.8996
pm1	Passengers on Renfe and FGC trains (thousands)	0.9399
pu2	Volume of water billed in the household sector in Catalonia (m^3)	0.9568
pi3	GHG emissions from the industrial sector (thousands of t of CO_2-eq)	0.9756
pm2	Goods on Renfe and FGC trains (thousands of tonnes)	0.9765
pt5	Foreign tourists' reasons for travelling to Catalonia (% professional tourism)	0.9858
pg1	Household consumption (l/inhab./day): Catalonia	0.9919
pe6	Primary energy intensity (energy content of GDP) (toe/€million in the year 2000)	0.9919
pg2	Household consumption (l/inhab./day): Metropolitan Area of Barcelona	0.9929

Table 10.3 Relative contribution of each indicator to factor 2, which evaluates environmental quality (primarily atmospheric emissions)

Code	Indicator	Factor 2
ps9	Catalan air quality index ICQA (% (satisfactory + excellent))	-0.9972
pi4	Imports of oil extraction and refining, coal (millions of euros)	-0.9645
pm5	GHG emissions from transport (kt CO_2-eq)	-0.9362
pm4	Energy consumed by transport (ktoe)	-0.8921
ps8	Green area per inhabitant in the city of Barcelona (m^2/inhabitant)	-0.6848
pi2	Final energy consumption of the industrial and service sectors (ktoe)	-0.5850
pi1	Water consumption: amount billed to industry and services (m^3)	-0.4366
ps6	At-risk-of-poverty rate: after social transfers in Catalonia	-0.4277
pm1	Passengers on Renfe and FGC trains (thousands)	-0.3416
pu2	Volume of water billed in the household sector in Catalonia (m^3)	-0.2909
pi3	GHG emissions from the industrial sector (thousands of t of CO_2-eq)	-0.2194
prd1	Domestic expenditure on R&D/GDP (%)	0.0728
pg2	Household consumption (l/inhab./day): Metropolitan Area of Barcelona	0.1186
pg1	Household consumption (l/inhab./day): Catalonia	0.1266
pe6	Primary energy intensity (energy content of GDP) (toe/€million in the year 2000)	0.1266
pe2	Number of special regime facilities in Catalonia	0.1413
pt5	Foreign tourists' reasons for travelling to Catalonia (% professional tourism)	0.1676
pm2	Goods on Renfe and FGC trains (thousands of tonnes)	0.2155
pa1	Total agricultural output of dryland crops with added value (t): olives+grapes	0.2812
pm3	Passengers on buses (thousands)	0.3711
pe1	% consumption of electricity from renewable sources	0.4554
pgf2	Timber harvesting for firewood and biomass in Catalonia (t)	0.5081
pgf3	Production of forest products (other than timber and firewood) (t): cork+truffles and other fungi+pine nuts	0.6596
pt14	Snow cannons on Catalan ski resort (km skislope/cannon)	0.7091
pu3	Final energy consumption of the household sector in Catalonia (ktoe)	0.8971
ps10	Maximum value of ozone immissions ($\mu g/m^3$)	0.9099
pgf5	Hectares burned per fire (%)	0.9447
pu4	GHG emissions from the residential sector (t CO_2-eq)	0.9822
pt2	Total overnight stays in hotels in Catalonia (% 3rd quarter/year total)	0.9972

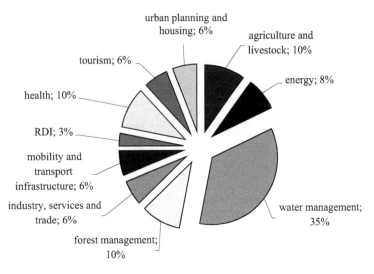

Fig. 10.1 Weight percentages of the indicators by sector

Table 10.4 Calculation of the final indicators based on the information for 2005 and 2011

	Use of resources 2005	Use of resources 2011	Environmental quality 2005	Environmental quality 2011
Value between −1 and 1	0.0118	−0.0065	0.0254	0.0115
Value between 0 and 10	5.0589	4.9674	5.1269	5.0576

diagnosis) resulted in the indicators being divided into the following five groups, from most to least importance (see Fig. 10.1):

1. Water management (35 %)
2. Agriculture and livestock; Forest management; Health (30 %, i.e. 10 % each)
3. Energy (8 %)
4. Industry, services and trade; Tourism; Urban planning and housing; Mobility and transport infrastructure (24 %, i.e. 6 % each)
5. Research, development and innovation (3 %)

Water management has been weighted with a 35 % because, according to ESCACC, water is the most vulnerable element with direct consequences for the other sectors.

Lastly, within each factor, the weighted value of the indicator is multiplied by the indicator's contribution to the factor and by the value (between 0 and 1) of the indicator during the selected time period (years). By performing this calculation for both factors and for 2005 and 2011, the results indicated in Table 10.4 are obtained.

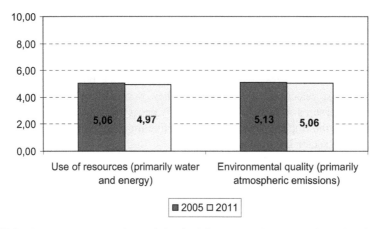

Fig. 10.2 Year-on-year comparison of the final factors based on the information for 2005 and 2011

As can be seen, both factors have a medium value (around 5) (see Fig. 10.2). In both cases, there was a slight decrease in the year 2011 compared with 2005.

Conclusions

As a result of applying the principal component analysis, **a synthetic adaptation indicator** was obtained. This will enable us to monitor the development of Catalonia's capacity to adapt to the impacts of climate change. This synthetic adaptation indicator is determined by two factors that explain 100 % of the variability of the original information contained in 29 indicators. Each of these factors corresponds to a different aspect: **(1) use of resources and (2) environmental quality**. Since having data with annual variability is essential, indicators for which there was no variability (qualitative indicators) or for which the information was not annual, based on a minimum of 10 consecutive years, were discarded.

The synthetic adaptation indicator, expressed as the result of both factors, shows a medium level in terms of the capacity to adapt to climate change impacts; just a pass. The evolution of this capacity has been decreasing slightly in recent years (2011 versus 2005). It is important to bear in mind that in order to monitor the synthetic indicators properly, rapid access to the information relating to the original indicators is required. These indicators should be reviewed every 5 or 10 years based on new information available (in order to include more aspects in the synthetic indicator).

It should be noted that biodiversity is the primary source of environmental services, so its effective or poor adaptation to climate change impacts will directly affect the other natural systems and many, if not all, economic systems.

Biodiversity has thus far not been included in this quantitative analysis of the adaptation, but its key importance means that a more qualitative evaluation is also needed.

The fishing industry was also omitted from the analysis, but for a different reason. In this case, there were initially three indicators, but they were rejected during the first selection process because they were largely indirect, since the adaptation measures were highly general. It is necessary to wait until more basic knowledge of the impacts and the most effective measures for combating climate change in this sector is available.

References

Catalan Office for Climate Change (2012) Strategy for adapting to climate change (ESCACC, horizon 2013–2020). Catalan Office for Climate Change, Barcelona (Catalonia). http://canviclimatic.gencat.cat/web/.content/home/campanyes_i_comunicacio/publicacions/els_papers_de_l_occc/resum_executiu_escacc_angles.pdf

Catalan Office for Climate Change (2014) Global indicator of climate change adaptation in Catalonia. Catalan Office for Climate Change, Barcelona (Catalonia). http://canviclimatic.gencat.cat/web/.content/home/actualitat/docs/Doc-Index-complet_ENG.pdf

Commission of the European Communities (2013) COM(2013) 216 Final Communication from the Commission to the European Parliament, the Council, the European Economic and Social Committee and the Committee of the Regions. An EU Strategy on adaptation to climate change. Commission of the European Communities, Brussels. http://eur-lex.europa.eu/legal-content/EN/TXT/PDF/?uri=CELEX:52013DC0216&from=EN

MEDACC (2013) Adapting the Mediterranean to climate change, LIFE 12 ENV/ES/000536 Demonstration and validation of innovative methodology for regional climate change adaptation in the Mediterranean area, July 2013–July 2018. http://medacc-life.eu/medacc-adapting-mediterranean-climate-change

Note to the Working Group 6 on adaptation under the Climate Change Committee and to the national contact points on adaptation: The adaptation preparedness scoreboard. http://webcache.googleusercontent.com/search?q=cache:yZ8tfwcRerIJ:ftp://ftp.boos.org/klimatkurs/Note1_scoreboard%2520(8).doc+&cd=1&hl=ca&ct=clnk&gl=es

Part II
Integrating Adaptation Strategies and Educational Approaches

Part II

Chapter 11
Facilitating Climate Change Adaptation on Smallholder Farms Through Farmers' Collective Led On-Farm Adaptive Research: The SAF-BIN Project

Romana Roschinsky, Sunil Simon, Pranab Ranjan Choudhury, Augustine Baroi, Manindra Malla, Sukleash George Costa, Valentine Denis Pankaj, Chintan Manandhar, Manfred Aichinger, and Maria Wurzinger

Abstract In India, Bangladesh and Nepal 70 % of farms are less than 2 ha in size. These subsistence oriented, rain-fed farming systems are highly vulnerable to climate variability. Climate change challenges local food and nutritional security. Adaptation is the key to address these vulnerabilities. Agricultural research and extension systems in the region ignore traditional food systems and culturally accepted food baskets. The EU funded project "Strengthening Adaptive Farming in Bangladesh, India and Nepal (SAF-BIN)", implemented by Caritas organisations, is building resilience to climate change through strengthening adaptive small scale farming systems in rain-fed areas. In a multi-sectoral collaboration a farmers' collective-led approach has been implemented with smallholder farmers at the centre. A major tool has been on-farm adaptive research trials in which farmers'

R. Roschinsky (✉) • M. Wurzinger
Centre for Development Research, BOKU University of Natural Resources and Life Sciences, Peter Jordan Strasse 82, 1190 Vienna, Austria
e-mail: romana.roschinsky@boku.ac.at

S. Simon
SACU SAF-BIN, Caritas India, New Delhi, India

P.R. Choudhury
Independent Consultant, New Delhi, India

A. Baroi • S.G. Costa
SAF-BIN, Caritas Bangladesh, Dhaka, Bangladesh

M. Malla • C. Manandhar
SAF-BIN, Caritas Nepal, Kathmandu, Nepal

V.D. Pankaj
SAF-BIN, Caritas India, New Delhi, India

M. Aichinger
SAF-BIN, Caritas Austria, Vienna, Austria

© Springer International Publishing Switzerland 2016
W. Leal Filho et al. (eds.), *Implementing Climate Change Adaptation in Cities and Communities*, Climate Change Management, DOI 10.1007/978-3-319-28591-7_11

collectives have developed and implemented on-farm trials in ten districts in Bangladesh, India and Nepal with the active involvement of civil society, researchers and government officials. Results are documentation and increased adoption of locally appropriate farming practises achieved through blending traditional and modern practices with awareness on organic, sustainable production. Farmers' collectives successfully built resilience to climate change, increased yields, improved nutritional security, reduced external dependency, and reduced input costs. SAF-BIN is evolving as a successful model for strengthening adaptive capacities of smallholders. Lessons on integration of a diverse set of stakeholders are transferable and applicable to similar initiatives.

Keywords Climate change adaptation • Farmers' collective-led approach • Food security on-farm adaptive research • Smallholders • South Asia

Introduction

In 2050 the global population is projected to exceed 9.1 billion (FAO 2009) which will mainly be fed by smallholders—the backbone of global food security who already produce over 80 % of food globally and especially in developing countries (IFAD & UNEP 2013). These smallholders are among those most affected by climate change and adaptation is a serious challenge for them.

The situation is especially worrying in South Asia. The region is extremely vulnerable to climate change (CDKN & ODI 2014). Effects include gradual shifts, such as temperature changes, rising sea level, and decreasing permafrost in the Himalayas (The World Bank 2006; IFAD 2011; CDKN & ODI 2014). The variability of weather extremes also increases including floods, storms, droughts, and the break of the monsoon cycle, which is extremely important for rain-fed agricultural and ecological systems in the region (The World Bank 2006; IFAD 2009). These changes have an impact on regional economies and the livelihoods of millions (The World Bank 2006), where the rural population, with low incomes and high reliance on traditional agriculture, is especially vulnerable (IFAD 2009).

South Asian farming is characterized by small farms, with the vast majority being less than 1 ha of land (Bangladesh 87 %, India: 62 % and Nepal: 75 %) (Joshi et al. 2007). Access to extension and public food distribution networks[1] is difficult due to physical and political remoteness affecting choices and production from local agriculture. Nutritional intake is reducing with decreasing purchasing ability, changing cropping patterns and decreasing production. Regional agriculture is constrained by reduced crop diversity, ill-suited technologies and a lack of climate change awareness. Effective strategies, developed over generations, to respond to climatic challenges of the South Asian context are no longer effective. Innovations

---

[1] Public Distribution System in India and Public Food Distribution System in Nepal and Bangladesh.

are needed to respond to new challenges. The increase of the adaptive potential of smallholders is a key means for global food security (FAO 2009).

In a region that made such advances in agricultural technologies in past decades (e.g. Green Revolution) the National Agricultural Research Systems (NARS) should produce outputs that benefit smallholders. But on small and highly diverse farms, a top-down, researcher-led approach has traditionally not led to farming output improvements (Goma et al. 2001). There are limitations to the locally applicable solutions that NARS can produce (Hazell 2008). Extension systems lack the capacity to effectively support small farmers. Mruthyunjaya and Ranjitha (1998)) have pointed out that for India, the extension system, apart from a lack of operating resources, also has a narrow focus, lack of technical capacity and inadequate cooperation with farmers.

For some time, donors, policymakers and civil society organisations (CSOs) have been demanding a transformation of formal agricultural research and development towards more relevant outputs for smallholders (McNie 2007; Waters-Bayer et al. 2015). In this context, the advantages of participatory research methods, in which different stakeholders participate and benefit jointly, come to mind (Goma et al. 2001). Farmer-led research produces relevant outcomes by bringing together farmers and support agents (e.g. researchers, extension) to improve the livelihoods of local people in agriculture or natural resource management (Waters-Bayer et al. 2015). As one alternative, Farmer Field School (FFS) were established in the 1980s reacting to detrimental effects of abundant chemical fertilizer application in Indonesia (Braun and Duveskog 2008).

The project presented here intends to produce outcomes that matter to smallholders through a multi-stakeholder approach that brings together smallholders, CSOs, research, extension agencies and other sector stakeholders.

SAF-BIN Project

"Strengthening Adaptive Farming in Bangladesh, India and Nepal (SAF-BIN)" is a research and development project funded through the EU Global Programme on Agricultural Research for Development implemented from 2011–2016. SAF-BIN is building resilience to climate change through strengthening adaptive small scale farming systems in rain-fed areas in Bangladesh, India and Nepal.

Project Aims and Objectives

The overall aim of SAF-BIN is to promote local food and nutritional security through adaptive small scale farming in four rain-fed agro-ecological zones in South Asia in the context of climate change. Specific objectives include:

1. Screen and document innovations in traditional food production, distribution and consumption systems (FPDCS) of smallholders (SHF) with respect to climate change adaptation and nutritional security.
2. Collectivize and strengthen smallholder farmers' institutions to achieve an organized and sustainable approach.
3. Test the potential of model strategies for FPDCS, defined by blending of traditional knowledge and modern technologies to enhance SHF adaptive potential.
4. Develop multi-stakeholder monitoring mechanisms to enhance the efficiency of these model strategies.
5. Influence national research and policy agendas for the promotion of collectivized FPDCS of SHFs to adapt to climate change effects and ensure nutritional security.

Project Implementation

Project Sites

SAF-BIN works with smallholder farmers' collectives (SHFCs) and has the ambitious aim to produce knowledge, processes and outputs relevant for 10 million smallholders in South Asia. The project sites were therefore selected within four agro-ecological zones (AEZ) that represent a considerable proportion of the South Asian agricultural landscape:

- Bangladesh: sub-humid flood plain and semi-arid upland agro-ecological system (AES) (FAO AEZ 3)
- India: sub-humid tropical hilly AES (FAO AEZ 5) and sub-humid tropical plateau AES (FAO AEZ 5)
- Nepal: humid sub-tropical montane AES (FAO AEZ 8).

The project districts within these AEZs were then selected by a stratified, purposive sampling technique based on the following criteria:

- climate change vulnerability
- traditional cropping practises
- large traditional agro bio-diversity
- large indigenous population
- small farm-holdings
- accessibility by road
- prior exposure to (adaptive) research
- absence of intra-village conflicts
- presence of partner organisations
- food insecurity: 4–8 months/household and year,
- malnutrition,
- maternal/infant mortality rate: 54–63

Fig. 11.1 SAF-BIN project districts in Bangladesh (adapted from Mapsopensource.com 2015) and Nepal (adapted from Traveltrendnepal.com 2015)

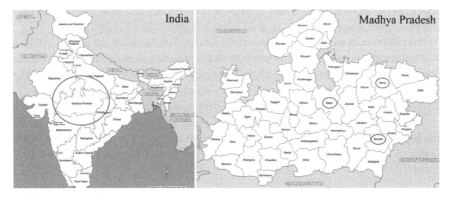

Fig. 11.2 SAF-BIN project districts in Madhya Pradesh, India (adapted from Dmaps.com 2015a, b)

- Human Development Index: 0.39–0.59

90 villages were targeted in three districts in Bangladesh (Natore, Naogoan and Rajshahi; Fig. 11.1), three districts in India (Mandla, Satna and Sagar in Madhya Pradesh state; Fig. 11.2) and four districts in Nepal (Kaski, Nawalparasi, Bardiya and Surkhet; Fig. 11.1).

Multi-stakeholder Cooperation

The South Asia Coordination Unit (SACU), at Caritas India, coordinates the project. Partners involved in SAF-BIN, including Caritas organisations (of Austria, Bangladesh, India and Nepal) and researchers (BOKU University of

Natural Resources and Life Sciences Vienna), enable a multi-disciplinary approach. Associated partners from the project countries like Action for Food Production (AFPRO), Local Initiatives for Biodiversity, Research and Development (LI-BIRD), Sam Higginbottom Institute of Agriculture, Technology and Sciences (SHIATS) and Bangladesh Rice Research Institute (BRRI) complement the consortium.

Methodological Design and Instruments

Of the multitude of activities within SAF-BIN, this publication focuses on the on-farm adaptive research activities.

SAF-BIN applied a farmers' collective-led approach combining elements of Participatory Rural Appraisal, FFS methodology and participatory on-farm adaptive research. Smallholder farmers' collectives (SHFCs) are the main drivers in all steps of the research process. The involvement of national and international researchers, decision makers and extension agents in a multi-stakeholder monitoring process ensured constant value addition and enriched the approach.

SAF-BIN selected Village Research Assistants (VRAs) from the project villages and trained them to successfully implement the project and build capacities at local level. VRAs assisted in the formation (or revived/selected if farmer initiatives were already present before) of 2–3 SHFCs per village, comprising of 10–15 farmers each, following selection criteria and feedback of all village members.

Initial assessment of the agro-ecological, social and climate change situation by the SAF-BIN team through a literature review was followed by a participatory rural appraisal conducted in all project villages to ascertain the local contexts. This included:

- mapping exercises on resources, social context, village landscape transact, agro-ecology and food-scape, technology adoption, and seasonality
- wealth ranking Venn Diagram (c.f. Sontheimer et al. 1999)
- time trend (c.f. Jain and Polman 2003)
- bio-resource flow diagram (c.f. Lightfoot et al. 1991)
- problem matrix tree and analysis (c.f. MDF 2005), and
- Food In and Food Out (FiFo) analysis.

A structured baseline survey was also conducted mainly focused on food security and climate change context issues at household level.

Screening and scouting exercises collected local challenges and solutions and were seasonally repeated to support farmers in reflecting and realizing possible causes and solutions for climate change related problems. Farmers identified traditional practises as well as food production and storage innovations for climate change adaptation which were documented by the SAF-BIN team. Recommendations concerning challenges were made by district level technical experts and research partners and shared with the SHFC.

Traditional practices, technologies based on formal research, and local innovations were analysed and blended into models for on farm adaptive trials suitable for the local vulnerability context by SHFCs, SAF-BIN team and other stakeholders at village level. These models, centred on specific crops, were further developed in a multi-stakeholder review process by SHFC, SAF-BIN field staff, government officials, technical and scientific advisers into trial designs and tested by the SHFC in on-farm adaptive research trials. The blending process was taken up as an integral part of the trial design. During regular meetings, SHFCs were responsible for planning, establishment, monitoring, data collection and initial analysis of farmer-led on-farm research trials. They collectively decided all matters related to the on-farm adaptive trials.

After each season, based on the data collected by the farmers themselves and the joint analysis with research and extension stakeholders, each SHFC decided if and how they wanted to adapt or refine on-farm trials and to what extent they wanted to be involved in the following season. They were supported by the project in form of capacity building and some initial inputs (for example new seed varieties) for experimental purposes. Data collected throughout the on-farm adaptive trials (e.g. crop performance, climatic data, yields, cost of production, etc.) was statistically analysed by the SAF-BIN research staff to create an empirical result base for recommendations to SHFCs as well as reporting and publication purposes. Additionally master students collected and analysed empirical data from selected on-farm trials.

To ensure that the project activities contributed to the fulfilment of the project objectives, regular exchanges between SHFCs and the national SAF-BIN project teams were organised (e.g. after the completion of a trial season). These opportunities were used to check the project progress by collecting information on key project parameters defined in the initial project phase (e.g. number of food secure days, number of items in the local food basket, number of functional SHFCs, etc.). Together with documentation of project activities these meetings provide data for regular reporting activities and strategic project decisions.

Results and Discussion

Results concentrate on the main insights gained to date from the collaboration with the SHFC during on-farm adaptive research trials.

Resilience Is Putting Farmers First and Believing in Their Practices and Capacity

Although the cooperation of a diverse set of stakeholders is challenging and structural changes can make the implementation of activities difficult, SAF-BIN has contributed to the spread of locally appropriate solutions for climate change adaptation that result from blending of traditional and improved practises. A number of activities were needed to achieve this result.

Information collected during the initial PRA exercises and the baseline household survey, as well as local practise and innovation screening, enabled the project team and participating farmers to critically analyse local climate change challenges. This improved their analytical and decision making skills, a reported effect achieved in FFS approaches (Pontius et al. 2002). Special care has been taken by the communities to select the most vulnerable groups (single women headed household, lower social classes) as participants. This made sure that SAF-BIN benefited those most vulnerable to climate change unlike the majority of FFS approaches that reportedly benefit better off farmers (Waddington and White 2014).

The local knowledge collected represents a valuable resource for further activities. In the first year alone, 297 local practises and innovations in response to climate change were documented (SAF-BIN 2013). Smallholders apply these practises to combat the diverse effects of climate change e.g. drought (through moisture conservation in seed bed preparation), low germination rates induced by changing climatic conditions (Petrů and Tielbörger 2008) and increased pest occurrence (Kocmánková et al. 2009). Some practises were selected as treatments for on-farm research trials (e.g. crushed larvae application as insect repellent).

The documentation serves as starting point to give recognition and value to local innovations described as crucial to institutionalise them in formal research and development systems to contribute to community empowerment and rural development (Okry and Mele 2006). Technologies developed at research stations often do not suit real life farm conditions due to the limited financial resources of smallholders (van de Fliert et al. 2007).

Knowledge has to be relevant to farmers for them to adopt it (Haug 1999). Chances of adoption increase if a technology is already applied by farmers from a similar agro-ecological and social setting. This process also created awareness on the detrimental effect of some local practises (e.g. termite control by mixing using engine oil into the irrigation water). By enriching local knowledge with modern technologies through the multi-stakeholder evaluation process, smallholders increased their access to technologies for the mitigation of climate change effects.

Table 11.1 provides details on SHFC and on-farm adaptive trials conducted in 2013 to illustrate the scope of on-farm research activities in SAF-BIN.

Table 11.1 Overview about the participating SHFCS, smallholders and on-farm adaptive trials conducted in SAF-BIN project during the trial period of 2013

Project country	Number of SHFCs	Number of smallholders active in SHFCs	On-farm adaptive trials conducted	Crop varieties used in on-farm adaptive trials
Bangladesh	90	1400	894	23
India	94	1433	543	11
Nepal	89	1335	302	34
Total	273	4386	1739	68

As representative examples, the details of selected trials are discussed below, where given the scale of the SAF-BIN project a full discussion of all trials is beyond the scope of this paper.

The first example was a trial to assess control mechanisms for sheath blight disease of rice. The disease, cause by *Rhizoctonia solani (IRRI* 2015*) is affected by climate change as it increases in occurrence as a result of higher temperatures and CO_2 concentrations (Gautam et al.* 2013*). In 2013 SAF-BIN* on-farm adaptive trials smallholders compared a farmers' practise[2] and the locally common dual fungicide application[3] with improved practises.[4]

Although, in terms of yield, improved practices were the most promising (4.43 t/ha), the simple additional control mechanism of debris collection in combination with fungicides resulted in higher yields compared to the locally practised dual fungicide application alone (4.4 vs 4.39 t/ha). Also the trial revealed that environmentally sustainable floating debris collection alone yielded a respectable 3.77 t/ha but reduced fungicide usage and input costs for farmers. The trial was up-scaled trough project intervention from 45 plots in 2012 to 93 in 2013 increasing awareness of a larger group of smallholders.

The adaptive manner of SAF-BIN on-farm trials contributed outcomes relevant for the participating farmers as the SHFC had the decision power over the trial design. An example of this is the adaptation of on-farm trials on Yellow Mosaic Virus in black gram in India (Table 11.2). Such changes were made according to the analysis of trial results and critical observation from the farmers (SAF-BIN 2014).

Throughout the project duration the SAF-BIN team has conducted workshops to sensitize farmers to organic production methods. The project has advocated for organic practises which were selected for many trials by SHFCs. Especially in the case of Nepal all trials were "by default" on organic production methods as neither mineral fertilizers nor chemical pest control methods were applied to trial plots.

In India, for example, SAF-BIN provided training on the formulation and application of organic, botanical pest repellents that could be made from locally

---

[2] Collection of floating debris (F.D.) (farmers practise); F.D. + fungicide (introduced farmers practise): yield 4.4 t/ha.

[3] Dual fungicide application ("Folicur") (farmers practice): yield 4.39 t/ha.

[4] Chemicals + MOP + urea (introduced).

Table 11.2 Trial trend of black gram trial in India as example for the development of trials according to the wishes of SHFCs

	Trial design adapted across different years		
Trial topic	2012	2013	2014
Yellow mosaic Virus in Black Gram due to abnormal rain and fluctuating temperature	2 local cultivars (*Khajua, Chikna*) 2 sowing methods (broadcasting, line-sowing) 2 management practises (organic, conventional)	1 local variety (*Khajua*), 1 improved variety (*Shikhar 3*) 2 sowing methods (2 sowing dates) 2 management practises (*Matka Khad*,[a] fish tonic[b])	3 improved varieties (*PSU 94-1, Jawahar Urd-2, Shikhar 3*) 1 sowing method 2 management practises (neem oil, fish tonic)

[a] See footnote 6
[b] Fish tonic is an organic pest repellent made of raw fish and jaggery

available resources of which local farmers had not been aware before: *Lamit Ark*[5] and *Matka Khad*.[6] A solution of Lamit Ark and water (ration 1:10) twice a week and a solution of Matka Khad and water (ratio of 1:10) once in ten days for three week successfully prevented crop failure due to Chilli Leaf Curl Virus (*Bemisia tabaci*). Organic practises and seed treatment with locally available materials have led to a reduction of production costs and external dependencies for smallholders.

The spread of successful technologies has been documented (Fig. 11.3) and shows that unlike FFS, that often fail to have an impact on neighbouring communities (Waddington and White 2014), SAF-BIN participants have been able to spread their knowledge and accessible technologies to relatives and friends. Innovations requiring no, or only a few external inputs have earlier been recognized to spread quickly (Waters-Bayer et al. 2015).

This is further supported by the upscaling of area planted in Nepal with rice and wheat trial combinations.[7] Rice combinations went from the trial plots to 49.6 ha in 2013 and 81.3 ha in 2014. Wheat combinations were increased from 18.8 in 2013 to 51.8 ha in 2014.

---

[5] Lamit Ark contains chillies, ginger and garlic and acts as pest repellent.

[6] Matka Khad contains cowdung, cow urine and jaggery and acts as nutrient supplement.

[7] Combination of varieties and cultivation practises tested in on farm trials.

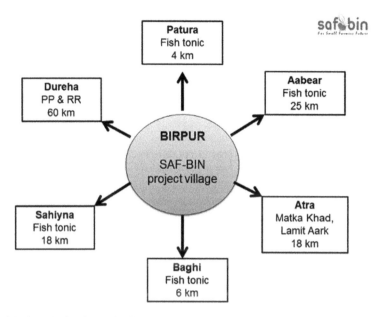

Fig. 11.3 Spread of technologies from SAF-BIN on farm trials in Birpur, Satna district, India to neighbouring villages. Indicated are village names (*bold*), the technology (Lamit Ark: an organic pest repellent containing chillies, ginger and garlic and fish tonic: a pest repellent made of raw fish and jaggery); PP and RR: plant spacing plant to plant and row to row and MathaKhad) and the distance from the source village (adapted from Pankaj 2015)

Increase in Yields, Nutritional Security and Resilience Towards Climate Change

In varietal trials new crop varieties, better adapted to the changing climatic conditions, were made available to SHFC through SAF-BIN and the associated partners (e.g. NARC: Nepal Agricultural Research Council, BRRI). A few examples of the new varieties tested are wheat cultivars (*GW273, GW366*) in India, drought tolerant rice (*Sukkha dhan 1, Sukkha dhan 2, Sukkha dhan 3*) and rust resistant wheat (*Vijay, Gautam; Gaura, Dhaulagiri, WK1204*) in Nepal and rice (*BRRIdhan56, BRRIdhan57, BINA7*), wheat (*BARIwheat-27*) and mungbean (*BINA mungbean-5, BINA mungbean-8*) varietiesin Bangladesh.

Under the same cultivation practises, improved varieties generally achieved higher yields than the local varieties although a statistical analysis did not reveal significant differences in all cases. Yet the improvement of crop yields has been a reported impact of FFS activities (e.g. Bekele et al. 2013).

As an illustrative example for yield increases, in Nepal SAF-BIN farmers conducted varietal trials of wheat and rice. In these trials, wheat yields increased

Table 11.3 Food secure days and food basked changes in SAF-BIN sites (results from the initial household survey and follow up assessment through group discussions with SHFCs after activity implementation) (values are rounded to the next full number)

	Food secure days/year before SAF-BIN	Food secure days/year after SAF-BIN	Change of food secure days/year through SAF-BIN (%)
Bangladesh	178	211	+19
India	200	330	+65
Nepal	150	240	+60
Average	179	260	+45
	Number of items in local food basket before SAF-BIN	Number of items in local food basket after SAF-BIN	Change of items in local food basket through SAF-BIN (number of items)
Bangladesh	9	10	+1
India	5	10	+5
Nepal	18	30	+12

(lowland varieties[8] from 3.9 t/ha to 4.2–4.5 t/ha; upland varieties[9] from 0.9 t/ha to 1.2–1.4 t/ha) and new rice varieties performed better with respect to both grain yield (local variety: 3.3 t/ha vs. new varieties 3.4–4.1 t/ha) and straw yield. Farmers decided to cultivate both the new and the local varieties, a diversification contributing to their food security.

In most project villages, the established SHFC agreed on a seed bank system for sustainable access to new seeds,[10] a risk-minimizing factor also reducing agricultural input costs. Established links with research institutions and extension agents opened the door to future collaboration. If these linkages can be sustained, SHFC members and their communities will continue to benefit from these contacts.

Participants reported that through the SAF-BIN their food security has increased and the local food baskets have diversified (Table 11.3). Diversification of production has been recognized as important in managing adaptation to climate change, especially for smallholders (IFAD 2010). The activities SHFC have realized, supported by SAF-BIN, contribute to diversified agricultural production systems in the target areas. Supporting activities were the establishment of kitchen gardens in all project countries, the revival of traditional food basket items (e.g. minor millets in tribal regions of India, snails and oysters in Bangladesh) and the introduction of new nutritional crops (mainly vegetables).

Through the initial discussions during PRA, the household survey, and the screening exercise, as well as through sensitization workshops on climate change effects, farmers capacity to link their local challenges with the global climate change phenomenon has been enhanced. This was further enhanced through the

---

[8] *Vijay, Gautam*

[9] *Gaura, Dhaulagiri, WK1204.*

[10] Communally managed seedbanks were established in all villages in Nepal and Bangladesh. In India 80 % of participating smallholders have access to communally or individually safely stored seed.

formation of SHFC as communication platforms in which farmers can share their experiences and interact with sector stakeholders to develop mitigation mechanisms that work for them.

Conclusions

While the SAF-BIN project is still ongoing, the project is already on a good path to promote local food and nutritional security through adaptive small scale farming. Farmer innovations in traditional food production, distribution and consumption systems have been screened and are being documented. Through the farmers' collectives approach, smallholder institutions have been created or strengthened entailing positive developments such as formalisation of organisations or creation of communal farming input management. The farmers' collective-led on farm adaptive research trials have produced outcomes that are relevant for farmers' adaptive capacity, nutritional security and enhanced their analytical and organisational skills. Food security has also been improved. Practices and technologies tested provide additional options for smallholders to mitigate the effects of climate change in a locally appropriate and acceptable manner.

Through the multi-stakeholder collaboration within SAF-BIN, the capacity of all actors to approach food production, distribution and consumption systems of the region in a multidisciplinary and participatory manner has been enhanced. The variety of collaborative engagements and communication strategies enables the sharing of emerging research results among scientists, extension workers, policy makers and especially local smallholders. The improved linkages between SHFC and development actors increased skills and information access of smallholders concerning the mitigation of climate change effects.

Challenges remain for the final stage, where the sustainability of the smallholder institutions beyond the project duration needs to be ensured. Experiences on the processes and technologies implemented in SAF-BIN, as well as challenges met, need to be documented and communicated effectively. Only then will national research and policy agendas for the promotion of collectivized approaches to adapt to climate change effects and ensure nutritional security be influenced, and the project achieve long term sustainability.

Acknowledgements SAF-BIN is a joint effort of many stakeholders. Without the willingness of all participating smallholders to collaborate none of the project activities could have been implemented so the authors wish to express their deepest gratitude. The SAF-BIN team members of the implementing Caritas Organisations in Bangladesh, India and Nepal were tirelessly working in the field (Village Research Assistants, District Officers) to contributed to the data base on which this publication is built. Research Officers are acknowledged for their direct and indirect contributions to this publication. The associated partners of SAF-BIN (AFRPO, SHIATS, LI-BIRD) contributed by their expertise as advisers for research matters throughout the implementation of SAF-BIN. SACU facilitated the design and implementation of the project and provided necessary inputs.

References

Bekele N, Obare G, Mithöfer D, Amudavi D (2013) The impact of group based training approaches on crop yield, household income and adoption of pest management practices in the smallholder horticultural subsector of Kenya. J Sust Dev Afr 15(1):117–140

Braun A, Duveskog D (2008) The Farmer Field School approach—history, global assessment and success stories. IFAD Rural Poverty Report 2011. Rome

CDKN & ODI (2014) The IPCC's Fifth Assessment Report. What's in it for South Asia? London

Dmaps.com (2015a) Map of India. http://tinyurl.com/lhydd7e. Accessed 8 Mar 2015

Dmaps.com (2015b) Map of Madhya Pradesh. http://tinyurl.com/q2epcta. Accessed 8 Mar 2015

FAO (2009) Food security and agricultural mitigation in developing countries: options for capturing synergies. October. Rome

Gautam HR, Bhardwaj ML, Kumar R (2013) Climate change and its impact on plant diseases. Curr Sci 105:12

Goma HC, Rahim K, Nangendo G, Riley J, Stein A (2001) Participatory studies for agro-ecosystem evaluation. Agric Ecosyst Environ 87:179–190

Haug R (1999) Some leading issues in international agricultural extension, a literature review. J Agric Educ Ext 5:263–274

Hazell PB (2008) An assessment of the impact of agricultural research in South Asia since the Green Revolution. Rome

IFAD (2009) Climate Change Impacts – South Asia. In: The global mechanism. United Nations convention to combat desertification. International Fund for Agricultural Development. Rome

IFAD (2010) IFAD's response to climate change through support to adaptation and related actions Comprehensive report: Final version. International Fund for Agricultural Development. Rome

IFAD (2011) Addressing climate change in Asia and the Pacific. International Fund for Agricultural Development. Rome

IFAD & UNEP (2013) Smallholders, food security and the environment. International Fund for Agricultural Development, Rome, pp 1–54

IRRI (2015) Sheath blight—IRRI Rice Knowledge Bank Training fact sheet. http://www.knowledgebank.irri.org/training/fact-sheets/pest-management/diseases/item/sheath-blight. Accessed 20 Mar 2015

Jain SP, Polman W (2003) A handbook for trainers on participatory local development. Food and Agriculture Organization of the United Nations/Regional Office for Asia and the Pacific: Bangkok

Joshi P, Gulati A, Cummings R (2007) Agricultural diversification in South Asia: beyond food security. In: Joshi P, Gulati A, Cummings R (eds) Agricultural diversification and smallholders in South Asia. Academic Foundation, New Delhi

Kocmánková E, Trnka M, Juroch J, Dubrovský M, Semerádová D, Možný M, Žalud Z (2009) Impact of climate change on the occurrence and activity of harmful organisms. Plant Prot Sci 45:48–52

Lightfoot C, Axinn N, John KC, Chambers R, Singh RK, Garrity D ... Salman A (1991) Training resource book for participatory experimental design. Report of a research design workshop on participatory design of on-farm experiments of the ICAR/IRRI collaborative Rice Research Project. Kumarganj, Faizabad, Uttar Pradesh, India

Mapsopensource.com. (2015). Map of Bangladesh. http://tinyurl.com/mjfkt94. Accessed 8 Mar 2015

McNie EC (2007) Reconciling the supply of scientific information with user demands: an analysis of the problem and review of the literature. Environ Sci Pol 10:17–38. doi:10.1016/j.envsci.2006.10.004

MDF (2005) Problem tree analysis. MDF Training and Consulting. http://www.toolkitsportdevelopment.org/html/resources/91/910EE48E-350A-47FB-953B-374221B375CE/03 Problem tree analysis.pdf

Mruthyunjaya S, Ranjitha P (1998) The Indian agricultural research system: structure, current policy issues, and future orientation. World Dev 26(6):1089–1101

Okry F, Van Mele P (2006) Documenting, validating and scaling-up local innovations. LEISA Mag 22(3):14–15

Pankaj VD (2015) SAF-BIN experiences from India. Presentation during the Conference on smallholders, New Delhi, 10–12 Mar 2015

Petrů M, Tielbörger K (2008) Germination behaviour of annual plants under changing climatic conditions: separating local and regional environmental effects. Oecologia 155:717–728. doi:10.1007/s00442-007-0955-0

Pontius J, Dilts R, Bartlett A (eds) (2002) From farmer field school to community IPM: ten years of IPM training in Asia. Food and Agriculture Organization of the United Nations, Bangkok

SAF-BIN (2013) Building resilience to climate change through strengthening adaptive small farming systems in Bangladesh, India and Nepal (SAF-BIN) project: narrative progress report May 2012–May 2013 (year 2). New Delhi/Vienna

SAF-BIN (2014) Building resilience to climate change through strengthening adaptive small farming systems in Bangladesh, India and Nepal (SAF-BIN) project: interim narrative report 01 June 2013–30 June 2014. New Delhi/Vienna

Sontheimer S, Callens K, Seiffert B (1999) Conducting a PRA training and modifying PRA tools to your needs. An example from a participatory household food security and nutrition project in Ethiopia. Rome

The World Bank (2006) Managing climate risk. Integrating adaptation into World Bank Group operations. Washington, DC

Traveltrendnepal.com (2015) Map of Nepal. http://tinyurl.com/lmbmw46. Accessed 8 Mar 2015

Van de Fliert E, Dung NT, Henriksen O, Dalsgaard JPT (2007) From collectives to collective decision-making and action: farmer field schools in Vietnam. J Agric Educ Ext 13(3):245–256. doi:10.1080/13892240701427706

Waddington H, White H (2014) Farmer field schools from agricultural extension to adult education. 3ie Systematic Review Summary No. 1. International Initiative for Impact Evaluation. London

Waters-Bayer A, Kristjanson P, Wettasinha C, van Veldhuizen L, Quiroga G, Swaans K, Douthwaite B (2015) Exploring the impact of farmer-led research supported by civil society organisations. Agric Food Secur 4:1–7. doi:10.1186/s40066-015-0023-7

Chapter 12
Assessing Student Perceptions and Comprehension of Climate Change in Portuguese Higher Education Institutions

P.T. Santos, P. Bacelar-Nicolau, M.A. Pardal, L. Bacelar-Nicolau, and U.M. Azeiteiro

Abstract The higher education system has a critical role to play in educating environmentally aware and participant citizens about global climate change. Yet, few studies have focused on higher education students' knowledge and attitudes about this issue. This study aims to contribute to a comprehensive understanding of views and attitudes about climate change issues, across the postgraduate student population in three universities—the on Campus University of Porto and University of Coimbra, and the distance learning Universidade Aberta, Portugal. We surveyed university students and graduates from three master programs in environmental sciences targeting their knowledge, attitudes and behaviour on climate change issues, and their views of the role that their master degree had on it. A majority of the respondents believed that climate change is factual, and is largely human-induced; and a majority expressed concerns about climate change. Still, the surveyed students hold some misconceptions about basic causes and consequences of climate change. Further research is necessary to comprehend the university post-graduate students' population, so that curricula programs can be adapted to grant consensus on scientific knowledge about climate change, and an active engagement of the graduate citizens, as part of the solution for climate change problems.

P.T. Santos (✉)
Faculdade de Ciências, Universidade do Porto, Porto, Portugal
e-mail: ptsantos@fc.up.pt

P. Bacelar-Nicolau · U.M. Azeiteiro
Departamento de Ciências e Tecnologia, Universidade Aberta, Lisbon, Portugal

Centre for Functional Ecology, Universidade de Coimbra, Coimbra, Portugal

M.A. Pardal
Centre for Functional Ecology, Universidade de Coimbra, Coimbra, Portugal

Faculdade de Ciências e Tecnologia, Universidade de Coimbra, Coimbra, Portugal

L. Bacelar-Nicolau
Institute of Preventive Medicine and Public Health & ISAMB, Faculdade de Medicina, Universidade de Lisboa, Lisbon, Portugal

Keywords Climate change • University students • Curriculum • Knowledge • Attitudes and behaviours

Introduction

Climate Change (CC) is one of the five priority areas of the Strategy 2020 of the European Commission and it constitutes a major global challenge needing local responses (always articulated at the macro, meso and micro levels) as *"Human influence on the climate system is clear, and recent anthropogenic emissions of greenhouse gases are the highest in history. Recent climate changes have had widespread impacts on human and natural systems."* (IPCC 2014). CC understanding entails identifying mutually influential relationships between Nature, Society, Culture, Education and Science, and any intervention, either to mitigate or adapt to CC, necessarily involves the full complexity of its environmental, sociocultural, educational and science dimensions (Viegas et al. 2014). The difference between climate change and global warming should be clear for the purpose of this work, being that Climate Change is a broader term referring to changes (increases or decreases) to long-term weather patterns, while Global Warming refers to increases in the Earth's average temperature caused by greenhouse gas build up in the atmosphere. As stated by Carvalho et al. (2014) the way CC is viewed and addressed in a given country depends on its cultural (including historical and social contexts), political, and scientific backgrounds and also, to a large extent, on the specific aspects of the changing climate in that country and on its actual and perceived future impacts. This work addressed CC research in Portugal focusing on its climate research history, on the observed climate trends, on assessments of CC in Portugal, and on the aspects of Portuguese CC policy (mitigation policies, adaptation policies and stakeholder engagement), together with the Portuguese public engagement with CC (concerns and knowledge, behavioral engagement, information sources and media reconstructions of CC (Carvalho et al. (2014)).

Lorenzoni and Pidgeon (2006) states that responsibility for CC action, risk communication and the implementation of actions to address CC will have to account for perceptions (and science comprehension). As most individuals relate to CC through their personal experience knowledge, and the balance of benefits and costs, as well as trust in other societal actors, the discourses about CC are situated in people's locality and in what people know, think, and believe relating to CC. Perceptions/Lay rationalities focus on lay knowledge that integrates individual understandings, explanations, life experiences, perceptions and intervention on climate change (Alves et al. 2014). The knowledge of the agents that serve social interaction does not correspond at all to the scientific hegemony of power and knowledge, being a different kind of knowledge where the need to produce senses require models much closer to the local and cultural symbolic universes (Alves et al. 2014).

Today's generation of students belong to the "*climate change generation*" (grew up with more information and less scientific uncertainty about CC) and Higher Education Institutions (HEI) have a critical role to play in educating them about global CC (Wachholz et al. 2014). CC is a matter of global concern, being a significant challenge facing society today and becoming a central issue to society (Leal Filho et al. 2014), and specific sectors such as universities need to engage and be active in the search for regional and local solutions for what is a global problem (Leal Filho 2010).

Research has found no correlation between levels of knowledge and behavioral change to address CC, and there is indication that behavioral engagement can occur in the absence of a complete understanding of the problem (e.g. Ortega-Egea et al. 2014; Rhodes et al. 2014), and studies in Portugal suggest that most people feel a significant lack of information and that this may be impacting the ways they relate to CC (Carvalho et al. 2014) following the same pattern in other countries (Anable et al. 2006; CIMC 2012). A better understanding of differences between specific publics is important in order to fully understand and act in this CC dimension. The purpose of this study was to investigate HEI students' perceptions of CC/global warming (GW). This work is the first one conducted within Portuguese university students and the results in this context should have meaningful implications for both national and international CC policies in the future (Jamelske et al. 2013) as well as university curricula adaptation to the students' perceptions.

Research Methodology

Surveyed Institutions and Courses

Research was conducted at three Portuguese Universities, Porto,[1] Coimbra[2] and the Portuguese Distance Learning University[3] (Azeiteiro et al. 2015). The survey was applied to students off our 2nd cycle programmes in the areas of Ecology, Environment, Natural Resources and Sustainability, during the evaluation period of January 2015.

The Master degree in Ecology, Environment and Territory, at Porto University, is aimed at the acquisition of knowledge to understand the biological aspects that are essential to the applied fields of ecology and biodiversity, landscape ecology and planning, environmental education and nature tourism, technology of environmental management, and the environmental characterisation, monitoring and recovery. Also at Porto University, the Master in Biological Aquatic Resources is a post-graduation designed for the management, sustainable exploitation and

---

[1] www.up.pt
[2] www.uc.pt
[3] www.uab.pt

preservation of the aquatic ecosystem. The Master in Ecology offered at the University of Coimbra, aims to provide specialised inter-and multidisciplinary training at an advanced level in different areas of Ecology and Environmental Sciences. It is intended to train highly qualified professionals with a profile adequate to the labour market in the area of environment, including environmental biomonitoring, or with a profile adequate to fundamental and applied research. Finally, the Master degree in Environmental Citizenship and Participation, offered at the Portuguese Distance Learning University, is designed to prepare individuals to work in environmental policy making, as well as to improve their environmental citizenship, participation and planning abilities (e.g. individuals working within government, environmental advisors, members of environmental NGOs, teachers, researchers and individuals involved in environmental practices, policies, planning, training, participation and citizenship) (Azeiteiro et al. 2015; Bacelar-Nicolau et al. 2009). It is taught in e-learning regime, and the pedagogical model underlying the learning process was developed specifically for e-learning 2nd cycle degrees at the UAb (Pereira et al. 2008).

The first three programmes are full-time, daytime attendance degrees with 120 ECTS and the duration of 2 years; the latter programme at the Portuguese Distance Learning University is 100 ECTS and three semesters. Their first year is dedicated to the curricular units (60 ECTS) and the second year is dedicated to the planning, developing, writing and defending of the Dissertation/Project (60 ECTS for the day time attendance degrees and 40 ECTS for the e-learning degree).

Questionnaire

The data were collected through self-completion questionnaire consisting of 30 - closed-ended questions and two open questions aiming at characterizing students from the socio-demographic, and from their perceptions, beliefs, motivations, attitudes, knowledge relating to the topic of CC. The questionnaire was adapted from Wachholz et al. (2014) and Manolas et al. (2010) and following studies by other authors (e.g. Leal Filho et al. 2014; TNS Opinion & Social 2014). Google Drive was used to host the questionnaire and allow for the online survey, which address was disseminated to students by e-mail (69 in the Porto University, 35 in Coimbra University and 20 in The Portuguese Open University). The questionnaire was pre-tested among faculty students not enrolled in the questionnaire application and results of the pilot study were used to refine the questionnaire. Students took part in the survey willingly and the time required for filling in the questionnaire was ca. 10 min. Respondents were assured that their responses remain anonymous.

Statistical Analysis

Data were downloaded into MS Excel and then exported to IBM SPSS Statistics for Windows, version 21®, for statistical analysis. Statistical exploratory univariate and bivariate analyses were performed on the data collected (frequencies, total and column percentages, adjusted residuals).

Findings

The questionnaire had a response rate of 60 %. Respondents were mostly full-time students (75 %) and less (25 %) were working-students, registered on the four different master degree courses. A few (15 %) were completing their curricular year, 64 % had just completed the curricular year and 21 % were developing research for their Master's dissertation. These students were 61 % female and 39 % male. Most (78 %) were aged between 21 and 30 years old; a few were in the age group 31–40 (6 %), in the age group 41–50 (11 %) and over 50 years old (6 %).

Nearly all respondents (99 %) believed that CC is happening, only 1 % being not sure about it. In what concerns their interest, recognition of importance and level of concern with the topic of CC these three aspects were differently perceived among the students. Most students were very much interested in the topic of CC (57 % versus 43 % who answered 'somewhat interested'), and a greater number (65 %) thought it is a very important topic (versus 25 % who expressed it is somewhat important), although their degree of concern was less: 44 %, were "very concerned", 54 %, were "somewhat concerned", and 1 % "had little concern with climate change".

Also, the perception of the impacts of CC on biota communities and human communities, either in one's country (Portugal) or abroad (as a whole) was sensed with different time scales, depending on the community and on the country. There was a generalised perception that biotic communities were already impacted by CC, particularly abroad (94 and 96 %, respectively for biotic communities in Portugal and abroad). The majority of respondents (92 %) also thought that CC impacts were already felt in human communities abroad, but less (76 %) in Portuguese human communities. Still, a minority though CC impacts would only be felt between 10 and 50 years from now (8, 6 and 4 %, respectively in human communities abroad, biotic communities in Portugal and abroad), and up to a 100 years from now in the Portuguese human communities.

The majority of students (75 %) expressed that the scientific community assumed that CC was happening, less (24 %) felt that there was much disagreement amidst the scientists relating to the subject, and a minority (1 %) felt not knowing enough to form an opinion.

The majority of students (96 %) also perceived that human activities were an important cause of CC (versus 4 % who expressed they were not). When inquired if humans could mitigate their effect on CC, most respondents shared the opinion that they could mitigate effects of CC, although 33 % thought that people did not have the will to change their behaviour in order to achieve mitigation, and the majority, 60 %, expressed it was not clear if that would ever be achieved; only 4 % believed that humans would manage successfully to mitigate the effect of CC. Lastly, 3 % thought that humans could not mitigate CC.

Almost half of the students (44 %) felt that their Master academic training had focused enough on the topic of CC, and a few (4 %) felt that they had too much training on that subject. Still 22 % felt the subject had not had enough focus; and 2 % had not attended enough classes to form an opinion. When latter inquired if they felt the need for more information on the topic to form an opinion, most students expressed the need for a little more of information (46 %), or more information (40 %), or even a lot more of information (7 %). Only a small number (7 %) felt they did not need more information in order to form an opinion on the topic of CC.

Furthermore, when inquired about their knowledge on the subject, most respondents (78 %) felt they had a moderate technical knowledge about the topic of CC, while some (17 %) perceived their knowledge as extended, and very few (1 %) as professional knowledge. Only a minority felt they had a minimum knowledge on the topic (4 %). Students knowledge of how their behaviour influenced CC followed a similar pattern, most (68 %) felling a moderate knowledge, some (26 %) perceiving it as extended, and very few (1 %) at a professional level; a minority (4 %) felt a minimum knowledge of how their behaviour influenced CC.

When asked about the effect of the ozone hole on CC, only 36 % acknowledged that the ozone hole was not a main cause of CC. Most students, 57 %, responded that it was a main cause of CC and 7 % didn't know whether it was or not a cause of CC.

In what concerns their ecological footprint, most students, 41 %, had not estimated it. For the students who had calculated their ecological footprint, most (26 %) had done so outside the scope of their academic degree and a little less (17 %) had calculated it as part of the training.

There was a large conviction that the main actions in mitigating CC effects should be taken by governments (74 %), although some respondents (18 %) were of the opinion that individuals/people should be the main responsible for taking actions, and a small number (8 %) thought that the responsibility should lay upon corporations. It was interesting to notice that non-profit organisations were not seen as being main actors in mitigating CC effects. Also, without an international agreement, e.g. Kyoto's protocol successor, most (51 %) felt that there would be still a way to mitigate CC, although many (42 %) felt that such an international agreement was essential, and a few students (7 %) had no opinion on the matter.

Regarding the behaviour of relational communities that surrounds each respondent, only 8 % thought that most people in their relational communities acted to mitigate climate effects. Nearly half, 49 %, perceived that most people in their relational community took some actions to mitigate CC effects, and 43 % perceived that they did not act in order to mitigate CC effects.

As individuals, 33 % of the respondents had already taken some actions to mitigate the causes of CC, 22 % had taken some actions but felt they were difficult to apply, 13 % had considered but never taken any actions, and 21 % had not considered taking actions even though they believed CC was happening. Still, 13 % believed that an individual's actions were not going to have any effect on mitigating CC.

When inquired if they were to reduce their contribution to CC most students felt that they would be increasing their life quality by a little (40 %) or a lot (25 %). Still, some (21 %) felt that reducing their contribution to CC would not change their life quality, and a few felt that it would decrease their life quality by a little (10 %) or a lot (4 %). As individuals, most respondents (57 %) also expressed that they would support actions to reduce greenhouse gases including measures which implied e.g. paying more for fuel and electric energy, while a lesser number (31 %) was not sure about it, and 13 % would not support such actions. The mitigation actions taken by the respondents were reducing the use of cars and fossil fuels, or "sharing the car of friends and colleagues" (24 %), increasing recycling and reutilization of materials, "namely the use of plastic bags" (18 %), reducing electricity consumption and water (18 and 10 %). Also, referred by less people, but important in this context was "raising environmental awareness and change attitudes and behaviour of people around oneself" (8 %), reducing meat consumption or "having a vegan diet" (7 %), consuming local products (4 %) and conserving nature and replanting trees (3 %).

The bivariate statistical analysis performed on the inquiry data allowed to have a deeper understanding of the analysed sample of students.

In this context, the question relating to ozone which may be seen as a knowledge indicator of the CC topic, allowed to note that the more (significantly) knowledgeable age group was the 31–40 years (75 % responded correctly) and the less knowledgeable were over 41 years old (25 % responded correctly). In the age group 21–30 years, only 36 % responded correctly. Even though the type of study regime—full–time and working-student—did not influence the percentage of knowledgeable respondents to this question; the only significant difference being the fact that students in full-time (10 %) admitted their ignorance on the matter (while working-students did not).

The self-perceived knowledge on the topic of CC did partially mirror the effective knowledge of the students, as indicated by the fact that only 31 and 58 % of self-perceived "moderate knowledge" and "extended knowledge" responded correctly to this question.

It was interesting to note that the respondents who are developing their dissertation are (significantly) more knowledgeable in this aspect (67 % respond correctly). Also, among the respondents who acknowledged that ozone was not the main responsible for CC, the perception of the change in one's quality of life relating to future contributions to mitigate CC was mainly "it won't change …" or "will improve a little my life quality". Other respondents' perception of change was more diversified from "will improve a lot ..." to "will worsen a lot" (data not shown).

In what relates to the bivariate analysis of the full-time vs. working-students profiles, the interest, importance and concern with the topic of CC, as well as self-perception of knowledge on the subject, was not affected by the type of study profile. Full-time students were mostly (96 %) aged 21–30, while working-students were distributed in 21–30 (41 %), 31–40 years (14 %), 41–50 years (27 %) and above 50 years (18 %). There were, however, several differences, some of which are worth pointing. The perceived time scale for impacts of CC on biotic and on human communities was longer among full-time students (up to 50 years and up to 100 year from now, respectively) than among working-students (up to 10 years from now for both biotic and human communities), both in Portugal and abroad. The working-students also perceived that the time scale for impact of CC on human communities was more immediate that full-time students did. The group of working-students also differed from full-time students in their greater perception (100 %) that human activities are an important cause of CC (100 vs. 94 % respectively).

Working-students had a more moderate belief that "humans can mitigate CC even though it is not clear that they want or will do so" (100 %), whilst full-time students had more diversified opinions, ranging from "humans won't be able" (4 %) to "humans will do it" (6 %) (data not shown). Interestingly, it is among the working-students that there was a (significantly) greater conviction that their contribution for mitigating CC will "improve a lot their life quality" (46 %) or "... a little" (18 %) but not "worsen a lot" (0 %), as opposed to the full time students (16 and 49 % and 6 %, respectively; data not shown).

It was also the group of working-students that assumed in (significantly) in greater number to have already taken actions to mitigate CC (59 vs. 20 %). Even though they attribute governments the major responsibility in setting action against CC, as full-time students (73 %) they differ from them by also giving individuals a greater role in mitigation actions of CC (27 vs. 14 %, respectively) (Table 12.1).

When analysing the age groups, students aged above 50 had globally (significantly) greater interest, perception of the importance and concern with the topic of CC, relatively to others (100 %, 100 %, 75 %). Students aged 31–40 years also had greatest interest on the topic of CC (100 %), the lowest being observed in the age groups 41–50 years (14 %) and 21–30 years (55 %). The perception of importance and concern did follow different profiles (Table 12.2).

In what relates to the perception of CC impacts on communities, the younger 21–30 years group perceived them as having a longer time scale for impact in human communities (particularly in Portugal; up to 100 years) and shorter in biotic communities, in opposition to students in age groups above 31 years, who perceived them has immediate. The students aged 41–50 are particularly interesting, as in opposition to their (lowest) interest in the topic, they perceived an immediate impacts in human communities but the longest in biotic communities, compared to other age groups.

Table 12.1 Respondents perception as a function of their regime of study (full-time students vs. working-students)

			Full-time student (%)	Working-student (%)	Total % (n)
Perception of climate change impacts on ...	Human communities Portugal	Now	69.4	90.9	76.1 (54)
		In 10 years	12.2	9.1	11.3 (8)
		In 25 years	8.2	0.0	11.3 (8)
		In 50 years	8.2	0.0	5.6 (4)
		In 100 years	–	–	1.4 (1)
	Human communities other countries	Now	87.8	100.0	91.5 (65)
		In 10 years	4.1	0.0	2.8 (2)
		In 25 years	6.1	0.0	4.2 (3)
		In 50 years	2.0	0.0	1.4 (1)
	Biotic communities Portugal	Now	93.9	95.5	94.4 (67)
		In 10 years	2.0	4.5	2.8 (2)
		In 25 years	–	–	–
		In 50 years	4.1	0.0	2.8 (2)
	Biotic communities other countries	Now	95.9	95.5	95.8 (68)
		In 10 years	0.0	4.5	1.4 (1)
		In 25 years	2.0	0.0	1.4 (1)
		In 50 years	2.0	0.0	1.4 (1)
Total % (n)			100.0 (56)	100.0 (4)	100.0 (7)

Gender was not, in most cases, a differentiating factor of the respondent's answers. The respondent's gender did not (significantly) influence the interest, perception of degree of importance or knowledge on the CC topic (e.g. ozone layer, human influence on CC). However, female respondents showed a (significantly) greater degree of concern with CC than male respondents (54 vs. 32 %, respectively) and they perceived impacts of CC, both on biotic and human communities, as immediate, while men tended to perceive them more in the medium term (up to 50 years from now, Table 12.3).

Discussion

Individuals' perceptions about CC are clearly complex and multidimensional. This paper provides the first results obtained from 2nd cycle Portuguese university students, towards a deeper understanding of these perceptions, which should be considered in national policies and university curricula adaptation taking into account that the sample size is not high and that its representativeness is limited.

Most students believe that CC is happening, that their main causes are anthropogenic, and they feel well informed on the topic of CC. However, when asked about particular concepts and processes of CC, their answers do not reflect their

Table 12.2 Respondents perceptions as a function of their age group

		21–30 years (%)	31–40 years (%)	41–50 years (%)	>50 years (%)	Total % (n)
The topic of CC has …	Interest	A lot of	55.4	100.0	14.3	100.0	56.3 (40)
		Some	44.6	0.0	85.7	0.0	43.7 (31)
	Importance	A lot of	62.5	75.0	71.4	100.0	66.2 (47)
		Some	37.5	25.0	28.6	0.0	33.8 (24)
	Concern	A lot of	44.6	25.0	42.9	75.0	45.1 (32)
		Some	53.6	75.0	57.1	25.0	53.5 (38)
Perception of climate change impacts on …	Human communities Portugal	Now	69.6	100.0	100.0	100.0	76.1 (54)
		In 10 years	14.3	0.0	0.0	0.0	11.3 (8)
		In 25 years	14.3	0.0	0.0	0.0	11.3 (8)
		In 50 years	7.1	0.0	0.0	0.0	5.6 (4)
		In 100 years	1.8	0.0	0.0	0.0	1.4 (1)
	Human communities other countries	Now	89.3	100.0	100.0	100.0	91.5 (65)
		In 10 years	3.6	0.0	0.0	0.0	2.8 (2)
		In 25 years	5.4	0.0	0.0	0.0	4.2 (3)
		In 50 years	1.8	0.0	0.0	0.0	1.4 (1)
	Biotic communities Portugal	Now	94.6	100.0	85.7	100.0	94.4 (67)
		In 10 years	1.8	0.0	14.3	0.0	2.8 (2)
		In 25 years	–	–	–	–	–
		In 50 years	3.6	0.0	0.0	0.0	2.8 (2)
	Biotic communities other countries	Now	96.4	100.0	85.7	100.0	95.8 (68)
		In 10 years	0.0	0.0	14.3	0.0	1.4 (1)
		In 25 years	1.8	0.0	0.0	0.0	1.4 (1)
		In 50 years	1.8	0.0	0.0	0.0	1.4 (1)
Total % (n)			100.0 (56)	100.0 (4)	100.0 (7)	100.0 (4)	

Table 12.3 Respondents perceptions as a function of gender

			Female (%)	Male (%)	Total % (n)
The topic of CC has ...	Interest	A lot of	51.2	64.3	56.3 (40)
		Some	48.8	35.7	43.7 (31)
	Importance	A lot of	65.1	67.9	66.2 (47)
		Some	34.9	32.1	33.8 (24)
	Concern	A lot of	53.5	32.1	45.1 (32)
		Some	46.5	64.3	53.5 (38)
Perception of climate change impacts on ...	Human communities Portugal	Now	81.4	67.9	76.1 (54)
		In 10 years	14.0	7.1	11.3 (8)
		In 25 years	4.7	7.1	11.3 (8)
		In 50 years	0.0	14.3	5.6 (4)
		In 100 years	0.0	3.6	1.4 (1)
	Human communities other countries	Now	95.3	85.7	91.5 (65)
		In 10 years	4.7	0.0	2.8 (2)
		In 25 years	0.0	10.7	4.2 (3)
		In 50 years	0.0	3.6	1.4 (1)
	Biotic communities Portugal	Now	100.0	85.7	94.4 (67)
		In 10 years	0.0	7.1	2.8 (2)
		In 25 years	–	–	–
		In 50 years	0.0	7.1	2.8 (2)
	Biotic communities other countries	Now	100.0	89.3	95.8 (68)
		In 10 years	0.0	3.6	1.4 (1)
		In 25 years	0.0	3.6	1.4 (1)
		In 50 years	0.0	3.6	1.4 (1)
Total % (n)			100.0 (56)	100.0 (4)	

self-perceived knowledge. Similar results were obtained in other studies (e.g. the study of Löfstedt 1991), where the greenhouse effect was confused with the issue of ozone depletion and other misconceptions can be found in the in other European (Manolas et al. 2010) and American (Wachholz et al. 2014 and references cited herein) university students. The erroneous beliefs held by university students about this environmental and socio environmental and political problem is a concern however, and like in other published works students believe that CC is real and largely human-induced with the majority expressing concerns about CC (Wachholz et al. 2014 and references cited herein).

CC has been viewed as a serious personal threat in previous studies (such as the Eurobarometer surveys) directed to the general Portuguese population, (Cabecinhas et al. 2008; Lazaro et al. 2011; TNS Opinion & Social 2011). In these earlier studies, respondents have consistently shown levels of concern above the European average. However, when asked in 2011 to mention the most serious problem that the world was facing, only 6 % of Portuguese respondents indicated CC (versus 20 % of Europeans) and were least likely to see CC as a serious issue (with only

28 %) (TNS Opinion & Social 2011). So, the outcomes of the survey must be placed in the framework of current social/political/historical conditions in Portugal and all the other European countries.

In the present study, and contrasting with the earlier results, the inquired sample showed higher levels of interest and concern about CC, which may be explained by their academic level and area of training (2nd cycle degree in ecological and environmental sciences), i.e. representing a specific social segment with more access to knowledge. It would, thus, be more likely that these individuals acted in an ecologically compatible manner and eventually had an increased sensitivity to CC problems. However, when asked about their actions to mitigate the causes of CC these were "short", both on the percentage that responded positively and on number of actions with effective contribution in CC mitigation. This result shows that environmental thinking, consciousness, attitudes and behaviours are not always related. However, as stated by Ortega-Egea et al. (2014) a profound shift in personal behaviour is needed to respond to the urgency of CC mitigation together with the need to actively engage future graduates and postgraduates as part of the solutions in their public and private roles (Wachholz et al. 2014).

The general public conceptions and knowledge on the topic of CC are known to be informally influenced by the various public media approaches (see De Kraker et al. 2014). For example, in our study, CC was felt by the Portuguese students as mainly happening in biotic communities and outside Portugal (with the exception of the respondents age 41–50); also, the main responsibility for mitigation was seen to lay upon governmental organisations. In fact, this may well be a reflection of the way CC is approached by the Portuguese media, which generally focuses in exotic geographies (e.g. polar geographies, small islands and biodiversity hotspots) and gets to citizens in the context of intergovernmental panels and in an institutionalised form (UN, IPCC, UNEP, UNESCO and WMO). Therefore, traditional media have an important role in the social construction of CC (Carvalho et al. 2014).

We can ask under which circumstances people initiate changes in their practices, as well as in the dynamics of climate governance and adaptation strategies to CC of their communities (Carvalho et al. 2014), and how can we manage to surpass the confusion reigning among media or detractors of the human influence on CC hypothesis (McCright et al. 2013), in order to raise the public awareness and deal with the fact that most researchers disagree not about the influence but on the magnitude of the negative impacts (Cook et al. 2013). We think that acting in the educational system providing information and new ways to improve climate literacy will be the future. Also beliefs and concerns about CC are affected in very important ways by short-term economic and weather conditions. These changes are consistent with over-weighting of recent events, and of shifting beliefs to reduce cognitive dissonance about short term needs and long term problems (Scruggs and Benegal 2010; TNS Opinion & Social 2014) and here too is fundamental to provide new ways of CC teaching in our education systems for a better informed generation of decision makers. The fact that respondents who were developing their Master's dissertations are better acknowledged on CC, and that they have taken more actions to mitigate CC, than students who were on their first curricular year, gives credit to the importance of their Master degree training in what concerns knowledge, attitudes and behaviours for CC.

The Wachholz et al. (2014) also reports differences by academic division/scientific discipline or field of study.

As any other social-environmental issue, perceptions and expectations of CC, as well as good knowledge on the topic, are important in the actions taken relating to this multidimensional subject. Hence, it is important that humans acknowledge that anthropogenic actions are a main cause of CC (which the majority of our student's sample did), but also that humans are a key factor in mitigating CC, which in our study was perceived with reserve, in what concerns "the will to change their behaviour in order to achieve mitigation", or by the fact that the main responsibility for CC mitigation is seen as falling upon governments, rather than in individuals. Also important to this issue are the perceptions that one's "relational community only take some actions to mitigate CC", and that these "actions are difficult to implement", which may have negative results on triggering behavioural changes for CC mitigation.

As most students lack adequate knowledge on CC causality factors, curricula (syllabi) and teaching methods should adapt accordingly, and should demonstrate that personal relational communities and attitudes may have positive results in reducing the CC drivers. Formal and informal education should centre the action not only in governments but in the individual. Moreover, education and policies should provide for a shift from the current "paying more for" to a "changing behaviour to mitigate" paradigm. Altogether these should foster the necessary behavioural changes.

Another result from our data relates to the role that some socio-demographic factors appear to have in CC issue, particularly student's occupation (full-time vs. working), gender and age group. For example, for the working-students (who are older students, even though not all age groups behaved in the same manner) the social dimension is a more important aspect of the CC impacts, which may be seen by their greater awareness of CC impacts in human communities, and by the perceived greater role of individuals as actors in the CC mitigation process. The age group of the respondent was also an important factor, as shown by the fact that younger students were more concerned with the biological dimensions (e.g. biodiversity) of CC, and were more open to admit their lack of knowledge and need of information than the older age groups. Socio-environmental issues seem to have a maturity dimension associated with knowledge acquisition, either formal, informal or professionally-linked. Adult learners have their own personal biography, views of the world, and experienced knowledge associated to CC. Gender was another interesting factor affecting CC views, which showed that female respondents were more concerned and aware of the social dimensions of CC than male respondents; eventhough their knowledge and global perceptions on what concerned CC were similar to the male respondents, women were also more willing to acknowledge the need for more information on CC (48.8 vs. 28.6 %, respectively) and more positive about increasing their life quality as a result of actions taken to mitigate CC. This findings mirror the ones of Wachholz et al. (2014) where women demonstrated greater levels of concern.

Conclusions

There is still much public debate and uncertainty regarding the reality of CC and the degree to which human activities are responsible but CC is a Global problem and we are educating a "climate change generation". Thus, Higher Education has a critical role to play in educating Higher University Students about Global CC. One of the major implications of this study is that higher education needs to expand its educational efforts to ensure that all university postgraduates (and graduates) understand scientific consensus about CC.

Our students hold misconceptions about CC and some of our students are not aware that we are already experiencing the consequences of CC. They generally believe in—and are concerned about, but they are not to be consequential in their actions to mitigate the problem. Different perceptions on CC evidenced by gender, age group, or occupation of students (full-time vs. working) are aspects which should have more insight and promote a discussion in the higher education system about the contents and methodologies to teach and improve climate literacy. In this same framework, formal education at all levels but especially at university must be called to adapt curricula, general teaching and evaluation approaches to foster a desirable growing of the willingness to change behaviour related to CC and it's impacts on the environment and socio-environmental and climate governance. We think universities are part of the solution, not only part of the problem and hope our results help to accelerate the rhythm of change.

References

Alves F, Caeiro S, Azeiteiro UM (2014) Lay rationalities of climate change. Int J Clim Chang Str Manage 6(1):1756–8692, http://www.emeraldinsight.com/journals.htm?issn=1756-8692&volume=6&issue=1

Anable J, Lane B, Kelay T (2006) An evidence base review of public attitudes to climate change and transport behaviour. Department for Transport, London, 228p. http://webarchive.nationalarchives.gov.uk/+/http://www.dft.gov.uk/pgr/sustainable/climatechange/iewofpublicattitudestocl5730.pdf. Accessed 29 Jan 2015

Azeiteiro UM, Bacelar-Nicolau P, Caetano F, Caeiro S (2015) Education for sustainable development through e-learning in higher education: experiences from Portugal. J Clean Prod. doi:10.1016/j.jclepro.2014.11.056

Bacelar-Nicolau P, Caeiro S, Martinho A, Azeiteiro UM, Amador F (2009) E-learning for the environment. The Universidade Aberta (Portuguese Open Distance University) experience in the environmental sciences post-graduate courses. Int J Sust Higher Ed 10(4):354–367. doi:10.1108/14676370910990701

Cabecinhas R, Lazaro A, Carvalho A (2008) Media uses and social representations of climate change. In: Carvalho A (eds) Communicating climate change: discourses, mediations and perceptions. Centro de Estudos de Comunicação e Sociedade, Universidade do Minho, Braga, pp 170–189. http://www.lasics.uminho.pt/ojs/index.php/climate_change. Accessed 29 Jan 2015

Carvalho A, Schmidt L, Santos FD, Delicado A (2014) Climate change research and policy in Portugal. WIREs Clim Chan 5:199–217. doi:10.1002/wcc.258

CIMC, Caribbean Institute of Media and Communication (2012) Report on climate change knowledge, attitude and behavioural practice survey. University of the West Indies, Mona Campus, Planning Institute of Jamaica: 148 p. http://www.climateinvestmentfunds.org/cif/node/12515. Accessed 29 Jan 2015

Cook J, Nuccitelli D, Green S, Richardson M, Winkler B, Painting R, Jacobs R, Skuce A (2013) Quantifying the consensus on anthropogenic global warming in the scientific literature. Environ Res Lett (IOP Publishing) 8(2). doi:10.1088/1748-9326/8/2/024024

De Kraker J, Kuijs S, Cörvers R, Offermans A (2014) Internet public opinion on climate change: a world views analysis of online reader comments. Int J Clim Chang Str Manage 6(1):19–33, doi:10.1108/IJCCSM-09-2013-0109

IPCC (2014) Climate change 2014: Mitigation of climate change. In: Edenhofer O, Pichs-Madruga R, Sokona Y, Farahani E, Kadner S, Seyboth K, Adler A, Baum I, Brunner S, Eickemeier P, Kriemann B, Savolainen J, Schlömer S, von Stechow C, Zwickel T, Minx JC (eds) Contribution of Working Group III to the Fifth Assessment Report of the Intergovernmental Panel on Climate Change. Cambridge University Press, Cambridge, UK

Jamelske E, Barrett J, Boulter J (2013) Comparing climate change awareness, perceptions, and beliefs of college students in the United States and China. J Environ Stud Sci 3(3):269–278. doi:10.1007/s13412-013-0144

Lazaro A, Cabecinhas R, Carvalho A (2011) Uso dos media e envolvimento com as alterações climáticas. In: Carvalho A (ed) As Alterações Climáticas, os Media e os Cidadaos. Gracio Editor: Coimbra. pp 195–222

Leal Filho W (ed) (2010) Universities and climate change: introducing climate change at University programmes. Springer, Berlin. ISBN 978-3-642-10750-4

Leal Filho W, Alves F, Caeiro S, Azeiteiro UM (eds) (2014) International perspectives on climate change: Latin America and beyond. Springer, 316 p 69 illus ISSN 1610-2010 ISSN 1610-2002 (electronic) ISBN 978-3-319-04488-0 ISBN 978-3-319-04489-7 (eBook). doi:10.1007/978-3-319-04489-7 Springer: Cham. Library of Congress Control Number: 2014934670

Löfstedt R (1991) Climate change perceptions and energy-use decisions in northern Sweden. Glob Environ Chang 1(4):321–324

Lorenzoni I, Pidgeon NF (2006) Public views on climate change: European and USA perspectives. Clim Chan 77:73–95

Manolas EI, Tampakis SA, Karanikola PP (2010) Climate change: the views of forestry students in a Greek university. Int J Environ Stud 67(4):599–609, http://dx.doi.org/10.1080/00207233.2010.499208

McCright A, Dunlap R, Xiao C (2013) Perceived scientific agreement and support for government action on climate change in the USA. Clim Change 119:511–518. doi:10.1007/s10584-013-0704-9

Ortega-Egea JM, García-de-Frutos N, Antolín-López R (2014) Why do some people do "more" to mitigate climate change than others? Exploring heterogeneity in psycho-social associations. PLoS One 9(9), e106645. doi:10.1371/journal.pone.0106645

Pereira A, Mendes AQ, Morgado L, Amante L, Bidarra J (2008) Universidade Aberta's pedagogical model for distance education. Universidade Aberta, Lisbon, https://repositorioaberto.uab.pt/bitstream/10400.2/2388/1/MPV_uaberta_english.pdf. Accessed 29 January 2015

Rhodes E, Axsen J, Jaccard M (2014) Does effective climate policy require well-informed citizen support? Glob Environ Chang 29:92–104

Scruggs L, Benegal S (2010) Declining public concern about climate change: can we blame the great recession. Glob Environ Chang 22(2). doi:10.1016/j.gloenvcha.2012.01.002

TNS Opinion & Social (2011) Climate change. Special Eurobarometer 372, Wave EB75.4. Brussels: European Commission. http://ec.europa.eu/public_opinion/archives/ebs/ebs_372_en.pdf. Accessed 29 Jan 2015

TNS Opinion & Social (2014) Climate change. Special Eurobarometer 409/Wave EB80.2. http://ec.europa.eu/clima/citizens/support/docs/report_2014_en.pdf. Accessed 29 Jan 2015

Viegas VM, Azeiteiro UM, Dias A, Alves F (2014) Alterações Climáticas, Perceções e Racionalidades [Climate change, perceptions and rationalities]. RGCI—Revista de Gestão Costeira Integrada. J Integr Coast Zone Manage 14(3):347–363. doi:10.5894/rgci456, http://www.aprh.pt/rgci/pdf/rgci-456_Viegas.pdf

Wachholz S, Artz N, Chene D (2014) Warming to the idea: university students' knowledge and attitudes about climate change. Int J Sust Higher Ed 15(2):128–141. doi:10.1108/IJSHE-03-2012-0025

Chapter 13
A Decade of Capacity Building Through Roving Seminars on Agro-Meteorology/-Climatology in Africa, Asia and Latin America: From Agrometeorological Services via Climate Change to Agroforestry and Other Climate-Smart Agricultural Practices

C. (Kees) J. Stigter

Abstract Climate change hits agricultural production areas hard. There is no knowledge base to counter its effects and this makes education and other capacity building in adapting to climate change imperative.

In the course of the last quarter of the twentieth century, applied agrometeorology/agroclimatology started to focus on traditional knowledge and agrometeorological services in agriculture. Since 2005, "Agromet Vision" offers Roving Seminars (RSs) of 2–5 days for university staff, professional agrometeorologists and extension intermediaries. They are particularly useful for training of extension trainers. To date, 37 RSs have been successfully delivered in 13 countries.

The first RSs offered were "Agrometeorological Services: Theory and Practice" and "Agrometeorology and Sustainable Development", followed from 2011 by "Reaching Farmers in a Changing Climate". In 2013 "What Climate Change Means for Farmers in Africa" was added, while in 2015 "Agroforestry and Climate Change" was included. This paper wants to review the need for and the contents of these RSs and reports on local evaluation by the institutes involved. Applied agrometeorology should not start with agrometeorology but with the conditions

---

C.(K).J. Stigter (✉)
Group Agrometeorology, Faculty of Agriculture, Department of Soil, Crop and Climate Sciences, University of the Free State (UFS), Agriculture Building, Campus UFS, Bloemfontein 9301, South Africa

Cluster Response Farming to Climate Change, Faculty of Social and Political Sciences, Department of Anthropology, Centre for Anthropological Studies, University of Indonesia (UI), Fl.6/Building H—Selo Soemardjan Room, Kampus UI, Depok 16424, Indonesia

Agromet Vision, Groenestraat 13, 5314 AJ Bruchem, The Netherlands
e-mail: cjstigter@usa.net

of where it should be applied, the livelihood of farmers. In the development of such RSs elsewhere, our experience could be of much value.

Keywords Agroclimatology • Capacity building • Roving Seminars • Sustainable development • Climate change

Introduction: Education and Capacity Building

Climate change as an environmental disaster, upsetting among others all farming communities, must be considered through its three sides from where the danger comes: (i) global warming; (ii) increasing climate variability; (iii) more (and often more severe) weather and climate extreme events (Stigter and Ofori 2014a). Experience in and development of doing so, must be transferred via new educational and capacity building commitments, among others in agrometeorology and agroclimatology (Stigter 2011). RSs are one way to do this very explicitly related to farmers' livelihoods (Stigter 2006a, 2015a, b).

This paper wants to review the need for and the contents of these RSs and it reports on their local evaluation by the institutes that were involved. The approach developed emphasizes that applied agrometeorology should not start with agrometeorology but with the conditions of where it should be applied, the livelihoods of farmers (Stigter 2010, 2015a, b). In the development of such RSs elsewhere, our experience could be of much value.

RSs in the form of very limited single "farmer days" were held by the World Meteorological Organization (WMO) since 2008, but they were never evaluated (Tall et al. 2014). Farmers have valuable traditional and more recent empirical knowledge but are unfamiliar with much new environmental knowledge. Our experience in Indonesia (e.g. Winarto and Stigter 2011; Stigter et al. 2013) learns that long periods of regular contacts have to be organized to get this new knowledge understood and applied. Farmers need strong support in this agrometeorological learning process.

This means that the new knowledge must first and for all go to university staff, professional agrometeorologists and extension intermediaries in (rural) supporting organizations. This was also the idea behind the author's book "Applied Agrometeorology" (Stigter 2010), of which parts are strongly related to the contents of the first two RSs dealt with below. These RSs are at a level that trainers of extension intermediaries would particularly benefit because the material shows what can be offered that can reach and change the livelihoods of farmers and how.

Adopting a new approach, since 2005, the authors' one man consultancy bureau "Agromet Vision" has developed and delivered RSs of 2–5 days with local evaluations. To date, Agromet Vision has delivered 37 RSs in 13 countries (including four trials and three mixes). The first two subjects were "Agrometeorological Services: Theory and Practice" and "Agrometeorology and Sustainable Development". The complete list of which RSs were held where and when is in Table 13.1.

Table 13.1 Agromet Vision Roving Seminars (chronologically for each seminar)

	Roving Seminar	Period of delivery	Countries of delivery	Number of times delivered
1	Agrometeorological Services, Theory and Practice	2005–2012	Iran (Tehran) India (mix with 2, trial) Brazil/Venezuela (id., trial) South Africa (Bloemfontein) Indonesia (Yogyakarta) Lesotho (Maseru) Swaziland (Mbabane) Argentina (Buenos Aires) South Africa (Bloemfontein) Lesotho (Roma) (id., trial) Zambia (Monze) Zimbabwe (Harare) Sudan (Gedaref)	13 (of which three mixed trials)
2	Agrometeorology and Sustainable Development	2005–2012	Iran (Gorgan) South Africa (Bloemfontein) Indonesia (Yogyakarta) Brazil (Piracicaba) South Africa (Bloemfontein) Zimbabwe (Harare) Sudan (Khartoum)	7 (not counting three mixed trials already mentioned under 1)
3	Extension Agrometeorology (developed for Iran)	2011 to the present	Qazvin Shiraz	2 sofar
4	Reaching Farmers in a Changing Climate	2011 to the present	Indonesia (Depok) South Africa (Bloemfontein) Iran (Esfahan) Iran (Gorgan) Zimbabwe (Harare) Sudan (Wad Medani) Cuba (Havana) Indonesia (Bogor) South Africa (Bloemfontein)	9 sofar
5	What Climate Change Means for Farmers in Africa (developed for Africa)	2013 to the present	Zimbabwe (Harare) South Africa (Pretoria) Sudan (Wad Medani)	3 sofar
6	Agroforestry and Climate Change	2015 to the present	Sudan (Khartoum, trial) Sudan (Khartoum) Zimbabwe (Harare)	3 sofar (of which one trial)

This includes from 2011 onwards a third Roving Seminar "Extension Agrometeorology", with material that was partly derived from the former two and partly from additional sources, developed for training extension intermediaries in Iran and delivered twice so far. In 2011 also the RS "Reaching Farmers in a Changing Climate" was added, while there are two more recently developed RSs as well that have been started in 2013 (What Climate Change Means for Farmers in

Africa) and 2015 (Agrometeorology and Climate Change) respectively (Table 13.1).

The conceptual framework was given by Stigter (2006b, 2007a, 2008) where he defined, between the two domains of (i) farmers' livelihoods and (ii) the scientific support systems of available knowledge and understanding, a third domain in which applied scientists, product intermediaries and farmer facilitators work on solutions of agricultural production problems as suffered by farmers. In this domain of the use of products of for example applied agrometeorology and climatology, these products are jointly made operational for establishment as climate services with farmers in their fields (Stigter 2007b, c). The extension behind these processes was described in Stigter and Winarto (2013) and Stigter et al. (2013). The Roving Seminars deal for their audience with describing the conditions under which science can be used that way, from the point of view of making applied science operational and from the angle of farmers' livelihood and its conditions that determine absorption and use of services as well as blending of the new knowledge with their traditional knowledge (e.g. Stigter et al. 2005; Zuma-Netshiukhwi et al. 2013). The RSs make use of a lot of case studies that illustrate difficulties with, consequences of and successes with these approaches.

This paper provides an overview of the "Agromet Vision" RS series. The early RSs, with a focus on agrometeorological services and development, are first briefly reviewed before discussing the more recent training that focuses on adapting to climate change. Experience with such knowledge transfer will be particularly of use in training of extension officers everywhere in Africa, Asia and Latin America in developing and establishing similar education and capacity building programmes.

All RSs, when given in 4 or 5 days, train their audience in discussions in small groups to answer specific questions after each presentation. The audience is distributed over groups of four to six participants and each group discusses the same questions. Then, in a plenary reporting on the proposed answers, conclusions are drawn on establishing services with farmers in their fields and problem solving in their livelihoods. It is our intention that these extra-curricular efforts to participate in the RSs bring applied science and its products much closer to problem solving with smallholders. Whether the audiences are related to extension training or to extension policy and practice issues in ministries, institutes and/or organizations, the local evaluations learn that reaching farmers is considered more within one's power than before the RSs.

Early Agrometeorological Roving Seminars

Applied agrometeorology/agroclimatology started to focus on traditional knowledge and climate services in agriculture in the last two decades of the twentieth century (Stigter 2006a). The Commission of Agricultural Meteorology (CAgM) of the WMO became instrumental in advocating capacity building in these directions (WMO 2006). The first RS was for the first time given as a Training Course in November 2005 in Tehran, Iran (Table 13.1). Table 13.2 shortly hints at the

Table 13.2 Overview of Roving Seminar 1 "Agrometeorological Services: Theory and Practice" (Stigter 2006a, 2015a)

Presentation	Synopsis/rationale
Intro: Some history	The history of why and how my opinions got formed that I want to get across in this course
1. Zoning & mapping as agrometeorological services in developing countries	From examples of zoning and mapping in among others China, Portugal, Sudan and Ecuador, as climate services for agriculture, preconditions and requirements were derived for a checklist for action on assisting farmers with such services
2. Farming systems, agrometeorology and agrometeorological services	An introductional survey to define and connect these action fields illustrated with case studies from Africa (slides) with as issue: designs of protective structures as agrometeorological services
3. The place of agrometeorological services in the livelihood of farmers	Use of a diagnostic and conceptual framework to explain the lack of operational agrometeorological services and knowledge that can make a difference in the livelihood of African farmers
4. Agrometeorological services for user communities, some lessons learned	Interpretation of case studies from Brazil and China showing conditions to be met for serving communities with climate knowledge effectively and what may prevent it
5. Using traditional methods and indigenous technologies for coping with climate variability	Options that Low External Input Sustainable Agriculture (LEISA) farmers have, what they do and may do with water, heat and crops combining traditional and scientific knowledge
6. Research and reality	Considering time scales and spatial scales with services to help reduce the impacts of natural disasters, with questions that have to cover the reality of research in funding research (education) proposals
7. Policies and preparedness	Beyond climate forecasting of flood disasters. Using an end to end framework to understand agrometeorological components of coping with flood disasters and policy implications
8. Agrometeorological services making a difference for poor farmers. I. Why it does not happen. II. How it can be done	No extension agroclimatology was developed to assist the majority of marginal farmers in the design of their production systems. With examples from Africa and China it is illustrated how to solve problems with agrometeorological components in the livelihood of farmers

The full presentations have recently been made available on ResearchGate under contributions from the author composition of its presentations. See Rahimi (2005) for a report and evaluation. The history of its development was given in Stigter (2015a), where also the literature supporting its presentations was reviewed.

The second RS was given for the first time in a Provincial Workshop in the same month in Gorgan, Iran (Table 13.1). Table 13.3 hints again at the composition of its

Table 13.3 Overview of Roving Seminar 2 "Agrometeorology and Sustainable Development" (Stigter 2006a, 2015b)

Presentation	Synopsis/rationale
1. Introduction to the approach	Agrometeorological services to prepare farmers for climate extremes and climate use. Examples from Africa, China, India, Indonesia and Vietnam how not to do it and how it should be done
2. What is sustainable development?	Means of communication & education are part of sustainable development. Developing a response farming approach, with forecasting capabilities that change and improve in the course of time, is a condition for sustainable development
3. The role of agricultural research in establishing agromet services	We need globalization as to the availability of methods, tools of research, but we need localization as to strategies, adaptation of research to local realities, problem identification and local innovations
4. Examples of agrometeorological services in the literature	African case studies confirmed that traditional adaptation strategies can be insufficient or may have degraded, that contemporary science is very often available on-shelf and that inappropriate policy environments do often exist
5. Preparation of farmers for climate extremes and climate use	Case studies confirm that (i) traditional adaptation strategies may fail, (ii) contemporary science is very often available on-shelf and (iii) inappropriate policy environments often prevent decisions and dissemination of locally obtained successes
6. Actual needs of farming systems and their farmers: some case studies	For other developing countries than China a similar farmer differentiation will be valid, but the stories that belong to each of their income groups and rural occupations will differ and the implications also
7. The role of civil servants and NGOs in preparing farmers	For actual agrometeorological services you need applied scientists to develop them, and extension or NGO intermediaries with sufficient knowledge to establish them with farmers in their fields
8. Training of agromet intermediaries to prepare farmers as end users	The education and in service training of agrometeorological extension intermediaries is an essential part of the new approach, that appears necessary in education, training and extension in agricultural meteorology
9. Conflicts of interests in a bottom-up approach in agrometeorology	Examples from Sudan illustrate that the use of science (agrometeorology) is not neutral but a matter of policies that can lead to conflicts. What is a disaster in one place may be a blessing elsewhere (*Eucalyptus, Acacia tortilis, Prosopis juliflora*)

The full presentations have recently been made available on ResearchGate under contributions from the author presentations. See Asadi (2005) for a report and evaluation. The history of its development was given in Stigter (2015b), where also the literature supporting its presentations was reviewed.

The conclusions on the above early "Agromet Vision" RSs in local evaluations have been that these RSs as educational commitments were suitable to get extension planners and trainers aware of the necessities of further establishment and use of climate services for agriculture (e.g. Kadi et al. 2011; Stigter 2011; Tall et al. 2014; WMO 2015). This was making use of policy trends of improved services climates in rural areas for which funds appear to become available, at least in some countries (e.g. Donnges 2003). Particularly due to the sufferings related to climate change, both background RSs had relevancy to all emerging countries (Stigter 2008, 2010, 2011; Stigter et al. 2007). However, during RS question and discussion periods, gradually the need was expressed for an agrometeorological/-climatological extension approach more directly related to adaptation to climate change in a second set of RSs to be developed (e.g. Winarto et al. 2008, 2010; Stigter 2011).

Roving Seminars Focused on Climate Change

In many countries or regions of Africa, Asia and Latin America, agriculture is still the backbone of societies, economically and in creating work. Climate change hits these production areas hard (e.g. Stigter 2010, 2011) and this creates new livelihood conditions in which climate services for agriculture need a new approach (Stigter et al. 2014a, b, c, 2015). Because of this and the above mentioned developments, the early RSs gradually got company from new RSs from 2011 till 2015 that were even more directly related to the needs for adaptations to climate change.

Reaching Farmers in a Changing Climate

Our early work on awareness raising and resilience improvement related to climate change with (largely female) farmers in Gunungkidul, Indonesia (e.g. Winarto et al. 2008, 2010, 2011; Winarto and Stigter 2011; Stigter and Winarto 2012a, b, c; Stigter 2012), added a strong extension flavor to the new RS "Reaching farmers in a changing climate". The target groups are the same as for the earlier Seminars, for the same reasons of existing extension needs as explained there, and it is again not country-specific. There is more emphasis on extension agrometeorology. Stigter and Winarto (2012c) pictured extension agrometeorology as a contribution to sustainable development and defined it as agrometeorology that attends to (i) local suffering from weather and climate and persistent ways to diminish it, and (ii) windows of opportunity that (micro)climate offers "on farm". Because this fourth RS (Table 13.1) is still currently delivered, we have given the information on the contents of this RS below in a somewhat more extensive form (Table 13.4) than that of Tables 13.2 and 13.3.

Table 13.4 Some representative conclusions/recommendations in presentations of the Roving Seminar "Reaching Farmers in a Changing Climate"

Introduction to the third Roving Seminar

The introduction is based on a review of the experience from the Mali agrometeorological pilot projects on serving farmers with agrometeorological knowledge. This project concluded that the Sahelian farmer can be technically assisted to reduce climate risk for his/her production. This project has after all made the activities of the National Meteorological Service better visible. It has strengthened the credibility of meteorology with the public at large as well as with the Government Services and the political authorities of Mali (e.g. Stigter 2010). It serves as example elsewhere (e.g. Winarto and Stigter 2011).

Presentation 1. *Applied agrometeorology: addressing the livelihood crises of farmers*

Science, and not only climate change science, has a role to play. But only when acknowledging the present livelihood crises of the poor and giving priority to policy preparations and policy mandate matters related to those crises; and to science only in that context. Especially in-service trained extension intermediaries are needed between the weather products (maps, forecasts, warnings, response proposals) as well as design rules (advisories on coping with weather and climate impacts) and their rural potential clients that are vulnerable and mostly have relatively low formal education (Stigter 2010).

Presentation 2. *Coping with climate change and disasters: "Intelligence does not solve problems" (Tagore)*

What the mainstream misses is in the undercurrent of applied agrometeorology: data, research, education/training/extension and policies, used/carried out in action as priorities in the undercurrent of applied agricultural meteorology. The developments in Low External Input Sustainable Agriculture (LEISA) research of the last 20 years show what is possible if norms and values in science show a paradigm shift. A shift towards valuing the basic issues in the undercurrent: realistic assessments of the environment and considerations of the plights of poor people (Stigter 2010).

Presentation 3. *New agrometeorological services: (i) (2006–2008)*

Services based on products generated by operational support systems in which understanding of farmer livelihood conditions and farmer innovations have been used. What we need is institutionalization of science supported establishment and validation of such services. We want to get into a situation in which, in a "farmer first paradigm", livelihood problems and farmer decision-making need to actually guide the bottom-up design of actual services (Stigter 2010).

Presentation 4. *New agrometeorological services: (ii) China (2007–2010)*

A comparison of a "Climate Field Classes" approach with the "cascade" down coming of extension information in China would be a great last phase of pilot projects started. Instead of this top down teaching, bringing new knowledge to farmers should preferably be done in dialogues between scientists, extension and farmers (e.g. Stigter and Winarto 2012d).

Presentation 5. *Agrometeorology of low inputs and scarce resources: Farmer differentiation*

(continued)

Table 13.4 (continued)

A massive investment in agriculture is indeed required. This should be primarily focused on the creation of knowledge that does justice to the local variation in water and nutrient availability. It should aim to empower farmers to experiment and be innovative, and remake agricultural extension and agricultural engineering (Stigter et al. 2014b). These adaptation solutions must be distinguished for the following main categories: type of farmer (farming system); natural resource management; markets and opportunities of economic activities; institutional opportunities; additional aspects.
Presentation 6. *First connecting principle: Agroforestry & other multiple cropping as multifunctional agriculture*
Agroforestry (+ other multi-functional agriculture) connects water and fertility issues (when introduced in monocropping and multiple cropping without trees); it does so by restoring (i) biological resources and natural capital (soil fertility, water, forests, etc.); (ii) livelihoods (nutrition, health, culture, equity, income); and (iii) agroecological processes (nutrient and water cycles, pest and disease control, etc.) (Stigter et al. 2014c).
Presentation 7. *Second connecting principle: Communication*
Much of the research and development needed for less-favoured lands does not involve high science. But rather the spread and adaptation of indigenous knowledge and practical innovations. NGOs have been very successful in pursuing this agenda. They work with local communities to overcome social and institutional constraints. There is a need for more participatory ways of innovative communication to test new technologies which shall be adopted by the small farmers. Climate services for agriculture! (Stigter et al. 2014c).
Presentation 8. *Science field shops and extension agrometeorology*
There are a number of important social outcomes from extension trainings. Farmers gain self-confidence, they start to work together to solve community problems, and they develop a different relationship with local government. As scientists we are supposed to propose and prepare policies. So in agricultural and social sciences we should among others care for policies of managing the rural response to climate change and of institutionalization of that response (Stigter and Winarto 2013).

What Climate Change Means for Farmers in Africa

An even more recent RS (5, Table 13.1) is based on three recent papers (Stigter and Ofori 2014a, b, c), runs since 2013 and has only been given three times as yet (Table 13.1). It was developed specifically for Africa (Table 13.5).

Table 13.5 Some representative conclusions/recommendations in presentations of the Roving Seminar "What Climate Change Means for Farmers in Africa: Facts, Impacts/Consequences and Possible Approaches Towards Adaptation"

Presentation 1. *Part One: Introductional Facts and a Start with Looking at Consequences of Global Warming for African Farmers A: Some Introductional Issues*

The issues are: (i) global warming, (ii) increasing climate variability, (iii) more (and possibly more severe) meteorological and climatological extreme events. The basis of our approach should be: listening to the farmers concerned in a "farmer first" paradigm in a participatory approach. Scientists should basically be the connection between applied science and the actual production environment.

Presentation 2. *Part One: Introductional Facts and a Start with Looking at Consequences of Global Warming for African Farmers B: Issues of Reality*

Once the period of a few years of "Science Field Shops" is over, scientists may return to their back-up functions. Coping with climate change adaptation must this way be seen as a matter of farmers and their communities, extension (so government) and applied scientists alike. We have to look beyond "adaptation to current climate variability". We must target the basic vulnerability factors of communities.

Presentation 3. *Part One: Introductional Facts and a Start with Looking at Consequences of Global Warming for African Farmers C: Talking Consequences*

Increases in maximum temperatures can lead to severe yield reductions and reproductive failure in many crops. Maize germplasm presently available in Africa is not suitable for projected climate change conditions. The economic and nutritional arguments for the diversification of agricultural production in Africa are now joined by climatological arguments.

Presentation 4. *Part Two: Increasing Climate Variability and a Response Approach for African Farmers A: El-Niño Southern Oscillation*

The combined forces of ENSO and global warming are likely to have dramatic, and currently largely unforeseen, effects on agricultural production and food security in Sub-Saharan African countries. We need an assessment of climate-related uncertainties associated with global warming and ENSO dynamics. Difficulties in making reliable climate predictions should resort to more response farming to climate but in the end they should be combined.

Presentation 5. *Part Two: Increasing Climate Variability and a Response Approach for African Farmers B: Response Farming*

Intercropping, adapting crops, crop varieties and crop densities to the expected season, and the traditional use of trees are examples of response farming developed by farmers. To get optimal preparations, farmers get new knowledge, through extension intermediaries, backed by scientists. New response farming approaches must be built on traditional knowledge and indigenous technology, using climate science as connections.

Presentation 6. *Part Three: Climate Extremes and Society's Responses, Including Mitigation Attempts as Part of Preparedness of African Farmers A: Extremes, Responses, Preparedness*

There is projected to be an increase in the number of days with heavy precipitation, but the number of dry days will also increase. New funding should enable delivery of enough drought tolerant maize seed to benefit 30–40 million people in sub-Saharan Africa. Farmer preparedness for extremes must be raised and established on-farm in a permanent way.

Presentation 7. *Part Three: Climate Extremes and Society's Responses, Including Mitigation Attempts as Part of Preparedness of African Farmers B:Research and Policy Responses; Contributions from Agriculture in Diminishing Greenhouse Gases*

What climate change brings is not new but it is more serious, so we should respond more seriously as well. Policies aimed at promoting farm-level adaptations need to emphasize the critical role of farmers' education, provision of improved climate, production and market knowledge and the means to implement adaptations through affordable credit facilities. Soil

(continued)

Table 13.5 (continued)

carbon sequestration has a higher mitigation potential than emission reductions in African agriculture, although both may be important.
Presentation 8. *Maize and Climate Change. A Choice From What a Recent Summary Says and Some Critical Additions*
Its high yields (relative to other cereals) make maize particularly attractive to farmers in areas with land scarcity and high population pressure. The adverse effects on maize production in southern Africa by the 2030s are projected to reach 50 % of the average yield levels in 2000. The world's many cultures must adapt to the changing dinner menu forced upon them by climate change.

Three times is only a start. Several countries are in the programme outlook for the coming years. So far invariably the reaction has been that the future looked much worse than the audience expected. The author recently participated in a meeting in Nairobi of the Food Security, Agriculture and Land Section of the United Nations Economic Commission for Africa (UNECA) with a group of scientists of the World Agroforestry Centre (ICRAF) working on a new book on "Farming systems and food security in Africa: Priorities for science and policy under global change". It appears that many of such farming systems have trees that particularly combine protective and productive functions (e.g. Stigter 2015c). Knowing of these trends I last year produced the latest RS as below.

Agroforestry and Climate Change

Most recently a new RS (6, Table 13.1), "Agroforestry and climate change", was developed, based on Stigter (2015c), and tried out and then fully delivered in Khartoum (Sudan) in February 2015 and in Harare (Zimbabwe) in March 2015 (Table 13.1). Because this RS is based on a book chapter not yet published, although I handled the printing proofs, and was so far only given in 2 days RSs, Table 13.6 takes again the abbreviated form of the Tables 13.2 and 13.3.

Table 13.6 Overview of Roving Seminar 6 "Agroforestry and Climate Change" (Stigter 2015c)

Presentation	Synopsis/rationale
1. Trees outside forests	Integrating all existing and new landscape eco-systems into a complex climate adaptation-oriented resilience approach appears highly promising
2. Trees and what they change at the micro-level	A participatory approach now must supplement the more traditional aspects of tree improvement and is seen as an important strategy towards the Millennium Development Goals
3. Agroforestry and agriculture I. Interactions; II. What agroforestry provides	Quantity of nutrients provided by tree prunings is determined by production rate & nutrient concentrations of biomass, which depend on climate, soil type, tree species, plant part, tree density and pruning regime
4. What we already know about agroforestry and climate change I. Implications; II. Details	The way forward is through focusing on the ability of agroforestry to boost food production and provide benefits for adaptation to climate change
5. Recent progress and additions in agroforestry; A. Foundations to build on; B. Strategic use of climate services for agroforestry	We know that trees can: (i) enhance understorey and improve water use efficiency; (ii) increase rainfall utilisation compared to annual crops; and (iii) have a direct impact on local and regional rainfall patterns
6. C. Coping with increasing climatic variability using agroforestry; both people and trees can adapt to change on various time scales	Increasing variability in rainfall is associated with lower tree cover in moist tropical forests. Effects of climatic variability in tropical dry lands appear to depend on the balance between extreme wet and dry events, and the opportunities for trees to grow during rainy periods
7: D. Coping with extreme weather and climate events using agroforestry; disaster risk reduction can lead the way	Agroforestry improves the quality of life of farmers by increasing real-time income due to the multiple harvests and sale of products from the various components of the system so providing regular income throughout the year. Science contributes on all these fronts, particularly in its many applications
8: E. Meteorological advisories/services of weather forecasting in agroforestry; F. Developing strategies to cope with risks in and with agroforestry	It was concluded that agroforestry can be used as an effective component within a broader development strategy to help subsistence farmers reduce their vulnerability to climate-related hazards

Conclusions

Ten years of RSs organized by "Agromet Vision" for various Universities, Institutes and Organizations appear to be a recommendable training of trainers of extension intermediaries or farmers in agrometeorology and -climatology, in which long term contacts with trainers of farmers and farmers themselves is promoted. Climate services for agriculture of many kinds have been exemplified and the trainers have been comforted in their attempts to establish these services with farmers in their fields. Many examples of (agro)meteorological and (agro) climatological products have been defined and discussed for operational use by farmers under conditions of a changing climate. The future is in expansion of such approaches, but existing extension systems need to be completely overhauled and new systems need to be developed.

References

Asadi M (2005) Agrometeorology and sustainable development. http://www.agrometeorology.org/news/news-highlights/agrometeorology-and-sustainable-development/. Accessed 8 Mar 2015

Donnges C (2003) Improving access in rural areas. Guidelines for integrated rural accessibility planning. International Labour Office, Bangkok. http://www.ilo.org/wcmsp5/groups/public/-asia/-robangkok/documents/publication/wcms_bk_pb_216_en.pdf. Accessed 26 Jul 2015

Kadi M, Njau LN, Mwikya J, Kamga A (2011) The state of climate information services for agriculture and food security in East African countries. CCAFS Working Paper No. 5. https://ccafs.cgiar.org/sites/default/files/assets/docs/ccafs-wp-05-clim-info-eastafrica.pdf. Accessed 25 Jul 2015

Rahimi M (2005) Agrometeorological services: theory and practice. http://www.agrometeorology.org/topics/needs-for-agrometeorological-solutions-to-farming-problems/agrometeorological-services-theory-and-practice-final-report/. Accessed 8 Mar 2015

Stigter CJ (2006a) A contemporary history of a new approach to applied agrometeorology. http://www.agrometeorology.org/topics/history-of-agrometeorology/a-contemporary-history-of-a-new-approach-to-applied-agrometeorology/. Accessed 7 Mar 2015

Stigter CJ (2006b) No policies, no cure: why the marginal farmers that need our agrometeorological support most are nowhere getting it. Farewell lecture Wageningen University, 14 Apr 2005. Published in a modified form (as presented at the Institute for Studies of the Future in Khartoum, Sudan, on 23 Apr 2005) under the title "Scientific research in Africa in the 21st century, in need of a change of approach". Afr J Agric Res 1:4–8

Stigter CJ (2007a) From basic agrometeorological science to agrometeorological services and information for agricultural decision makers: a simple conceptual and diagnostic framework. Agric For Meteorol 142:91–95 (Guest Editorial)

Stigter, K. (2007b) Agrometeorological services to prepare farmers for climate extremes and climate use. Paper presented under "Agrometeorology and Sustainable Development" at the XVth Congress of the Brazilian Society for Agrometeorology, Aracaju, Brazil, July. Invited paper. In Revista Brasileira de Agrometeorologia, 15, pp 202–207

Stigter K (2007c) Coping with climate risk in agriculture needs farmer oriented research and extension policies. Invited paper presented under "Socialization of the contents of agrometeorology in Latin America" as the closure lecture on 30 November 2007 at the "First Venezuelan Congress of Agrometeorology" and the "Fifth Latin American Meeting on Agrometeorology", concurrently held in Maracay, Venezuela. Scientia Agricola (Piracicaba, Brazil), 65 (special issue): pp 108–115. http://www.scielo.br/pdf/sa/v65nspe/a16v65nsp.pdf. Accessed 15 Apr 2013

Stigter CJ (2008) Agrometeorology from science to extension: assessment of needs and provision of services. A review. Agric Ecosyst Environ 126:153–157

Stigter K (ed) (2010) Applied agrometeorology. Springer, Berlin

Stigter CJ (2011) Agrometeorological services: reaching all farmers with operational information products in new educational commitments. CAgM Report 104, WMO, Geneva, Switzerland

Stigter C(K)J (2012) Climate-smart agriculture can diminish plant hopper outbreaks but a number of bad habits are counterproductive. http://ricehoppers.net/2012/02/cimate-smart-agriculture-can-diminish-planthopper-outbreaks-but-a-number-of-bad-habits-are-counterproductive/. Accessed 19 Feb 2015

Stigter C(K)J (2015a) The history of capacity building through Roving Seminars on agro-meteorology/-climatology in Africa, Asia and Latin America by "Agromet Vision". I. Agrometeorological services: theory and practice. INSAM-website, under History of Agrometeorology. http://www.agrometeorology.org/topics/history-of-agrometeorology/the-history-of-capacity-building-through-roving-seminars-on-agro-meteorology-climatology-in-africa-asia-and-latin-america-by-201cagromet-vision201d.-i.-agrometeorological-services-theory-and-practice. Accessed 10 Jul 2015

Stigter C(K)J (2015b) The history of capacity building through Roving Seminars on agro-meteorology/-climatology in Africa, Asia and Latin America by "Agromet Vision". II. Agrometeorology and sustainable development. INSAM-website, under History of Agrometeorology. http://www.agrometeorology.org/topics/history-of-agrometeorology/the-history-of-capacity-building-through-roving-seminars-on-agro-meteorology-climatology-in-africa-asia-and-latin-america-by-201cagromet-vision201d.-ii.-agrometeorology-and-sustainable-development. Accessed 10 Jul 2015

Stigter K(CJ) (2015c) Agroforestry and (micro)climate change. In: Ong C, Black C, Wilson J (eds) Tree/crop interactions: agroforestry in a changing climate, 2nd edn. CABI, Wallingford, Chapter 5

Stigter C(K)J, Winarto YT (2012a) Considerations of climate and society in Asia (I). What climate change means for farmers in Asia. Earthzine 4(5). http://www.earthzine.org/2012/04/04/what-climate-change-means-for-farmers-in-asia/. Accessed 22 Nov 2014

Stigter C(K)J, Winarto YT (2012b) Considerations of climate and society in Asia (II): our work with farmers in Indonesia. Earthzine 4(6). http://www.earthzine.org/2012/04/17/considerations-of-climate-and-society-in-asia-farmers-in-indonesia/. Accessed 22 Nov 2014

Stigter K, Winarto YT (2012c) Extension agrometeorology as a contribution to sustainable agriculture. New Clues Sci 2(3):59–63

Stigter C(K)J, Winarto YT (2012d) Coping with climate change: an active agrometeorological learning approach to response farming. Invited opening keynote presentation on the first day of the APEC Climate Symposium 2012 "Harnessing and Using Climate Information for Decision Making: An In-Depth Look at the Agriculture Sector", St. Petersburg, Russia. Extended Abstract in Proceedings, 2 pp. http://www.apcc21.org/eng/acts/pastsym/japcc0202_viw.jsp. Accessed 25 Jul 2015

Stigter C(K)J, Winarto YT (2013) Science field shops in Indonesia. A start of improved agricultural extension that fits a rural response to climate change. J Agric Sci Appl 2(2):112–123

Stigter C(K)J, Ofori E (2014a) What climate change means for farmers in Africa. A triptych review. Middle panel: introductional matters and consequences of global warming for African farmers. Afr J Food Agric Nutr Dev 14(1):8420–8444

Stigter C(K)J, Ofori E (2014b) What climate change means for farmers in Africa. A triptych review. Left panel: increasing climate variability and a response approach for African farmers. Afr J Food Agric Nutr Dev 14(1):8445–8458

Stigter C(K)J, Ofori E (2014c) What climate change means for farmers in Africa. A triptych review. Right panel: climate extremes and society's responses, including mitigation attempts as part of preparedness of African farmers. Afr J Food Agric Nutr Dev 14(1):8459–8473

Stigter CJ, Zheng D, Onyewotu LOZ, Mei X (2005) Using traditional methods and indigenous technologies for coping with climate variability. Clim Change 70:255–271

Stigter CJ, Tan Y, Das HP, Zheng D, Rivero Vega RE, Van Viet N, Bakheit NI, Abdullahi YM (2007) Complying with farmers' conditions and needs using new weather and climate information approaches and technologies. In: Sivakumar MVK, Motha R (eds) Managing weather and climate risks in agriculture. Springer, Berlin

Stigter K, Winarto YT, Ofori E, Zuma-Netshiukhwi G, Nanja D, Walker S (2013) Extension agrometeorology as the answer to stakeholder realities: Response farming and the consequences of climate change. Special issue on agrometeorology: from scientific analysis to operational application. Atmosphere 4(3):237–253

Stigter K, Walker S, Das H, Dominguez-Hurtado IM, Nanja D (2014a) Meeting farmers' needs for agrometeorological services: a review with case studies. Part I: Introduction and history. Ital J Agrometeorol 1/2014:59–65

Stigter K, Walker S, Das H, Huda S, Haasbroek PD (2014b) Meeting farmers' needs for agrometeorological services: a review with case studies. Part II: Context 1, The existing situation. Ital J Agrometeorol 2/2014:51–60

Stigter K, Walker S, Das H (2014c) Meeting farmers' needs for agrometeorological services: a review with case studies. Part III: Context 2, The future. Ital J Agrometeorol 3/2014:45–52

Stigter K, Zheng D, Liu J, Li C, Dominguez-Hurtado IM, Mohammed AE, Abdalla AT, Bakheit NI, Al-Amin NKN, Wei Y, Kinama JM (2015) Meeting farmers' needs for agrometeorological services: a review with case studies. Part IV: Historical case studies. Ital J Agrometeorol 1/2015(2015):58–66

Tall A, Hansen J, Jay A, Campbell B, Kinyangi J, Aggarwal PK, Zougmoré R (2014) Scaling up climate services for farmers: mission possible. Learning from good practice in Africa and South Asia. CCAFS Report No. 13. CGIAR Research Program on Climate Change, Agriculture and Food Security (CCAFS), Copenhagen

Winarto YT, Stigter K (eds) (2011) Agrometeorological learning: coping better with climate change. LAP Lambert Academic Publishing, Saarbrucken

Winarto YT, Stigter K, Anantasari E, Hidayah SN (2008) Climate field schools in Indonesia: coping with climate change and beyond. Low External Input Sustainable Agriculture (LEISA) Magazine 24(4):16–18

Winarto YT, Stigter K, Anantasari E, Prahara H, Kristyanto (2010) "We'll continue with our observations": agro-meteorological learning in Indonesia. Farming Matters (formerly LEISA Magazine) 26(4): 12–15

Winarto YT, Stigter K, Anantasari E, Prahara H, Kristyanto (2011) Collaborating on establishing an agrometeorological learning situation among farmers in Java. Anthropol Forum 21(2): 175–197

WMO (2015) Climate services introduction. https://www.wmo.int/pages/themes/climate/climate_services.php. Accessed 15 Jul 2015

WMO (with Kees Stigter as lead author and integrator) (2006) Commission for Agricultural Meteorology (CAgM). The first fifty years. WMO-No. 999, WMO, Geneva, Switzerland

Zuma-Netshiukhwi G, Stigter K, Walker S (2013) Use of traditional weather/climate knowledge by farmers in the south-western free state of South Africa: agrometeorological learning by scientists. Special issue on agrometeorology: from scientific analysis to operational application. Atmosphere 4(3):383–410

Chapter 14
West African Farmers' Climate Change Adaptation: From Technological Change Towards Transforming Institutions

Daniel Callo-Concha

Abstract The effects of climate change are widely threatening West African farming. The case of the farmers in the West African Sudan savannah is particularly severe due to the specific ecological vulnerability and the political and socioeconomic instability in the region.

Since 2012, the West African Science Service on Climate Change and Adapted Land Use (WASCAL) has been leading a regional effort to enhance the understanding, data availability and building capabilities to strengthen the ability of agricultural systems in coping with the effects of climate change. One of the activities focuses on determining the drivers of farmers' adaptation.

The case studies in this research in Benin and Burkina Faso identified major climate hazards, farmers' perceptions and adaptation measures. Research methods included ethnography, quantitative and mixed analyses.

Results show mixed mainstream and specific perceptions of climate change and its effects. The mainly perceived impacts relate to yield decline, pest increase and water scarcity. Adaptation is widespread, but practices are diverse and often neither correlated with the type of hazards nor with impacts.

This indicates that it is necessary to acknowledge the role of adaptation to problems not only caused by climate change, and to consider adaptation as an enhancing factor of social-ecological resilience. Furthermore, the focus on technological changes in the local context should be directed towards enhancing adaptive and transforming institutions. Ongoing studies on agroforestry, subsidies provision and livestock demonstrate such strategies.

Keywords Climate change • West African Sudan Savannah • Technological innovation • Institutional adaptability • Agricultural land-use systems

D. Callo-Concha (✉)
Center for Development Research (ZEF), University of Bonn, Walter-Flex-Str. 3, 53113 Bonn, Germany
e-mail: d.callo-concha@uni-bonn.de

Introduction

Although the Intergovernmental Panel on Climate Change (IPCC) confirms global trends in the increase in temperature and greenhouse gases (GHG) concentration, declining snow and ice caps, and rising sea level (IPCC 2014), the results of Global Circulation Models(GCM) are uncertain with respect to regional changes, particularly in the tropics (Diallo et al. 2012). This is especially the case for West Africa due to its variable vegetation cover, oscillating rainfall regimes and periodic regional winds (Sylla et al. 2009).

This situation is particularly problematic for the agricultural sector, which depends heavily on rainfall and is strongly influenced by temperature. Since rainfall amounts have decreased by around 10 % in the past 60 years, and can vary up to 7 % between seasons and 20–30 % between years (Hulme et al. 2001; Nicholson 2001; Obeng-Asiedu 2004; Neumann et al. 2007), crop yields are reduced. Regional increases in temperature in West Africa are comparable with the global trend and estimated at 0.26 ± 0.05 °C per decade (Malhi and Wright 2004). Modeled yield losses suggest reductions from 5 to 20 % between 2008 and 2015 for the main crops, becoming higher in the north towards the Sahel (Paeth and Capo-Chichi 2008; Roudier et al. 2011).

Due to the inherent regional vulnerabilities, i.e., poor soil fertility and water scarcity, weak or absent infrastructure and institutions, and the farmers' limited technical, managerial and financial skills (Sanchez 2002; Challinor et al. 2007; Callo-Concha et al. 2013), the current scenario is cause for concern.

Farmers' interest in climate is pragmatic and refers mainly to the effects of the climate on natural cycles (Roncoli 2006). In the case of West Africa, empirical observations coincide with the general claim of the farmers of lower rainfall amounts and increasingly irregular rainfall periods and longer and drier spells (Saïdou et al. 2004; Yaro 2013). However, it is also acknowledged that farmers' perceptions may differ depending on their circumstances. Indicators of such differences are, for example, the degree of involvement in the market, gender and ethnicity, general political, economic and geographic settings, and even super natural and fatalistic beliefs (Patt and Schröter 2008; Bryan et al. 2009; Cuni Sanchez et al. 2012; Yaro 2013). The contrast between measured data and farmers' perceptions is not trivial. Social scientists suggest that isolating climate change from farmers' overall settings and other environmental changes is misleading, and instead their cultural backgrounds need to be considered to explain the changes and the responses of the farmers to these changes (Cannon and Müller-Mahn 2010; Mertz et al. 2010; Eguavoen and zur Heide 2012).

The factors cited above define farmers' adaptation strategies with respect to climate change in West Africa. Most adaptation measures deal with biophysical constraints, such as poor soil fertility and irregular provision of water. Hence, soil fertility can be improved through the addition of key nutrients and through soil conservation practices (Bationo and Vlek 1998; Schlesinger 1999; FAO 2008; Naab et al. 2012). The impact of water scarcity can be reduced through supplemental irrigation, better storage capacities, utilization of surface water and shallow groundwater, and strengthening of community-based management (Steenhuis et al. 2003; Sandwidi 2007; Mdemu et al. 2010).

Nevertheless, the success of these measures depends greatly on the local biophysical conditions (Stoorvogel and Smaling 1990; Giller et al. 2009), while the concurrent generation of goods and benefits for the households depends on broader social, commercial and political contexts (Ifejika Speranza 2010; Leakey 2012). This is why, despite the fact that the recommended measures are the result of sound scientific research, the implementation of such measures in farmers' fields for coping with climate change does not always mean an increase in adaptive capacities (Crane et al. 2011).Therefore, recommended measures need to be tested and validated in the local conditions they have been developed for Graef and Haigis (2001). In this regard, a number of studies on local adaptations have validated specific measures for specific settings (Paavola 2008; Barbier et al. 2009; Bryan et al. 2009; Ringler 2010; Fosu-Mensah et al. 2012; Yaro 2013).

In the above-depicted context, this paper provides insights into the adaptive behavior of West African farmers regarding climate change and climate variability. First, by identifying the farmers' perceptions and the local settings. Second, by determining the details, extent and importance of the coping measures. Finally, by cross-analysis of these factors to determine the underlying reasons for the selection of specific coping measures and shaping of the overall adaptation strategies, thus extracting lessons for the improvement of policy design.

Materials and Methods

The climate of the West African Sudanian Savanna his influenced by the Inter-Tropical Convergence Zone (ITCZ) and the constant exposure to high solar radiation, which lead to seasonal winds, unimodal rainy seasons, high and variable temperatures and low humidity (Ouédraogo 2004; Sandwidi 2007). The landscape is dominated by agricultural land use, grasslands, and scattered forest patches and trees in parklands. Streams and ponds are ephemeral, and the wild fauna is mainly restricted to conservation zones (Laube 2007).

Two catchments representative of the Sudanian Savannah and also within the sampling frame of the West African Science Service on Climate Change and Adapted Land Use (WASCAL) were selected for analysis: Dassari in Benin and Dano in Burkina Faso. Dassari belongs to the municipality of Materi, department of Atacora, located in the north–west of Benin near the borders of Burkina Faso and Togo. It has a population of more than 22,693 in about 15 villages (INSAE 2006). Annual rainfall exceeds 1000 mm and the annual average temperature is 28.7 °C with maximum peaks over 40 °C (Stechert 2011). Dano is the capital of the Ioba Province located in south-western Burkina Faso about 250 km from Ouagadougou. The community covers 669km^2 and counts 22 villages and one urban center. The population is estimated at43, 829; the area has the highest population density in

Burkina Faso (INSAE 2006; Gleisberg-Gerber 2012). Monthly rainfall peaks can reach 270 mm and humidity 75 %, while temperatures range from an average minimum of 20.1 °C in December to a maximum of 38.4 °C in March (Schmengler 2011). In both regions, the original dominant vegetation used to consist of scattered trees and perennial grasses, but is now dominated by annual crops and grassland (Stechert 2011; Schmengler 2011; Gleisberg-Gerber 2012). Dominant soils are lixisols with low nutrient content, low water-holding capacity and high erodibility due to their lateritic origin (Saïdou et al. 2004).

A mix of methods was applied for data collection. First, farmers' climate change perception and adaptation strategies were determined through exploratory focus groups(n = 4) to screen the responses of farmers, and through an extensive survey (n ≈ 260) covering household characterization, description of farming systems, farmers' perception of climate change and variability, and information about the implemented coping measures. Semi-structured interviews were conducted with elderly farmers (n = 11) focusing on their perceptions, and with representatives of local institutions (n = 10)focusing on the coping measures and adaptation strategies. Households and farming plots showing coping measures were geo-localized for forthcoming studies. Data analyses comprised discourse and statistical analyses, including measures of central tendency and dispersion, sampling distribution (Chi square test), and concordance and symmetry (Kappa and McNemar tests).

Results and Discussion

Climate Change Perception

The findings of the exploratory focus groups show that farmers identify as the main distressing climate elements rainfall, wind and heat. The major impacts are lack, scarcity and variability of water, increased pest attacks, soil loss through erosion, and lack of feed for cattle. When asked about solutions, these varied from the construction of dams, improvement of soil fertility by the use of fertilizers and physical conservation measures, reforestation as a multipurpose measure, and migration as the last option.

In the interviews with elderly farmers, these generally confirmed changes in most climate elements. Examples of the answers illustrate how these changes have affected the seasonality of agricultural practices:

> "*Everything changed. The heat. There is a lot of heat. Before there was not much heat (...), the wind, the wind is too much, too fast (...) brings the illnesses/diseases. The rain also (...). Everything has changed!*
>
> *(...) the nature has a calendar that was respected, now there is no longer such a calendar. It used to indicate when to start (agricultural activities), now things just happen like that*" (Mr. Dary Kiatti, Dassari, 20.04.2013)

In contrast, the reasons for those changes appear unclear to the farmers and range from fatalistic to supernatural. However, they also point out human activities as the main reason.

> *"It can be called: 'la vie d'aujourd'hui' (the life nowadays), this is the expression! The changes we can see. (...) Nowadays there areno trees. The rain is rare. Cannot say why it is like this nowadays. I do not know why. I have seen the changes but I do not know who is responsible"* (Dary Kiatti, Dassari, 20.04.2013).

> *"Before there were the totems who told us what was for bidden. But meanwhile the (fellow) Africans have started to do what is forbidden (...) and the change was sent. Voila!"* (Sumit Onviourou, Dano, 25.04.2013).

> *"The first issue is man. That is it. When my grandpa was here there were lots of trees, big trees. He did not chop them down. Now we have chopped them all down! It is man! (Arguing about other possibilities) No, no! It is man. Is man, he's the one that does it. In the time of my mom (...) she watched the foreigners who came, who confirmed that man was the problem, and now we gradually understand, and therefore have to change (...)"* (Anbene Quandi, Dano, 20.04.2013).

But when asked about measures to address the change, positions became concrete and operational, highlighting first the improvement of soil fertility as the major target, and second the improvement of community organization.

> *"In the past you could work the field (alluding to the natural regeneration of soil fertility) now you are obliged to use chemicals! To use organic manure and ask for fertilizer from the government. So we are using both! (...) If we were not able to use the fertilizer for farming, we would have to leave to look for new lands. So we need organic manure and the fertilizers from the government!"* (Dabire Kuntera, Dano, 20.04.2013).

> *"A way to address the change is: In the past, there were reunions and now when one calls the people to come, they do not come. Then the solution is to attend the meeting and find the solution. So there is (will not be) no problem (...)"* (Sumit Onviourou, village chef, Dano, 25.04.2013).

In the survey, the responses referring to being officially informed of climate change or not and measures to cope with it show no significant difference ($p < 0.05$). Asked about the changes during the last 10 years,[1] between 60 and 80 % of the farmers in Dassari and Dano believed that temperatures and winds had increased while rainfall had decreased ($p < 0.01$). Concerning the resulting hazards, 80 % of the farmers in Dassari mentioned droughts, and in Dano droughts and floods with 30 and 35 %, respectively; up to 40 % gave no answer ($p < 0.01$). When asked about the direct harm to their livelihoods, droughts and storms were significant for both catchments ($p < 0.05$), while floods ranked low in Dassari and were not significant for Dano.

However, when overlapping major hazards with harm to farmers' livelihoods, results are inconsistent. Droughts in Dassari and floods and droughts in Dano are

---

[1] In all subsequent questions on climate, the time frame was 10 years. For farming activities, e.g., crop selection, it was 5 years.

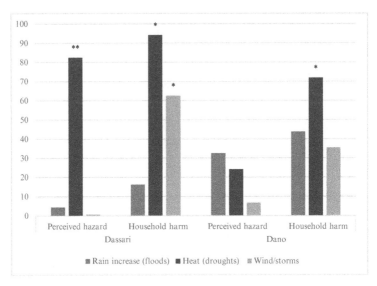

Fig. 14.1 Comparison of percentage of farmers' perception of climate change hazards and direct harm to livelihoods in Dassari, NW Benin, and Dano, SW Burkina Faso (* = p < 0.05, ** = p < 0.01)

stated as being responsible for household damage. Moreover, storms, although not considered a major hazard, are generally considered harmful (Fig. 14.1).

Similarly, when farmers were asked about specific occurrences of these events in the last 10 years (which? when? how many?) and the direct harm to their households, results became less categorical. Mean flood occurrences dropped to <1 per decade, and although occurrences were still high for droughts and storms (1.6–3.66), variance is high (2.96–3.83). Medians are below 2 for all climatic shocks, most observations lie in the lower quartiles, and a few out liers stand for frequent shock occurrences (Fig. 14.2).

Coping Measures and Adaptation Strategies

A broad spectrum of governmental and non-governmental organizations (NGOs) operate in both catchments. In general, all of them have embraced the mainstream discourse of climate change, i.e., "the triggering effects of global warming are heavily impacting rural householders' well-being, and there is an urgent need for action".

However, field activities do not correspond to this discourse, but instead activities are continued that are determined by institutional agendas and have not changed for decades, e.g., soil conservation practices, advice on the use of chemical fertilizers, etc., and even utilizing conventional extension approaches like demonstration plots. Government officials complain about the chronic lack of finances and

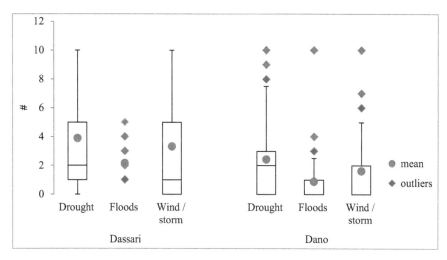

Fig. 14.2 Distribution of the perceived occurrences of climate shocks 2002–2012 in Dassari, NW Benin and Dano, SW Burkina Faso

material as the reason for their inactivity. In the case of NGOs, their focus varies according to the priorities established by their funders. In the past, they focused mainly on farming, but this has gradually been extended to other agricultural and non-agricultural activities, e.g., credit provision and marketing support, HIV prevention, urban sanitation, wildlife conservation, etc.

Although government officials and representatives of NGOs acknowledge the existence of National Adaptation Programs of Action (*Programme d'Action National d'Adaptation*; PANA), these still do not have a comprehensive and countrywide action plan, and there is a lack of logistic and financial capacities to lever actions on the ground. Instead, PANA tasks were re-addressed to existing organizations reframing and/or enlarging their portfolios. In Dano, PANA actions were taken over by the Program of Agricultural Development (*Programme Dévelopment de l'Agriculture*; PDA), which has operated since 2013 with area-specific agendas in selected parts of the country. In our research area, the program's focus is on soil conservation; however its activities restrict networking and capacity building. In Dassari, PANA is operated by the local office of the Agricultural Regional Agricultural Centers for Rural Development (*Centres Agricoles Régionaux pour le Développement Rural*; CARDER), which is involved in networking functions among several other conventional tasks, such as technical advice for farmers on soil and water conservation, and other less conventional tasks such as the provision of fertilizers and seeds for cotton farmers.

Nonetheless, regarding their impact on farmers' households, it is often said that the regular activities of these organizations, i.e., promotion of good agricultural practices, contribute indirectly to the increase in householders' ability to cope with hazards, including climate change.

The quantitative assessment of farmers' responses shows that virtually all farmers implement measures to cope with climate change ($p < 0.01$). These can be

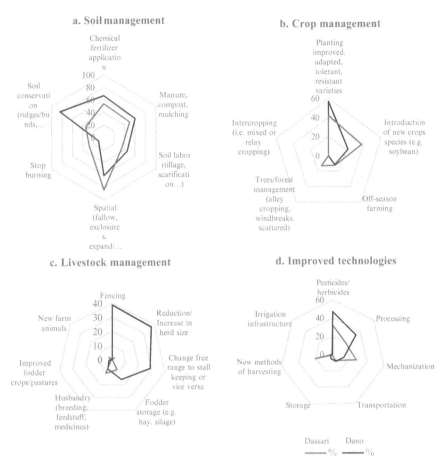

Fig. 14.3 Coping measures grouped by type of farming activities in Dassari, NW Benin (n = 136), and Dano, SW Burkina Faso (n = 121); %per activity

categorized in five groups: land/soil, cropping, livestock, modern technologies, and market related. On average, each farmer implements three soil-related measures, and one crop- and technology-related measure; livestock-related measures vary from 0.3 to 1, and market-related are below 0.3 per farmer.

Soil-related coping measures mainly include fertilization and soil conservation, which vary depending on the topography of the region, e.g., stone ridges in the hilly Dano and fenced enclosures in Dassari. Predominant crop-related measures are the change and introduction to adapted varieties and species, off-season farming, and agroforestry. Livestock-related measures are, for example, fencing, herd size control, stalling, etc. Regarding modern technologies, the use of agrochemicals and mechanization can be observed, and with respect to marketing, sales organization is the most important (Fig. 14.3a–d).

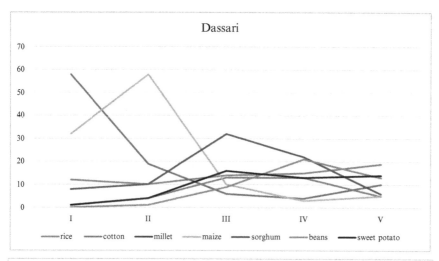

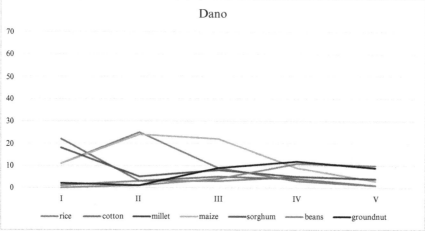

Fig. 14.4 Monocropping (%) in farmers' main five farm plots in Dassari, NWBenin and Dano, SW Burkina Faso 2008–2012

Broadly suggested as an adaptation measure, changes in crop species were tested. Farmers' crop selection from 2008 to 2012 in five different plots was documented, and inter-reliability tests applied (Armitage and Colton 2005). The working hypothesis was that crop species selection was due to climate change or other non-random reasons, and that farmers base their crop choice on previous and ongoing climate/weather changes. However, this does not occur. On the contrary, the applied tests show that the main determinant of crop selection is the crop previously cultivated, which leads to high rates of monocropping (Fig. 14.4).

The inconsistencies in responses regarding climate change (CC): 'have heard of CC', 'CC hazards have occurred', 'have experienced CC', and 'have been harmed by CC', discloses a discrepancy between what is framed within the climate change mainstream and what truly happens in the farmers' livelihoods. That is why although 'only' 80 % of the farmers acknowledge changes in climate elements ($p < 0.01$), all of them implement at least five measures to mitigate climate change impacts, and these often show no relationship with the hazards that threaten the household.

Hence, it is clear that climate change adaptation is not restricted to climate change only. Climate change does not seem to particularly determine farmers' adaptation/adaptive decisions. Under the highly unstable conditions in West Africa, these changes become just one more factor together with the already wide spectrum of factors that influence farmers in their attempt to sustain their livelihoods.

The evaluation of the adaptation measures carried out by the farmers shows that the similarities presumed by the comparable latitudes are overlain by the local geographic, ecological, and socioeconomic differences. For example, the selection of soil conservation measures is determined by slope and soil type, and livestock management by technological know-how and traditions.

Furthermore, when assessing the consistency of specific measures such as the crop choice supposedly dependent on climate, it is observed that these are ad-hoc measuring response to broader demands, which altogether may grant better coping opportunities to farmers. In general, it seems that the farmers' decisions tend to guarantee financial income, maintenance of and increase in soil fertility, follow traditional choices, and ensure food security. All these may refer well to climate change but not exclusively to it.

In this context, the major levering power of two institutions is identified. First, the market framed within government incentive programs such as for cultivation of cotton for export. These determine not only crop choice and the allocation of production area and sales organization, but also the type of practice, e.g., use of agrochemicals, mechanization and transport. Second, the support by the government and NGOs, identifiable by the ongoing agricultural practices implemented by the farmers. It is recommended that the above factors should be considered when planning dissemination and implementation of climate change coping measures.

Conclusions

The study investigates two components of the behavior of West African farmers: climate change perception and adaptation measures.

The detailed analysis on farmers' perception indicates that there is no direct connection between climate and weather variability and the identification of hazards and harm to their households. The repeatedly observed decline in yields, increase in pests and irregular provision or lack of water, although accurate, are not causally linked to climate change only. Climate change triggers assumed by the farmers vary from anthropogenic to supernatural.

However, the ongoing adaptation measures tend to be pragmatic and framed within a broader set of socio-ecological changes. They are in response to multiple demands grounded in the needs of households, e.g., food security, inflow of cash or livestock capitalization. Short-term economic benefit, local socio-ecological conditions, traditional practices and influence of extension institutions shape the selection of coping measures.

Against this background, efforts to encourage and support farmers' climate change adaptation need to avoid being limited to mainstream causal responses and unsound and inappropriate innovations, and instead should consider options able to fulfill the multiple demands framed in existing practices. They should also rely on the existing institutions for their implementation. In short, it is recommended to support multifunctional local technological options, institutions and organizations. Good examples are agroforestry and subsidized cotton, which are subjects of ongoing research.

Acknowledgements This research was carried out within the West African Science Service Center on Climate Change and Adapted Land Use (WASCAL) framework, financed by the German Federal Ministry of Research and Education.

References

Armitage P, Colton T (2005) Encyclopedia of biostatistics, vol 1–8. Wiley, New York

Barbier B, Yacouba H, Karambiri H, Zoromé M, Somé B (2009) Human vulnerability to climate variability in the Sahel: farmers' adaptation strategies in Northern Burkina Faso. Environ Manag 43(5):790–803

Bationo A, Vlek P (1998) The role of nitrogen fertilizers applied to food crops in the SudanoSahelian zone of West Africa. In: Renard G, Becker K, von Oppen M (eds) Soil fertility management in West African land use systems. Margaf Verlag, Weikersheim, Germany, pp 41–51

Bryan E, Deressa TT, Gbetibouo GA, Ringler C (2009) Adaptation to climate change in Ethiopia and South Africa: options and constraints. Environ Sci Pol 12(4):413–426

Callo-Concha D, Gaiser T, Webber H, Tischbein B, Müller M, Ewert F (2013) Farming in the West African Sudan Savanna: insights in the context of climate change. Afr J Agric Res 8 (38):4693–4705

Cannon T, Müller-Mahn D (2010) Vulnerability, resilience and development discourses in context of climate change. Nat Hazards 55(3):621–635

Challinor A, Wheeler T, Garforth C, Craufurd P, Kassam A (2007) Assessing the vulnerability of food crop systems in Africa to climate change. Clim Chang 83(3):381–399

Crane TA, Roncoli C, Hoogenboom G (2011) Adaptation to climate change and climate variability: the importance of understanding agriculture as performance. NJAS Wageningen J Life Sci 57(3–4):179–185

Cuni Sanchez A, Fandohan B, Assogbadjo AE, Sinsin B (2012) A countrywide multi-ethnic assessment of local communities' perception of climate change in Benin (West Africa). Clim Develop 4(2):114–128

Diallo I, Sylla MB, Giorgi F, Gaye AT, Camara M (2012) Multimodel GCM-RCM ensemble-based projections of temperature and precipitation over West Africa for the early 21st century. J Geophys. 2012:Article ID 972896

Eguavoen I, zur Heide F (2012) Klimawandel und Anpassungsforschung in Äthiopien. Z Ethnol 137(1):97–118

FAO (2008) Investing in sustainable agricultural intensification. The role of conservation agriculture. A framework for action. Food and Agriculture Organization of the United Nations, Rome

Fosu-Mensah BY, Vlek PLG, MacCarthy DS (2012) Farmers' perception and adaptation to climate change: a case study of Sekyedumase district in Ghana. Environ Develop Sustainability 14:495–505

Giller KE, Witter E, Corbeels M, Tittonell P (2009) Conservation agriculture and smallholder farming in Africa: the heretics' view. Field Crops Res 114(1):23–34

Gleisberg-Gerber K (2012) Livelihoods and land management in the Ioba Province in south western Burkina Faso. ZEF Working Paper 91. Center for Development Research, Bonn, Germany

Graef F, Haigis J (2001) Spatial and temporal rainfall variability in the Sahel and its effects on farmers' management strategies. J Arid Environ 48(2):221–231

Hulme M, Doherty R, Ngara T, New M, Lister D (2001) African Climate Change: 1900–2100. Clim Res 17:145–168

Ifejika Speranza C (2010) Resilient adaptation to climate change in African agriculture (Studies 54). Deutsches Institut für Entwicklungspolitik/German Development Institute, Bonn

Institute National de la Statistique et de l'Analyse Economique (INSAE) (2006) Population beninoise en age de voter en 2006. http://www.insae-bj.org/2012/doc/Population_beninoise_en_age_de_voter_en_2006.pdf

Intergovernmental Panel on Climate Change (IPCC) (2014) Climate change 2014: synthesis report. Contribution of Working Groups I, II and III to the Fifth assessment report of the intergovernmental panel on climate change. IPCC, Geneva, p 151

Laube W (2007) Changing natural resource regimes in northern Ghana. Actors, Structure and Institutions. In: Gerke S, Evers H-D (eds) ZEF development studies, vol 4. Lit Verlag Dr. W. Hopf, Münster, Germany

Leakey R (2012) Living with the trees of life: towards the transformation of tropical agriculture. CABI, Wallingford, UK

Malhi Y, Wright J (2004) Spatial patterns and recent trends in the climate of tropical forest regions. Philos Trans R Soc Lond Ser B 359:311–329

Mdemu M, Laube W, Barry B (2010) Temporal water productivity of tomato irrigated from a small reservoir and hand-dug wells in dry season cropping in the Upper East Region, Ghana. J Appl Irrig Sci 45(1):75–93

Mertz O, Mbow C, Østergaard Nielsen J, Maiga A, Diallo D, Reenberg A, Diouf A, Barbier B, Bouzou Moussa I, Zorom M, Ouattara I, Dabi D (2010) Climate factors play a limited role for past adaptation strategies in West Africa. Ecol Soc 15(4):25

Naab J, Bationo A, Wafula BM, Traore PS, Zougmore R, Ouattara M, Tabo R, Vlek PLG (2012) African perspectives on climate change and agriculture: impacts, adaptation, and mitigation potential. In Hillel D, Rosenzweig C (2012) Handbook of climate change and agroecosystems, global and regional aspects and implications. ICP series on climate change impacts. In Adaptation, and mitigation, vol 2. Imperial College Press, London, pp 85–106

Neumann R, Jung G, Laux P, Kunstmann H (2007) Climate trends of temperature, precipitation and river discharge in the Volta Basin of West Africa. Int J River Basin Manag 5(1):17–30

Nicholson SE (2001) Climatic and environmental change in Africa during the last two centuries. Clim Res 17:123–144

Obeng-Asiedu P (2004) Allocation water resources for agricultural and economic development in the Volta River Basin. PhD thesis, University of Bonn

Ouédraogo E (2004) Soil quality improvement for crop production in semi-arid West Africa. Ponsen & Looijen Bibliotheek Wageningen UR, Wageningen

Paavola J (2008) Livelihoods, vulnerability and adaptation to climate change in Morogoro, Tanzania. Environ Sci Pol 11(7):642–654

Paeth H, Capo-Chichi A (2008) Climate change and food security in tropical West Africa—A dynamic statistical modeling approach. Erdkunde 62(2):101–115

Patt AG, Schröter D (2008) Perceptions of climate risk in Mozambique: implications for the success of adaptation strategies. Glob Environ Change 18(3):458–467

Ringler C (2010) Climate change and hunger: Africa's smallholder farmers struggle to adapt. EuroChoices 9(3):16–21

Roncoli C (2006) Ethnographic and participatory approaches to research on farmers' responses to climate predictions. Clim Res 33:81–99

Roudier P, Sultan B, Quirion P, Berg A (2011) The impact of future climate change on West African crop yields: what does the recent literature say? Glob Environ Change 21:1073–1083

Saïdou A, Kuyper TW, Kossou DK, Tossou R, Richards P (2004) Sustainable soil fertility management in Benin: learning from farmers. NJAS Wageningen J Life Sci 52(3–4):349–369

Sanchez P (2002) Soil fertility and hunger in Africa. Science 295(5562):2019–2020

Sandwidi JP (2007) Groundwater potential to supply population demand within the Kompienga dam basin in Burkina Faso, Ecology and development series. PhD thesis. Center for Development Research, University of Bonn, Bonn, Germany, p 55. http://www.zef.de/fileadmin/template/Glowa/Downloads/Sandwidi_doc_thesis_2007.pdf.

Schlesinger WH (1999) Carbon sequestration in soils. Science 284(5423):2095

Schmengler AC (2011) Modeling soil erosion and reservoir sedimentation at hillslope and catchment scale in semi-arid Burkina Faso. http://www.zef.de/fileadmin/webfiles/downloads/zefc_ecology_development/eds_80_schmengler_text.pdf

Stechert C (2011) Einsatz von Insektiziden im Baumwollanbau in Benin und deren Auswirkung auf Nicht-Zielorganismen, pp 7–9. http://digisrv-1.biblio.etc.tu-bs.de:8080/docportal/receive/DocPortal_document_00041115

Steenhuis TS, Masiyandima MC, van de Giesen N, Diatta S, Windmeijer PN (2003) The hydrology of inland valleys in the subhumid zone of West Africa: rainfall-runoff processes in the M'beexperimental watershed. Hydrol Process 17(6):1213–1225

Stoorvogel J, Smaling EMA (1990) Assessment of soil nutrient depletion in Sub-Sahara: 1983–2000. Report 28, 4 volumes. The Winand Staring Centre for Integrated Land, Soil and Water Research. Winand Staring Centre, Wageningen

Sylla MB, Gaye AT, Pal JS, Jenkins GS, Bi XQ (2009) High-resolution simulations of West African climate using regional climate model (RegCM3) with different lateral boundary conditions. Theor Appl Climatol 98:293–314

Yaro JA (2013) The perception of and adaptation to climate variability/change in Ghana by small-scale and commercial farmers. Reg Environ Change 13:1259–1272

Chapter 15
Livelihood Options as Adaptation to Climate Variability Among Households in Rural Southwest Nigeria: Emerging Concerns and Reactions

Isaac B. Oluwatayo

Abstract Climate variability is no doubt one of the greatest challenges facing inhabitants of sub-Sahara African countries, and especially those relying on agriculture as their primary source of livelihood. This is not unconnected with the fact that agriculture, a climate-dependent and climate-controlled livelihood source, remains the largest employer of labour in these countries. This study examined the activities engaged in by farming households in rural southwest Nigeria as a way of mitigating climate variability shocks. Data was collected through a questionnaire administered to a random sample of 250 households in rural Nigeria. Analytical methods employed include descriptive statistics and tobit regression model. A descriptive analysis of respondents' socioeconomic characteristics revealed their average age to be 52 years with only about one-third having tertiary education. Results indicated a noticeable shift in farming household activities, where livelihood options embraced to cushion climate variability include 'okada' riding (motorcycle passenger transport), attending political rallies, taking up menial jobs in neighbouring communities, trading, and migrating to city centres in search of paid jobs. However, this development is already taking its toll on food security status of residents in terms of availability and affordability. Results of a Tobit model revealed a positive and statistically significant correlation between the livelihood options harnessed and age, years of formal education, income, membership of social group, as well as access to credit. However, negative and significant relationships were found between the livelihood options harnessed and household size and poverty status of respondents. Suggested policy prescriptions include investment in capacity building of farming households so as to enhance their earning potential and effort should be geared at encouraging cooperative activities since this can help in sharing and mitigating covariate shocks like climate change.

I.B. Oluwatayo, PhD (✉)
Department of Agricultural Economics and Animal Production, School of Agricultural and Environmental Sciences, University of Limpopo, Sovenga 0727, South Africa
e-mail: isaacoluwatayo@yahoo.com

Keywords Adaptation • Climate change • Food security • Livelihood options • Rural Nigeria

Introduction

Climate variability is one of the greatest challenges confronting Nigeria and other sub-Sahara African countries particularly those that depend on agriculture as a source of livelihood (Apata 2011). Agriculture accounts for about 30.9 % of Nigeria's Gross Domestic Product (GDP after rebasing) and employs more than 40 million people through the different nodes of its value chain (Ogundipe 2014). Meanwhile, the agricultural sector provides employment for around two-thirds (approximately 67 %) of the population and climate change is already taking its toll on this all-important sector. This is because agriculture in most African countries is rain-fed, and hence exposed to the vagaries of weather. African agriculture, Nigeria not being an exception, is characterized by undercapitalization, use of crude technology, small and fragmented farmlands and infrastructure deficits among others (Oluwatayo 2009a; Oluwatayo and Ojo 2014). This explains the high exposure to extreme weather events (flood and droughts) and the attendant welfare losses experienced by smallholder farmers (Abiodun et al. 2011).

The terms climate change and climate variability are often used interchangeably in the literature, and refer to a sustained fluctuation in the average weather conditions of a particular geographical location over time (Ademola and Oyesola 2012). Climate variability directly affects farm output, and thus the food security status of agrarian economies (Fatuase and Ajibefun 2013). These impacts of climate variability materialise through outright crop failure, climate-induced transportation problems, and post-harvest losses (Fatuase et al. 2015). However, climate variability also poses indirect challenges to the production of agricultural produce. At present, the majority of rural households in Nigeria are involved in different nodes of the agricultural value chain (Oluwatayo 2009b). However, productive resources are now flowing away from agriculture to other sectors, with young people now moving into sectors with less climate risk such as transportation, electoral fraud and "commercial" violence among others. These issues necessitated the need for this study to examine the effect of livelihood options (necessitated by climate change) harnessed in rural Southwest Nigeria on the food security status of households in the study area.

Literature Review

A number of studies have been conducted on adaptation strategies employed by diverse categories of people in Nigeria. Ademola and Oyesola (2012) examined the adaptation strategies of selected tree crop farmers to climate variation in Oyo State,

Nigeria. The challenges faced by tree crop farmers due to climate change were low yield, stunted growth, high sunlight intensity, pests and diseases, and prolonged dry seasons among others. They reported that about 60 % of the respondents were aware of climate change impacts. The adaptation strategies employed in the study area included mulching, increased irrigation and crop diversification.

In another study, Ogbo and Onyedinma (2012) investigated the prospects and problems of climate change adaptation in Nigeria. They used primary data and interviewed 109 respondents. The findings revealed a strong relationship between climate change adaptation and effective management technologies. Coexistence enhancement, promotion of preventive measures, expansion of awareness and human discipline, technology transfer, development, and enhancement of agricultural extension services were some of the benefits of climate change adaptation.

Adebayo et al. (2013) analysed farmers' awareness, vulnerability and adaptation to climate change in Adamawa State, Nigeria. They interviewed 340 farmers and found that adaptation strategies the farmers used were: planting tolerant crop varieties, altering planting schedules, using early maturing varieties, and crop diversification. However, according to the study, the constraints to adaptation were inadequate access to information, appropriate technology, necessary input and labour. Again, Farauta et al. (2012) carried out a study on farmers' adaptation initiatives to the impact of climate change and agriculture in Northern Nigeria using rapid rural appraisal, focus group discussions (FGDs) and semi-structured interviews. They reported planting of early maturing crops, use of chemicals, increased use of fertilizers, use of resistant varieties, processing to minimize postharvest losses, and afforestation were the adaptation strategies used by the farmers in Northern Nigeria.

Fatuase and Ajibefun (2013) carried out a study on adaptation to climate change in Nigeria using rural farming households in Ekiti State as a case study. They interviewed 80 respondents and analysed the data using descriptive statistics and a multinomial logit model. It was reported that educational level, farming experience, access to extension services, access to climate information, and access to credit were factors affecting adaptation choices of the respondents. They identified inadequate funds and climate information as the constraints to adaptation in the study area. Fatuase and Ajibefun (2013) investigated the determinants of adaptation measures to climate change by arable crop farmers in Owo local government area of Ondo State, Nigeria. The study drew a sample of 120 crop farmers using a multistage sampling technique and the data were analysed using descriptive statistics and tobit regression model. The study identified household size, education, farm size, income, experience and access to extension agents as the factors influencing the rate of utilization of adaptation strategies. However, inadequate funds and information on climate change undermined adaptation to climate change in the study area.

From the foregoing, it is very clear that none of these studies examined the implications of some of the strategies employed in addressing climate change/variability on food security of households.

Research Methodology

This study was carried out in Southwest Nigeria. Primary data were collected by questionnaire from 250 respondents using a multistage sampling technique. The first stage of data collection involved a random selection of two states (Ekiti and Oyo) out of the six states in the zone. The second stage was the selection of three and six local government areas (LGAs) respectively from the selected states. In Ekiti state, 3 LGAs were randomly selected out of the 16 LGAs in the state while in Oyo State, 6 LGAs were randomly selected out of the 33 LGAs in the state. The selection of these LGAs was done based on the number of LGAs in each state to make it representative. In the third stage, two rural communities/villages were randomly selected in each LGA, and the fourth stage was the random selection of respondents within each community/village based on probability proportionate to size, comprising 87 and 163 respondents (household heads) from Ekiti and Oyo States respectively. The questionnaire sought to elicit information regarding the socioeconomic characteristics of respondents, livelihood strategies harnessed, reasons for taking up such strategies, and food security challenges, among others. The data were analysed using descriptive statistics and a Tobit regression model.

Results and Discussion

Socioeconomic Characteristics of Respondents

A descriptive analysis of the respondents is presented in Table 15.1. The average age of respondents was 52 years indicating that respondents are more or less still active in their occupation. Over a fifth (22 %) had a household size ranging from 1 to 3 people, with more than a third (39.2 %) having a household size ranging from 4 to 6. This is comparable to the findings of Anyanwu (2013), where it was revealed that average household size in Nigeria is 5 persons with more in rural areas (5.2 persons) than in the urban areas (4.7 persons). This is not a particularly large household size, which could be due to the high level of education of the respondents, where 34.4 % of the respondents had tertiary education. Meanwhile the proportion of people with tertiary education is more in urban Nigeria than in rural Nigeria and this is due to the ease of getting white collar jobs in the cities. It should also be noted that excessively high household size reduces income per capita, which invariably leads to high poverty where poor households are not likely to be able to employ appropriate adaptation strategy to climate variability. For example, Oyekale (2009) and Owombo et al. (2014) found that farm households in Ondo State were unable to use some adaptation options because of their status and also because they do not possess the skills required for effective application. With respect to occupation, 57.2 % of the respondents were dependent on agriculture for their livelihood.

Table 15.1 Descriptive analysis of respondents' socioeconomic characteristics in Ekiti and Oyo states

Variable	Frequency	%
Age		
≤30	41	16.4
31–40	33	13.2
41–50	60	24.0
51–60	87	34.8
≥60	29	11.6
Household size		
1–3	55	22.0
4–6	91	39.2
7–9	78	28.4
10–12	20	8.0
>12	06	2.4
Education		
No formal	67	26.8
Primary	42	16.8
Secondary	55	22.0
Tertiary	86	34.4
Occupation		
Civil service	49	19.6
Trading	23	9.2
Farming	143	57.2
Artisans	35	14.0

Livelihood Options Harnessed by Respondents to Cushion Climate Variability

A number of livelihood strategies were being employed by respondents in Ekiti and Oyo States to cushion the effect of climate variability shocks. Responses from the households revealed that some of the livelihood activities currently being harnessed were necessitated by recurrent cases of climate variability in the study area and this has led them adopting a number of adaptation strategies. Examples of adaptation options harnessed by households in the study area include relocation to city centres, use of mulching (in the case of farming households), 'okada' riding, participation in political rallies, taking up menial jobs in neighbouring communities and trading to mention just a few as indicated in their responses.

With respect to agricultural adaptation, 8.4 %, 4.7 % and 2.5 % of respondents reported use of mulching, crop rotation and strip cropping respectively as measures employed to mitigate the impacts of climate variability on food production, and thus contribute to the food security of the study area.

However, a significant proportion of respondents were engaging in non-agricultural adaptation strategies. Notable among these strategies include alternative livelihood options such as 'okada' riding (motorbike passenger transport; 43.9 %), taking up menial jobs (9.3 %), and trading (18.3 %), alongside

Table 15.2 Livelihood option harnessed by respondents

Livelihood activity	Frequency	%
Okada riding	97	38.8
Participation in rallies	35	14.0
Doing menial job	23	9.2
Trading	41	16.4
Relocation to other places	15	6.0
Mulching	21	8.4
Other measures (crop rotation and strip cropping)	18	7.2
Total	250	100.0

participation in political rallies (14.2 %), and relocation to neighbouring communities (6.0 %). The likely consequence of taking up alternative livelihood options is increased food insecurity in the study area as the majority of the able-bodied residents (who ought to have been engaged in agricultural activities such as production, processing, marketing, value addition etc.) have migrated into either the cities or neighbouring communities to pursue other livelihood opportunities. This is corroborated by the findings of Adeoti et al. (2010) where flooding necessitated the migration of fishermen to other locations with significant negative implications for their welfare (Table 15.2).

Determinants of Livelihood Strategies Employed by Respondents

A number of socioeconomic factors were found to influence the adaptation/livelihood options harnessed by respondents in Ekiti and Oyo States, Nigeria and these are discussed below and presented in Table 15.3.

Age There exists a positive and statistically significant relationship between age and adaptation strategies harnessed, where an increase in the age of the respondent increased the likelihood of choosing an adaptation strategy to reduce climate variability impacts. This could be related to knowledge of climate variability increasing with age. This is consistent with the findings of Oyekale (2009) in his study on climate variability and its impact on income and welfare of households in Southern and Northern Nigeria.

Household Size Household size was found to be negatively related to adaptation strategies harnessed. In other words, an increase in the household size of the respondent decreased the likelihood of harnessing adaptation strategies. This could be attributed to the limited financial power that comes with large households, which inhibits adaptive capacities.

Table 15.3 Results of the tobit regression model showing determinants of livelihood/adaptation strategies harnessed

Variable	Coefficient	Standard error	t-Statistic
Age	0.237**	0.104	2.279
Gender	0.510	0.392	1.301
Household size	−0.4201***	0.108	−3.898
Years of formal education	0.722***	0.298	2.423
Income	0.169*	0.088	1.920
Season of the year	0.163	0.125	1.304
Member of social group	0.302***	0.107	2.822
Primary occupation	0.081	0.106	0.764
Years of experience	0.117*	0.062	1.887
Member of ruling political party	0.401*	0.203	1.975
Access to credit facility	0.555**	0.256	−2.168
Poverty status	−0.141	0.088	−1.602
Constant	1.571*	0.803	1.956
Number of observations	250.00		
Log likelihood	−15003.07		
LR(Chi^2(14))	151.37		
Prob > Chi^2	0.000		
Pseudo R^2	0.32		

***Significant at 1 %
**Significant at 5 %
*Significant at 10 %

Years of Formal Education An increased level of formal education increased the likelihood of harnessing adaptation/livelihood options/strategies. This is because a high level of education will influence the level of information and opportunities to adapt to climate variability. This is consistent with findings of Owombo et al. (2014) in their study on gender dimensions of adaptation to climate change in Ondo State, Nigeria.

Income An increase in income increased the likelihood of a respondent harnessing adaptation strategies. This is because a higher income will allow respondents to choose from available combinations of adaptation strategies/options.

Membership of Social Group Membership in social groups was found to have a positive relationship with harnessing adaptation strategies. This is because membership of social groups will increase access to information and productive resources (money, inputs etc.) which will in turn enhance adaptation of the respondents.

Years of Experience in Primary Occupation There is a positive and statistically significant relationship between years of experience and use of adaptation strategies. This is because the respondents must have witnessed climate variability over time and the experience acquired over the years will in no small measure assist in managing the situation or cope well with climate variability. Experience is often said to be a good teacher hence employing available adaptations become easy.

Membership of Ruling Political Party There exists a positive relationship between membership of ruling political party and harnessing adaptation strategies in the study area. This is because of the differences in resource allocations due to political considerations that characterise ruling party membership in Nigeria.

Access to Credit There is a positive relationship between access to credit and harnessing adaptation strategies. This is because access to credit increases economic access to adaptation strategies.

Conclusion and Recommendations

This study reveals that respondents with higher levels of education and active participation in social groups (cooperative activities) have a higher likelihood of harnessing available agricultural adaptation strategies, as opposed to taking up adaptation options that threatens the already precarious food security status in the study area. Suggested policy prescriptions based on the study findings include the following:

(i) Increased investment in capacity building of respondents through education to enhance their earning potential.
(ii) Efforts to encourage cooperative activities between farmers should be intensified by government and private organisations, since this can help in risk sharing and mitigating covariate risk like climate variability.

References

Abiodun BJ, Salami AT, Tadross M (2011) Climate change scenarios for Nigeria: understanding biophysical impacts. A report published by Building Nigeria's Response to Climate Change (BNRCC) project, pp 1–10

Adebayo AA, Mubi AM, Zemba AA, Umar AS (2013) Awareness of climate change impacts and adaptation in Adamawa state, Nigeria. Int J Environ Ecol Family Urban Stud 3(1):11–18, ISSN: 2250-0065

Ademola AO, Oyesola OB (2012) Adaptation strategies of selected tree crop farmers to climate variation in Oyo state, Nigeria. J Agric Ext Rural Dev 4(199):4–5, ISSN: 214-2154

Adeoti AI, Olayide OE, Coster AS (2010) Flooding and welfare of fishers' households in Lagos state, Nigeria. J Hum Ecol 32(3):161–167

Anyanwu JC (2013) Marital status, household size and poverty in Nigeria: evidence from the 2009/2010 survey data. African Development Bank (AfDB) Working Paper Series No. 180 – September, 2013. 28 pp

Apata TG (2011) Factors influencing the perception and choice of adaptation measures to climate change among farmers in Nigeria: evidence from farm households in South West, Nigeria. J Environ Econ 2(4):1–10

Farauta BK, Egbule CL, Agwu AE, Idrisa YL, Onyekuru NA (2012) Farmers' adaptation initiatives to the impacts of climate change on agriculture in northern Nigeria. J Agric Ext 16 (1):1–13, ISSN: 1119-94X

Fatuase AI, Ajibefun AI (2013) Adaptation to climate change. A case study of rural farming households in Ekiti state, Nigeria. A paper presented at the International Conference on Climate Change Effects, Postdam, 27–30 May 2013

Fatuase AI, Aborisade AS, Omisope ET (2015) Determinants of adaptation measures to climate change by arable crop farmers in Owo local government area of Ondo state, Nigeria. World Rural Observ 7(1):1–9

Ogbo AI, Onyedinma AC (2012) Climate change adaptation in Nigeria: problems and prospects. Sacha J Environ Stud 2(1):1–16

Ogundipe GAT (2014) Protecting animals, preserving our future. Inaugural lecture of the University of Ibadan delivered on the 4th of September, 2014, p 14

Oluwatayo IB (2009a) Poverty and income diversification among households in rural Nigeria: a gender analysis of livelihood patterns. Conference Paper Number 41 of *Instituto de Essudo Socias e Economicos (IESE)*, pp 8–9

Oluwatayo IB (2009b) Vulnerability and adaptive strategies of staple food crop farmers to seasonal fluctuations in production and marketing in southwest Nigeria contribution to Seasonality Revisited International Conference Institute of Development Studies, UK

Oluwatayo IB, Ojo AO (2014) Socioeconomic contributions of neglected and underutilized species to livelihood security in rural southwest Nigeria: *Thaumatococcus Danielli* as a test case. Mediterranean J Soc Sci 5(27):1–7

Owombo PT, Koledoye GF, Ogunjimi SI, Akinola AA, Deji OF, Bolarinwa O (2014) Farmers' adaptation to climate change in Ondo state, Nigeria: a gender analysis. J Geogr Reg Plann 7 (2):30–35, ISSN: 2070-1845

Oyekale SO (2009) Climate variability and its impact on agricultural income and households' welfare in southern and northern Nigeria. Electr J Environ Agr Food Chem 8(7):1–10

Chapter 16
Climate Change Projections for a Medium-Size Urban Area (Baia Mare Town, Romania): Local Awareness and Adaptation Constraints

Mihaela Sima, Dana Micu, Dan Bălteanu, Carmen Dragotă, and Sorin Mihalache

Abstract Numerous studies worldwide emphasize the impacts that cities now face due to increasing variability of weather extremes associated with climate change, exacerbating the urban heat island effect, air pollution and health impacts. In this respect, cities need to cope with these new threats and get prepared. One way of doing this is to develop climate change mitigation and adaptation strategies, focusing on local vulnerabilities in relation to the current societal development needs and adaptation options. There are many recent initiatives and examples for climate change adaptation strategies and plans for large cities, but only a few examples for small and medium-size cities. This paper investigates the mid-(2021–2050) and far-future (2071–2100) climate change signals in an urban area located in northern Romania (Baia Mare town), with around 200,000 inhabitants, exploring the results of some CORDEX GCMs under the new IPCC RCPs (RCP2.6, RCP4.5 and RCP8.5). The study analyzes these signals in relation to the local awareness of key institutional stakeholders, as well as to the current social and economic constraints towards considering the climate change adaptation. The findings, based on a survey, highlight how the local authorities perceive and include climate change aspects in their activities as an important step towards real implementation of specific climate-based decisions, as well as their needs in terms of climate information and data. The study provides useful scientific insights about future climate and expected impacts in the Baia Mare Urban System to stakeholders, which could increase their awareness and knowledge in terms of climate change.

Keywords Climate change adaptation • Urban area • Awareness • Baia Mare town

---

M. Sima (✉) • D. Micu • D. Bălteanu • C. Dragotă • S. Mihalache
Romanian Academy, Institute of Geography, 12 Dimitrie Racovita St., Sector 2, Bucharest 0233993, Romania
e-mail: simamik@yahoo.com

Introduction

Climate change is recognised as a serious threat for cities around the globe, their geographical location influencing the magnitude of impact as well as the choice of appropriate adaptation measures and strategies. Climate change is increasingly discussed as an emerging global security issue (UNEP 2012) and cities are becoming more aware of the need to prepare for coping with greater variability in temperature, precipitation, as well as an intensification of weather extremes (World Bank 2011). Rosenzweig et al. (2011) emphasized that worldwide many cities have already faced significant climatic and environmental challenges, related to the urban heat island effect, air pollution and existing disruptive climatic extremes (e.g. typhoons, hurricanes). As a consequence of increasing concern regarding climate change impacts on cities, there are several examples of how cities have considered the issue of climate change adaptation in their future plans, where these range from extensive adaptation plans (e.g. New York, London or Amsterdam), to various commitments towards adaptation or disparate actions and measures.

Generally, it is larger cities, with their larger financial and human resources, that have engaged with climate change adaptation initiatives. In comparison, small towns and rural areas are lagging behind, likely due to challenges such as financial constraints, lack of specialised human resources or awareness of the issue (Cohen 2011). Within these communities, the role of local institutions in adapting to climate change is higher (Agrawal 2008) and the need to adapt is just as great (if not greater) for these smaller population centres. For example, past hazardous events have proved disastrous for small local communities, mainly because they affected almost the entire population (Ford and Berrang Ford 2011). A first step towards better coping with such events is increasing awareness in terms of climate change information and potential impacts for local planners and municipality leaders (Ford and Berrang Ford 2011; Carmin et al. 2012).

At the European level, several barriers have been identified in terms of urban adaptation to climate change, lack of awareness, lack of local data and knowledge, and limited funding being the prominent ones (EU 2013). The adaptive capacity of a community in terms of climate change depends, among others, on the awareness, capacity building and willingness of stakeholders to respond and being prepared for future risk (Mehrotra et al. 2009; Hartman and Spit 2014). In this respect, the ability of the local stakeholders to act and work together to develop and implement adaptation measures is a key step towards building adaptive capacity (Carter et al. 2015). Raising awareness has proved to be a key step towards ensuring decision-makers commitment and responsibility in terms of climate change (Prutsch et al. 2010), which help in building a stronger adaptive capacity (World Bank 2011). For this, communication in terms of framing climate change and dialogues between scientists and stakeholders are important for long-term active engagement and motivation (de Boer et al. 2010). There is a need at local level for tailored climate data and information at fine spatial resolution as well as for impact and vulnerabilities studies at local scales (Rosenzweig et al 2011; IPCC 2012).

At a European level, the issue of climate adaptation is considered in the EU Strategy on Adaptation to Climate Change (2013), which explicitly states the aim to promote climate adaptation actions in European cities. Furthermore, the proposal of the EU Multi-annual Financial Framework for the period 2014–2020 foresees a much higher share (20 %) of the budget for climate change actions, including the support for both adaptation and mitigation of climate change impacts in urban areas. In support, the recently established European Climate Adaptation Platform CLIMATE-ADAPT (http://climate-adapt.eea.europa.eu) brings together adaptation case studies, the outputs of European complementary research projects (e.g. EU Cities Adapt project) and guidance on adaptation planning. Currently, a rather small number of European cities and urban regional governments have developed or are in the early stages of implementing local plans for climate change adaptation (e.g. Copenhagen, London) (EEA 2012).

In response to the EU Green Paper "Adapting to climate change in Europe - options for EU action", in Romania, the Ministry of Environment and Forests developed in 2008 the "Guide on the adaptation to the climate change effects" (GASC), approved by Ministerial Order (no. 1170/2008). The guideline provides recommendations on measures aimed to reduce the risk of the negative effects of climate change in 13 key sectors. GASC identified urban areas among the priority sectors in which adaptation is a necessity, in order to reduce the risks for population, enterprises and infrastructure to climate-change-related impacts. The National Strategy on Climate Change of Romania (2013–2020) addresses two main existing challenges: the reduction of greenhouse gas emissions and adapting to climate change. However, the adaptation component of the strategy has no special references for cities.

Climate adaptation in Romanian cities is still at an early stage. However, there are some actions toward adaptations worth mentioning. The Romanian Municipalities Association (RMA) supports the involvement of municipalities in actions, initiatives and projects in the field of climate change. In 2009, RMA launched and supported signing of the "Municipalities commitment to fight climate change", a partnership agreement with the EU General Directorate for Energy and Transport, aimed to contribute to the achievement of European 2020 targets. Currently the agreement is signed by 35 cities.[1] In March 2014, a new EU Adaptation Initiative for cities was launched, encouraging local authorities to adopt local adaptation strategies and participate in awareness raising activities. In this framework, the Mayors Adapt Initiative (The Covenant of Mayors Initiative on Adaptation to Climate Change), is a flagship European initiative for cities on taking action on climate change mitigation. This initiative was developed in the framework of EU Cities Adapt project aiming to strengthen capacity building and provide assistance in developing and implementing local adaptation strategies for 21 European cities.

---

[1] http://www.amr.ro

The Sfântu Gheorghe town (Covasna County, Romania) was selected among these cities and currently has a draft adaptation strategy prepared.

The latest EU survey on Europeans' perception of climate change (2013[2]) shows a progressive decline of population concern of climate change issue in Romania relative to the earlier surveys (50 % 2013–2008[3]; 32 % 2013–2009[4]; 6 % 2013–2011[5]). Climate change is a threat currently ranking the third among the most serious issues (10 % share in the total number of answers) behind poverty, hunger and lack of drinking water (42 %) and the economic situation (30 %). However, the role of climate action in responding to the climate change-related targets of the Europe 2020 Strategy is currently well recognized in the country Partnership Agreement, which is setting up the development investments in Romania for the 2014–2020 period. Generally, the awareness on climate change of decision-makers improved in respect to the previous financial programming period (Partnership Agreement 2007–2013), reflecting a better mainstreaming of mitigation and adaptation in most operational programmes of this country.

This study presents the results of regional climate projections available from the EU FP7 ECLISE project and aims to investigate the future climate change signals for a medium-size urban area, Baia Mare, located in northwestern Romania, in relation to understanding the local stakeholders needs and interest in climate change issues as a precondition towards enabling concrete adaptation measures. The paper is organized in two main sections: (i) evaluation of changes in the mean climate and expected driven extreme events for the Baia Mare urban area in forthcoming decades, under different scenarios; (ii) overview of the results of the survey on stakeholder perception about climate change, their interest in climate information and data, and their knowledge and awareness in terms of climate change.

Study Area

BaiaMare is a medium-size town (123,738 inhabitants at the 2011 Census), with a very high population density of 530 inhabitants/km^2, located in the BaiaMare intra-Carpathian depression (north-western Romania) at an average altitude of about 228 m a.s.l. The town is surrounded by mountains to the north and north-east and hill alignments to the west and south (Fig. 16.1). The town is crossed by the Sasar River and some of its tributaries, exposing parts of the city to floods due to the insufficient flood protection infrastructure.

The town is the most important polarizing centre of Maramureş county, and has a tertiary-secondary economic profile, where industrial and construction services

---

[2] http://ec.europa.eu/public_opinion/archives/ebs/ebs_409_en.pdf
[3] http://ec.europa.eu/public_opinion/archives/ebs/ebs_300_full_en.pdf
[4] http://ec.europa.eu/public_opinion/archives/ebs/ebs_322_en.pdf
[5] http://ec.europa.eu/public_opinion/archives/ebs/ebs_372_en.pdf

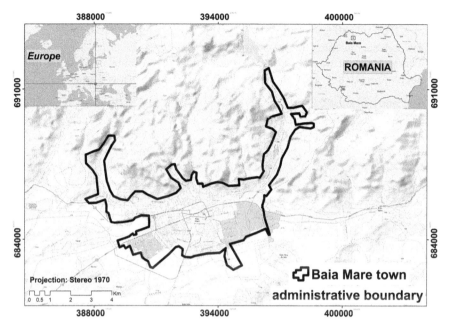

Fig. 16.1 Location of Baia Mare town

and activities prevail. Baia Mare plays a central role in the Baia Mare Urban System (established in 2006), stimulating the sustainable development in this intercommunity development association with 17 other bordering localities. It has a share of about 17 % in the total administrative surface of the Baia Mare Metropolitan area. The town is also a third order centre of regional importance in the North-West Development Region in terms of connectivity and accessibility.

The town lies in an area with poor agricultural resources, but with valuable non-ferrous ore deposits that made the mining industry a prominent economic activity. The environmental quality in the Baia Mare area has improved in recent years following the closure of mining operations in 1997 and the main sources of air pollution in the area (SC Romplumb, SA Baia Mare and SC Cuprom SA Baia Mare) in 2007–2008.

Dragotă et al. (2013) identified the major types of climatic hazards in the Baia Mare urban area (heavy rainfalls, hail storms, heat waves, fog, thermal inversions) and highlighted some relevant urban population exposure and social sensitivity indicators, as well the attributes of the existing limited adaptive capacity to these hazards. These included:

- the share of built-up areas in the total administrative Baia Mare Metropolitan area (30 %);
- a low share of urban green surfaces well below the EU recommendation (5.8 m^2/inhabitant, as compared to 26 m^2/inhabitant);

- low access to health infrastructure (3.02 doctors/1000 inhabitants, 0.38 pharmacies/1000 inhabitants);
- low average income of population;
- a large share of population groups vulnerable to warming in the total urban population (10.8 % aged above 60 years, 10.2 % aged below 10 years).

The study area has a annual mean temperature of 9.7 °C and a total precipitation of 873 mm, suggesting a cool and wet climate, specific to intra-Carpathian depression areas from northern Romanian Carpathians (ANM 2008). The instrumental measurements provide evidence that the climate of Romania grew warmer since the mid 1980s, with notable regional differentiation of magnitude (GASC 2008; Busuioc et al. 2010). Generally, the southern and south-eastern lowlands are the "hot spot" regions of this country, where the mean annual temperature increased with about 1 °C. According to GASC (2008), in Baia Mare the temperature increase was about 0.7 °C over the 1901–2000 period, with a great contribution of industrial activities and urban heat island effects, which accelerated the upward trends and exacerbated their effects. In spite of the clear ongoing warming and effects of associated extreme events in the urban area (e.g. heavy rainfalls of 2005; heat waves of 2012, 2015), actions to support climate change adaptation are not yet an investment priority in the sustainable development strategy of the Baia Mare town for the 2020 horizon.[6] However, the urban strategy foresees development investments to support climate mitigation e.g. development of green areas, increase of energy efficiency, improvement of transport infrastructure.

Local governance in terms of climate change-related issues is represented by several institutions subordinated to the national central administration: Baia Mare Municipality with a main responsibility in urban planning and development; Environmental Protection Agency in charge of environmental protection at county level, and monitoring of environmental quality; Civil Protection Inspectorate with a county-wide responsibility in prevention, monitoring and management of emergency situations, including the natural climate-driven hazards; Public Health Direction in charge of monitoring and protection of public health at county level, including the human health impacts associated to extreme weather events. Climate change-related aspects have not yet been considered in the local development plans promoted by these institutions, with the exception of the sustainable development strategy which includes some mitigation actions on energy efficiency and transport infrastructure.

---

[6] http://www.baiamare.ro/Baiamare/Strategia%20de%20dezvoltare/Strategia%20de%20Dezvoltare%20Baia%20Mare%202020.1.pdf

Future Climate in the Baia Mare Urban Area

Climate Data This study investigates the future trends in extreme temperature and precipitation over the twenty-first century at local scale, based on two multi-model ensembles available on a 25 km horizontal resolution. Four simulations from the EU FP6 Project ENSEMBLES (http://ensembles-eu.org), forced under the SRES A1B emission scenario, until the mid of the twenty-first century (2021–2050), have been used to detected the changes in temperature extremes. The second ensemble consists of three simulations from the CORDEX project (Coordinated Regional Climate Downscaling Experiment), under three new IPCC scenarios (RCP2.6, RCP4.5 and RCP8.5), covering two future time slices: 2021–2050 and 2071–2100. The CORDEX runs have been used to assess the changes in precipitation extremes. Table 16.1 summarizes the description of models participating in the two ensembles used in this study.

The performance of individual models was tested at monthly level against observation data from Baia Mare weather station (216 m a.s.l.) over the control climate period (1971–2000).

The change signals in climate extremes were relative to the 1971–2000 period and were addressed as differences of mean changes in annual and seasonal mean temperature (°C) and total precipitation (%) for the scenarios A1B, RCP4.5, RCP2.6 and RCP8.5.

Eleven climate extreme indices have been calculated on a daily maximum and minimum temperature and daily precipitation basis to assess the potential impacts of climate change in the Baia Mare urban area. A full description of the climate extreme indices used is presented in Table 16.2, where the selected indices can be broadly classified into four different categories using different calculation methods:

- *percentile-based indices* including the upper 95th percentile (e.g. HW),
- *threshold indices*, defined according to the national/international meteorological practice (e.g. Tmin0, Tmax30, Tmin20, R10mm),
- *absolute indices*, representing the maximum value within a month/year (e.g. R1day),
- *duration indices* define as intervals of extreme weather (e.g. HW, Rdry, Rwet).

The 95th percentiles used in the calculation of HW index were determined seasonally, using the daily maximum temperature values in the each time-series of the ENSEMBLES selected model simulations under the A1B, for both the baseline period and the 2021–2050 future interval. Two further indices were also studied to assess the frequency and magnitude of heat stress in relation to the health of the urban population:

- Tmax37, which is considering the 37 °C threshold in the maximum temperature values, according to the current Romanian legislation (Government Emergency Ordinance no. 99/2000);
- the combined heat stress index (CHT), which is a composite index between the number of tropical nights (Tmin $\geq$ 20 °C) and hot days (Tmax $\geq$ 35 °C), proposed by Ficher and Schar (2010).

Table 16.1 Overview of the simulations used to detect the future changes in climate extremes

Simulations	Short name	Spatial resolution	Scenarios	Time horizons	Climatic variables
ENSEMBLES					
CNRM_RM5.1_ARPEGE	CNRM-ARPEGE	25 km	A1B	1971–2000, 2021–2050	Maximum and minimum temperature
HAD_RM3Q0	HADRM				
SMHI_RCA3_BCM	SMHI-BCM				
SMHI_RCA3_ECHAM5	SMHI-ECHAM5				
CORDEX					
CNRM-CERFACS-CM5	CNRM-CM5	25 km	RCP4.5, 8.5	1971–2000, 2021–2050, 2071–2100	Mean temperature, precipitation
ICHER-EC-EARTH	ECE	25 km	RCP2.6, 4.5, 8.5		
MOHC-HadGEN2-ES	HC	25 km	RCP4.5, 8.5	1971–2000, 2021–2050, 2071–2095	

Table 16.2 Climate extreme indices for the Baia Mare urban area

Index	Short name	Definition	Relevance
Frost days	Tmin0	Annual number of days with Tmin ≤ 0 °C	Cold stress
Extremely hot days (summer)	Tmax37	Number of days with Tmax ≥ 37 °C in summer (cf. GO nr. 22/2000)	Heat stress
Tropical days (summer)	Tmax30	Number of summer days with Tmax ≥ 30 °C	
Tropical nights (summer)	Tmin20	Number of summer days with Tmin ≥ 20 °C	
Heat spells/ heat waves	HW	Spells of at least 3 (5) consecutive days when Tmax exceeds the 95th percentile	
Combined heat stress index (summer)	CHT	Summation of tropical nights (Tmin ≥ 20 °C) and hot days (Tmax ≥ 35 °C) in summer	
Cooling degree days	HDD	Seasonal summation of differences between the mean daily air temperature and the base temperature of 22 °C	Proxy for energy demand of urban population for adaptation to outdoor temperature variations
Dry days	Rdry	Number of days with daily precipitation <1.0 mm	Drying conditions
Wet days	Rwet	Number of days with daily precipitation >1.0 mm	Wetting conditions
Greatest 1 day precipitation	R1day	Maximum precipitation amount cumulated over a 24 h time span	Urban pluvial flooding
Heavy precipitation days	R10mm	Annual number of days with daily precipitation ≥ 10 mm	

Results of the Climate Simulations

Mean Changes in Temperature and Precipitation

For mean temperature, the CORDEX models exhibited a positive skill against observations at all time-scales (annual, seasonal, monthly), reproducing adequately the timing and magnitude of the annual cycle of mean temperature. The RCMs analyzed tended to underestimate the mean and variability of maximum temperatures from January to June, returning good results from July to December. Model simulations underestimated the minimum temperatures in winter, summer and autumn within acceptable limits (1.0 °C), but the cold bias exceeded 2.5 °C in spring, especially in April.

For mean precipitation, the models captured the prominent characteristics and the timing of the observed annual cycle, but tended to underestimate its magnitude. The models underestimated precipitation amounts from April to December (10–20 %) and particularly in summer (30–40 %), and showed excessive wetness in February and March (20 %). By model ensemble mean, there was a 15 % underestimation of the total annual precipitation amount in the area.

The projections showed a robust and statistically significant warming in the Baia Mare area by the end of 2100, continuing the general upward trends revealed from observations during the control climatology. Annually, the projected warming was in the range of 1.0–2.5 °C for RCP4.5, 1.6–4.2 °C for RCP8.5 and 1.0–1.3 °C for RCP2.6 (Fig. 16.2). Seasonally, warming was predicted to be stronger than at annual scale in winter and spring, particularly under RCP8.5 and RCP4.5, and weaker in autumn and spring, particularly under RCP2.6 (Table 16.3).

Simulations showed an asymmetric change in the variation of daily extreme temperatures by 2050 under the A1B scenario, suggesting that the urban area will exhibit a stronger warming in day-time (1.4 °C) than night-time (0.8 °C). Seasonally, the models project the greatest warming in summer (1.5 °C) and winter (1.4 °C) in terms of maximum temperature values, and in winter (0.9 °C) in terms of minimum temperature values (although this is similar to the warming of 0.7–0.8 °C projected for the other seasons).

By ensemble mean, the climate of Baia Mare urban area is expected to become wetter by the end of the twenty-first century under all the RCPs considered. This climate change signal is opposite to the observed trends over the control climate interval (1971–2000), which suggest a slight decrease in precipitation (not statistically significant). By 2050, the annual precipitation in the area is projected to increase more for RCP8.5 and RCP4.5 (21-22 %) and slightly less for RCP2.6 (19 %) (Fig. 16.3). By 2100, the increase of annual precipitation is up to 5 % larger

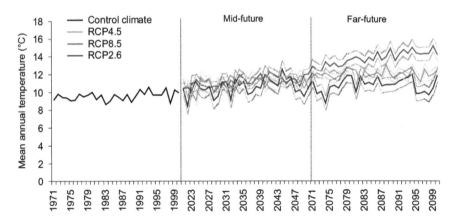

Fig. 16.2 Expected future variation of mean annual temperature in the Baia Mare area by 2100 under different RCPs. The solid lines depicts the ensemble mean, while the dashed lines mark the one standard deviation confidence interval

Table 16.3 Projected changes in seasonal mean temperature (°C) in the Baia Mare area

Future time slices	RCP2.6				RCP4.5				RCP8.5			
	DJF	MAM	JJA	SON	DJF	MAM	JJA	SON	DJF	MAM	JJA	SON
T2050	0.8	1.2	1.1	1.0	3.2	1.7	1.1	0.9	3.5	1.8	1.2	1.5
T2100	1.3	1.9	0.8	1.1	3.2	3	1.9	2	5.1	4.4	4.1	3.7

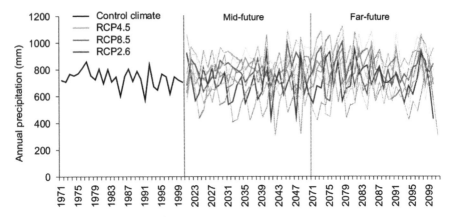

Fig. 16.3 Expected variation of annual precipitation in the Baia Mare area by 2100 under different RCPs. The solid lines depicts the ensemble mean, while the dashed ones mark the one standard deviation confidence interval

than the previous future period, ranging from 22.6 % (RCP8.5) to 26.1 % (RCP4.5). The upward trend in precipitation is generalised over the year for most of the selected model simulations. The positive precipitation changes are projected to be larger in the mid-future period (2021–2050) relative to the far-future ones (2071–2100), especially in winter (RCP4.5 and RCP8.5) and summer (RCP2.6) (Table 16.4). By 2100, precipitation is projected to increase with more than 25 % against the control climate in winter and autumn for RCP4.5 and RCP8.5, and in summer and autumn for RCP2.6.

Changes in Temperature and Precipitation Extremes

The analysis of the changes in the distribution of daily maximum and minimum temperature with respect to the control climatology suggest an enhanced daily temperature variability related to changes at both annual and seasonal time scales. The positive shifts in the upper tail of the underlying minimum and maximum temperature distribution indicate an increasing occurrence probability of warm extremes, especially in winter and summer. This future behaviour is also suggested by the changes in upper extreme quantile thresholds (P90 and P95), particularly of the maximum temperatures. The greatest shifts (2.3–2.6 °C) were detected in summer in July and August. For minimum temperature, the most important shifts do not exceed 1.6–1.7 °C (January).

The mid-term change signals suggest a decrease in the number of frost days (Tmin0), as well as a significant increase in the frequency of tropical nights (Tmin20), tropical days (Tmax30), extremely hot summer days (Tmax37) and heat waves (Table 16.5). This would result in a growing heat load of the Baia Mare town and an increasing exposure of the urban population to heat stress in the

Table 16.4 Projected changes in seasonal precipitation (%) in the Baia Mare area

Future time slices	RCP2.6				RCP4.5				RCP8.5			
	DJF	MAM	JJA	SON	DJF	MAM	JJA	SON	DJF	MAM	JJA	SON
P2050	21.5	13.6	36.1	43.8	23.8	14.5	19.7	20.3	26.8	8.2	20.9	24.2
P2100	23.8	12.2	53.6	45.3	28.0	15.9	22.4	32.3	28.7	19.5	11.3	26.2

Table 16.5 Key messages for urban adaptation to future changes in temperature and precipitation extremes in the Baia Mare town

Climate extreme indices	Projected change	Expected implications for urban community life and health	Other effects
Tmin0	Significant decrease. The decline is more evident in the spring (20–27 %) and autumn (20–40 %). In winter, the frequency of Tmin0 is likely to decrease with 7–14 % (up to 10 days)	Increase of winter bioclimatic comfort by the reduction of cold stress	Contribution to the reduction of snow cover duration, earlier snow-melt and runoff in the surrounding mountains
Tmin20	The area is not prone to tropical night occurrences in the current climatology (less than 1 day annual frequency). By 2050, a significant increase is likely under the projected summer warming trends (up to 7 days)	Increase of the night-time bioclimatic discomfort of urban population; Decisive impact on population health	Augmentation of heat island effects
Tmax30, Tmax37	Significant increase in the frequency of hot summer days in the area, from 1–3 days over 1971–2000 to 6–13 days by 2050 (70 %)	Increase of bioclimatic discomfort for outdoor activities; Aggravation of population health conditions, particularly of vulnerable groups (e.g. elderly people, people with chronic diseases, children)	Increase of thermo-convection associated atmospheric instability favouring the occurrence of torrential rainfall events and urban flooding; Augmentation of heat island effects; Contribution to air quality decrease
CHT	The area has a low exposure to heat stress and heat load due to its geographical location. Projections show a significant increase in heat stress exposure (80 %), with the greatest contribution on tropical nights. An expansion of the heat stress interval towards early autumn (September) is also expected in the area		
HW	Heat waves are sporadic in the control climate interval. A significant increase in heat wave duration (17 to 24 days) and frequency (3–21 days) at annual time-scale is expected in the area by 2050. Seasonally, the changes are statistically significant only in autumn		

CDD	The demand for cooling in the control climate period is rather low (centred only on July and August) due to the limitative effect of summer warming induced by the surrounding mountains. The annual CDD is rather low ranging between 3 and 6, with up to 100 % contribution of summer months. A significant increase in the annual CDD and an expansion of cooling period towards late spring (May) and early autumn (September) is expected by 2050, but mostly by 2100 in the area		Additional socio-economic challenges induced by a higher energy demand and consumption for urban cooling are likely to be related to lower productivity and failure of power transport networks,
Rdry	Low agreement between model simulations results: both increases (summer and autumn, under RCP4.5) and decreases are likely (winter and spring, under RCP8.5); most visible changes by 2021–2050 (decrease of 2.9-3.4 %)	Increase of bioclimatic discomfort during the warm periods associated with dry intervals	Intensification of thermo-convection associated atmospheric instability favouring torrential rainfalls in the months of the warm half of the year; Contribution to air quality decrease
Rwet	Increase under RCP4.5 in winter and spring (stronger by 2050) and a generalized decrease under RCP4.5 and RCP8.5, mostly in summer and autumn	Increase of bioclimatic comfort in the cold half of the year, by reduction of cold stress	Potential increase of urban flood and flash-flooding occurrence probability Increase in runoff from the surrounding mountains
R1day	Significant increase particularly under RCP8.5 (January, June, November) by 2071–2100; increase (March, November) and decrease (October, December) under both RCP4.5 and RCP8.5 by 2021–2050		
R10mm	Increase in most seasons under RCP4.5, stronger by 2021–2050; decrease in autumn and summer under RCP8.5, stronger by 2071–2100		

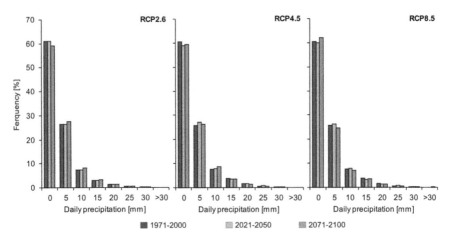

Fig. 16.4 Changes in the distribution of daily precipitation by the end of the 21st century under different RCPs in the Baia Mare area

future. The change signals in temperature extremes are particularly important for future climate change adaptation strategies as the town is located in an area less prone to severe heat stress due to the limiting effect on warming induced by the surrounding mountains.

The control climatology shows that the urban area exhibited a wet climate due to the prevailing westerly and north-westerly airflows. However, the occurrence probability of heavy precipitation events (>10 mm) was limited to less than 10 %. Figure 16.4 illustrates the changes in the distribution of daily precipitation for the two future periods under the selected RCPs. By 2100, the frequency of wet days (>1.0 mm) is expected to increase in both future periods, where this is accompanied by a decreasing frequency of dry days (<1.0 mm), except for summer. A slight increase in the occurrence probability of heavy precipitation days is likely by 2050. The projected changes in the distribution of daily precipitation by 2100 derived from CORDEX simulations suggest that the urban area will have to cope with a slight intensification of wet extremes under most RCPs, particularly in the mid-future term (Table 16.5).

Stakeholder Awareness of Climate Change and Implications for Effective Adaptation in the Baia Mare Urban Area

Stakeholder Interaction Approach Stakeholders are important actors in adaptation (Conde and Lonsdale 2005). There is a growing literature referring to stakeholders/users interaction in climate change adaptation (Shackley and Deanwood 2002; André et al. 2012), on the best methods for the participation process, or how the interaction with stakeholders should be planned or oriented. André et al. (2012)

developed a systematic method for identifying stakeholders using a variety of criteria, assigning them certain functions and capabilities in the climate change adaptation process. The basic questions in identifying relevant stakeholders can be reduced to "who", "what" and "how" in relation to climate change. "Who" refers to who are the representative groups at local and regional level who are either affected by climate change ("non-professional" group of stakeholders) or who make decisions regarding, or otherwise influence, the process of climate adaptation (professional stakeholders). "What" refers to what are these stakeholders needs and what is their adaptive capacity in the process. Finally, "how" refers to how the process and interaction should be oriented in order to be effective for both sides.

In 2012, several stakeholder-centred interviews were conducted to survey the perceptions on climate change held by some key local authorities, to understand the way in which they consider and include climate change aspects in their activities, as well as in the development strategies of the urban area, and to identify their expressed needs regarding climate change data and information. The study focused on professional stakeholders, either individuals or representatives of a group, that represent key local and regional institutions in a position to advance adaptation measures or to build capacity for urban adaptation. The selected stakeholders were involved in environmental and public health protection, urban development, disasters and risk management, and energy efficiency (Table 16.6). The stakeholders were identified using the "snowball method" starting from a key stakeholder (Baia

Table 16.6 Stakeholders in this study

Represented institutions	Sectors/professional activity field	Number of interviewed stakeholders	Contact
Baia Mare Municipality	Administration, urban planning and development	3	http://www.baiamare.ro/
Agency for Environmental Protection Maramureş	Protection of environment and monitoring of environmental quality	3	http://apmmm.anpm.ro/
County Inspectorate for Civil Protection	Prevention, monitoring and management of emergency situations	2	http://www.isumm.ro/
NGO "Baia Mare Metropolitan Area"	Promotion of investments and public services at a metropolitan level, for improving the quality of citizen's life and attenuating of social, economical and territorial disparities between member localities	1	http://www.zmbm.ro
County Energy Management Agency	Promotion of energy efficiency and renewable energy	1	http://www.amemm.ro/
Public Health	Monitoring and protection of public health	1	http://www.nodismed.ro/directia-de-sanatate-publica-dsp-maramures

Mare Municipality), who suggested other contacts that could be of interest for the study.

The main interaction tool was the individual or group interview. This tool was preferred instead of workshop meetings, as institutions proved more receptive after establishing a prior contact with their representatives and discussing relevant issues of concern for their field of activity. Moreover, previous experience with the other case studies of the FP7 ECLISE project showed that workshop discussions proved less successful in conducting perception surveys, and were more difficult to organize, as they required gathering the representatives of target institutions in a single day.

The semi-structured interview was organized in four main sections (Table 16.7) and designed to cover the following topics: identification of the local authorities perception on climate change, the way the institution views the importance of this topic and includes it in their activities, the perception regarding change in environmental quality in the urban area and its significance in terms of associated health impacts, the importance of climate services in their activities and the need for climate data and information.

Four interviewers participated in the discussions with the selected stakeholders. Around 1–3 representatives of each institution participated in the face-to-face interviews, the majority of the interviewees having a leading role within their organisations (head of departments, directors, technical experts). The respondents' answers were transcribed by the interviewers by filling in a pre-designed questionnaire form, following the four sections presented in Table 16.7. The answers were finally summarised and organised to draw broad conclusions of the survey.

During the interviews, tailored climate change data and information were presented and discussed with the stakeholders in respect to the observed climate trends and changes in the frequency and intensity of some climate extremes (e.g. storms, floods, heavy rainfall, heat waves), derived from the local and regional instrumental observations. The objectives, workpackage activities and an overview of the regional climate simulations of the FP7 ECLISE project were also presented. Communications continued after the interviews through telephone or email to

Table 16.7 General structure of the individual interviews organized with representative stakeholders in Baia Mare

I. General framework of climate change as perceived by institution's experts
– What is climate change, observed effects at local level, extreme events, how the town will be affected by climate change, opportunities/threats
II. Institutional responsibilities to tackle climate change
– How the institution consider climate change aspects in their activities, dedicated projects/programmes, the access to funds, the needed resources, specialised job positions
III. Current environmental state of the urban area
– The level of pollution, pollution sources, specific health problems associated to pollution
IV. Interest in terms of climate services
– What are "climate services", particular interest to climate data, the future time frame of interest, climate adaptation

exchange information and to receive feed-back from the participants regarding the climate data. At the end of the project, a synthesis report with relevant findings of the study was sent to each of the contacted stakeholders.

Results of Interaction with Stakeholders

A synthesis of the key issues derived from the discussions held with professional stakeholders in the Baia Mare Metropolitan Area is shown in Table 16.8. In general, all interviewees displayed a high interest in the topic of climate change and potential impacts on the urban area, although neither the consequences of climate change nor the way that adaptation measures should be orientated were exactly known. The representatives of the interviewed institutions perceived that climate had changed in recent years, in terms of increasing temperatures, different rainfall patterns, less snow in winter or season shrinking. Although climate change was considered relevant for the urban area (mainly in the last few years where the extremely hot 2012 summer was somehow a turning point in orienting perceptions towards the issue), at an institutional level climate change has not been considered a priority and there are not as yet specialized departments (job positions) to tackle this

Table 16.8 Key issues of the interviews with local stakeholders in the Baia Mare town

I. General framework of climate change as perceived by institution's experts
– Climate is perceived to have changed in the last 20–30 years (since 1985–1986): increase in temperatures during the summer, torrential rainfalls, seasonal shrinking, less solid precipitation, decrease in air moisture, drought, 2012 summer—extremely hot
II. Institutional responsibilities to tackle climate change
– Institutions are poorly aware of the potential impacts of climate change and associated extremes in the area, but show a real interest for future climate estimations necessary to undertake reliable adaptation measures. – At an institutional level, specialized departments (job positions) do not exist to tackle or manage climate information and climate-related aspects, where this has not been perceived as a pressing or priority issue for the town – There are several projects in implementation referring to greening the town, renewable energy, and waste management – A need for increasing the green spaces inside the town, for thermal building rehabilitation, promoting public transport, traffic decrease in the town centre, waste management, noise reduction
III. Current environmental state of the urban area
– A general improvement of air quality as a result of ceasing the mining industry, increasing the life expectancy, the most important problems: dust blown from the waste deposits, soil contamination, burning of fossil fuels for thermal heating, urban traffic, waste management; specific diseases trigger by the mining history of the town
IV. Interest in terms of climate services
– Usually short-term interest (next 5–10 years), territorial planning and strategies, driven by the need to consider climate change aspects at a regional level, otherwise local strategies would not tackle the issue; long-term energy strategy; hazard mapping

issue. A general improvement in the quality of the environmental was reported for Baia Mare, a fact outlined by all respondents, although some existing neighbourhoods (e.g. Ferneziu neighborhood) lack in appropriate urban infrastructure and green spaces. All the stakeholders perceived that increasing the green infrastructure in the town, promoting public transport, decreasing traffic in the city centre and associated noise reduction, and waste management are important actions that should be considered in the future development plans of the urban area, where these actions relate to both climate change adaptation and mitigation.

The interviewed organisations do not use raw climate data in their activities and they tend to be more concerned about the impact of some extreme events (floods and drought). Climate data and information are of interest to them in terms of short-medium term climate information (next 5–10 years), the supplied information having a potential use in the local planning strategies (Table 16.9). Although there was generally high interest in climate change information, some stakeholders

Table 16.9 User needs for Baia Mare case study

Users	Climate information needed	Time horizon	Potential climate actions based on the requested climate services
Baia Mare Municipality	General information on climate variability and change in air temperature and precipitation, with a focus on the associated extremes	Near-future (the next 5–10 years)	Short, medium and long-term urban planning
Agency for Environmental Protection Maramureş	General information on climate variability and change in air temperature and precipitation, with a focus on the associated extremes	Near-future (the next 5–10 years) and mid-future (2021–2050)	No actions
County Inspectorate for Emergency Situations	General information on climate variability and change and associated weather extremes	Near-future (the next 5–10 years) and mid-future (2021–2050)	Extreme events management (particularly floods)
Public Health Agency	Climate information on the projected range of change in the frequency, duration and intensity of some weather extreme events posing threats to population health	Near-future (the next 5–10 years) and mid-future (2021–2050)	No actions
Energy Efficiency Agency	Expected seasonal temperature change in the urban area in relation to the energy demand/consumption in the urban area	Near-future (the next 5–10 years) and mid-future (2021–2050)	Mid-term and long-term energy efficiency strategy
Metropolitan Association of Baia Mare (NGO)	General information on climate variability and change and associated weather extremes	Near-future (the next 5–10 years) and mid-future (2021–2050)	No actions

recognised that presented data may not find future applicability in their activities as they don't have specialised people to further use the data's results:

> Interviewee X: *"I don't know what to do with the data you are presenting, this is not my job in my organisation and I don't know any other person who can handle them......"*

As in other studies conducted in different countries (Shackley and Deanwood 2002), there was a clear demand from stakeholders for high resolution climate projections at the regional and even sub-regional level, with a focus on extreme events and their particular effects on the urban area. The strongest request for this was from the civil protection, who needs to elaborate detailed hazard and risk maps for preparedness activities in case of extreme events. This is to be expected, as these stakeholders have a responsibility for decision-making and implementation at local level, and wish to reduce uncertainties to a minimum for long-term planning. They see the scientific community as a support for that and are interested in being provided with tailored climate information, although not necessarily for well-defined purposes or to meet specific needs.

The interviewed potential users of climate data did not express specific needs in terms of climate data and information, as they did not have the requisite knowledge and expertise. In this respect, the involved scientists tried to provide a general background in terms of the climate change topic and also to discuss potential impacts for the urban area. On this subject, the researchers tried to provide some consistent messages in terms of climate change in the region based on existing literature as well as results of climate simulations of the project. Based on these sources, the main climate-related messages for the region presented to stakeholders referred to: the projected increase in mean, maximum and minimum temperatures, the projected increase in precipitation and change in rainfall patterns, and the projected increase in evaporation and specific humidity, as well as the potential implications for human health and urban management.

During the individual interviews, the discussions on the presented climate information were somewhat limited, with no discussion of model skills or uncertainty aspects. However, when presenting the data to a larger group of stakeholders, the discussions were more intense in terms of trying to understand the climate data and their reliability, but also in identifying some options and measures for adaptation, which could be implemented considering the institution's mission and expertise. Addressing local climate aspects is a volunteer task for the institutions and depends very much on the expertise and interest of people working in the institutions, thus in most cases, climate aspects can be ignored and it is less likely that the supplied climate data and information will have the impact to influence a decision at local level. Indeed, as long as the issue is not pressing, it is not deemed important, considering that institutions usually face other pressing problems (e.g. infrastructure development, social problems, road traffic etc.) that require almost immediate decision and action. A single initiative to provide climate data (and thereby promote action) is therefore unlikely to produce long-term effects and structural changes, unless the climate change issue becomes an obligatory aspect to be considered at local level. Financial constraints were also mentioned by all the involved organisations, with requests for a certain degree of financial help and guidance from the leading national institutions:

> Interviewee Y: *"To consider climate change issue is not compulsory in the annual reports we need to prepare for the central authority. However, we took the decision to include some information based on available literature. We will of course consider this more if it will be obligatory required from our leading institution"*.

However, an essential step that any organisation should take to respond to climate change effectively is to understand the problem and the potential implications in order to identify and engage with key people and organisations within the community (Defra 2010). There are numerous examples of what communities in other countries have already done, which can be used and adapted to the local particularities (Prutsch et al. 2010; Sharma and Tomar 2010; Hardoy and Ruete 2013). It is recognised that good practice examples help in raising awareness for adaptation planning and in motivating stakeholders towards action (Prutsch et al. 2010). During the discussions, the involved stakeholders recognised the need to exchange information and ideas on the topic with peer-towns from Europe, and to have some good practice examples in an advanced situation that had implemented specific adaptation actions. Currently, they have limited knowledge and interest in promoting climate change actions, although the municipality plans some actions in the sustainable development strategy that cover climate change mitigation and adaptation, promoted mainly as a response to increasing the environmental quality in the urban area (e.g. development of green areas, increase of energy efficiency, improvement of transport infrastructure).

In summary, the survey for the Baia Mare town proved to be successful in terms of changing and orienting participants' perception and awareness in terms of climate change and increasing their knowledge on the issue. A stakeholder mentioned that his institution will try to publish a summary of the ECLISE project results on their webpage or in their informative journal, although this has not happened as yet. However, it is less likely that further concrete action on the stakeholders part will follow after this survey, as none of them seemed to be pressed by any immediate decision or need. One of the most important outcomes of the interaction was that these organisations are now aware that there are research institutions in the country that study the topic and may offer them guidance and assistance, if needed.

Conclusions

Considering the cumulative effects of temperature and precipitation changes at annual and seasonal time-scales, particularly during the extreme seasons, the climate of the Baia Mare urban area is expected to become far warmer (especially on a far-future term) and wetter (mainly by 2050) than the control climate interval. In winter, the projected temperature and precipitation trends are expected to favour an increasing frequency of liquid and mixed precipitation falls to the detriment of solid precipitation falls. Spring climate change signals are expected to determine earlier snowmelts in the surrounding mountains, as well as an increasing frequency of 'rain-on-snow' events, contributing to early runoff and spring flooding, with potential impacts on urban flooding (the Săsar River is crossing the centre of the town and flooded parts of the urban area in April 2005). Despite the moderate magnitude of changes in summer mean temperature and precipitation compared to other seasons, the evaporation and specific humidity are expected to significantly increase in this season, posing threats to urban population health and adaptation challenges with respect to a lower bioclimatic comfort, particularly during outdoor activity intervals. The magnitude of the expected changes in temperature extremes are likely to be exacerbated locally by the effects of urbanization (e.g. the urban heat island).

Under a warmer and wetter urban climate, an increase in the frequency of river floods (the Săsar river), urban drainage flooding and flash flooding (on some streams flowing through the neighbourhoods of northern and south-eastern parts of the town) is expected. Moreover, considering the historical mining tradition of the area, future precipitation changes might have implications for the stability of the limitrophe mining sites (e.g. waste dumps, tailings impoundments).

Some local characteristics of the Baia Mare town (e.g. the large share of impervious areas due to urbanization) are expected to amplify the potential impacts from projected increases in the frequency of wet days and heavy precipitation events, particularly in the residential, commercial and public buildings areas. The projected changes in the frequency of extreme precipitation events (e.g. heavy precipitation days) could significantly contribute to the depletion of the urban drainage systems, in lower lying urban areas combined with sewer overflow (urban drainage flooding).

A first and important step towards adaptation of small urban areas is increasing the awareness of climate change among municipal leaders who may then drive structural changes at an institutional level to address climate change. The current level of awareness for the Baia Mare urban areas is relatively low, where the actions taken to date by the involved organisations cover mainly environmental quality aspects, and have been less directed towards climate change adaptation. Although municipal leaders showed a good interest in climate information and data, it is considered less likely that immediate future actions will be taken, as climate change is not yet considered a pressing issue for the town. Most interest was shown to the potential impacts of climate change at a local level, with a focus on floods and drought. The impacts of climate change are place-specific, thus involving local actors in the adaptation process requires tailored and downscaled information at a local level to meet, attract and inform stakeholder needs and lead to a better understanding of the future challenges and the need for action. As small communities are facing a wide range of social, economic and environmental stresses, climate change come as an additional pressure, and is usually seen as none critical and none immediate. A more careful consideration of the issue will likely require a certain degree of obligation or financial support from the government or other leading institutions. Effective adaptation needs to be supported by relevant policies and guidelines for local communities, which may identify opportunities to include climate issues in the existing local plans and developments with low financial and human efforts.

Acknowledgements This work was supported by the research project FP7 ECLISE (Enabling Climate Information Services for Europe), no. 265240 (2011–2014), coordinated by the Royal Netherlands Meteorological Institute (KNMI). The authors acknowledge the access to ENSEMBLES and CORDEX climate projection data kindly provided by the Swedish Meteorological and Hydrological Institute (SMHI).

References

Agrawal A (2008) The role of local institutions in adaptation to climate change, social dimensions of climate change. Social Development Department, The World Bank, Washington, DC
André K, Simonsson L, Swartling AG, Linnér B (2012) Method development for identifying and analysing stakeholders in climate change adaptation processes. J Environ Policy Plann 14(3):243–261
ANM (Administraţia Naţională de Meteorologie) (2008) Clima României. Editura Academiei Române, Bucureşti, p 365
Busuioc A, Caian M, Cheval S, Bojariu R, Boroneanţ C, Baciu M, Dumitrescu A (2010) Variabilitatea şi schimbarea climei în România. Editura Pro Universitaria, Bucureşti
Carmin J, Roberts D, Anguelovski I (2012) Preparing cities for climate change: early lessons from early adaptors. In: Hoornweg D, Freire M, Lee MJ, Bhada-Tata P, Yuen B (eds) Cities and climate change: responding to an urgent agenda, vol 2. World Bank, Washington, DC, pp 470–501
Carter J, Cavan G, Connelly A, Guy S, Handley J, Kazmierczak A (2015) Climate change and the city: building capacity for urban adaptation. Progr Plann 95:1–66
Cohen S (2011) Overview: climate change adaptation in rural and resource-dependent communities, climate change adaptation in developed nations. Adv Glob Change Res 42(2011):401–412
Conde C, Lonsdale K (2005) Engaging stakeholders in the adaptation process. In: Lim B, Spanger-Siegfried E (eds) Adaptation policy frameworks for climate change: developing strategies, policies and measures. UNDP, Cambridge University Press, Cambridge, pp 47–66
de Boer J, Wardekker JA, van der Sluijs JP (2010) Frame-based guide to situated decision-making on climate change. Glob Environ Change 20:502–510
Defra (2010) Adapting to climate change: a guide for local councils. Department for Environment, Food and Rural Affairs, UK
Dragotă C, Grigorescu I, Sima M, Kucsicsa G, Mihalache S (2013) The vulnerability of the Baia Mare Urban System (Romania) to extreme climate phenomena during the warm semester of the year. In: Proceeding of the international conference air and water. Enviromental Components, pp. 71–78
EEA (2012) Urban adaptation to climate change in Europe. Challenges and opportunities for cities together with supportive national and European policies, Report no. 2. Copenhagen, p 143
EU (European Union) (2013) Climate change adaptation: empowerment of local and regional authorities, with a focus on their involvement in monitoring and policy design, Local Governments for Sustainability, European Secretariat (ICLEI) and CEPS (Centre for European Policy Studies). doi:10.2863/92867
Ficher EM, Schar C (2010) Consistent geographical patterns of changes in high-impact European heat waves. Nat Geosci 3:398–403

Ford J, Berrang Ford L (eds) (2011) Climate change adaptation in developed nations from theory to practice. Springer Science and Business Media, Heidelberg

Guideline for Climate Change Adaptation (GASC) (2008) Ministry of Environment and Forests, p 40. http://www.meteoromania.ro/anm/images/clima/SSCGhidASC.pdf. Accessed 22 Jul 2015

Hardoy J, Ruete R (2013) Incorporating climate change adaptation into planning for a liveable city in Rosario, Argentina. Environ Urban 25:339–360

Hartmann T, Spit T (2014) Capacity building for the integration of climate adaptation into urban planning processes: the Dutch experience. Am J Clim Change 3:245–252

IPCC (2012) Managing the risks of extreme events and disasters to advance climate change adaptation. In: Field CB, Barros V, Stocker TF, Qin D, Dokken DJ, Ebi KL, Mastrandrea MD, Mach KJ, Plattner G-K, Allen SK, Tignor M, Midgley PM (eds) A special report of working groups I and II of the Intergovernmental Panel on Climate Change. Cambridge University Press, Cambridge, New York, NY, p 582

Mehrotra S, Natenzon CE, Omojola A et al (2009) Framework for city climate risk assessment. Commissioned research, World Bank Fifth Urban Research Symposium, Marseille, France

Prutsch A, Grothmann T, Schauser I, Otto S, McCallum S (2010) Guiding principles for adaptation to climate change in Europe, ETC/ACC (The European Topic Centre on Air and Climate Change) Technical Paper 2010/6

Rosenzweig C, Solecki WD, Hammer SA, Mehrotra S (2011) Climate change and cities, the first assessment report of the urban climate change research network. Cambridge University Press, Cambridge, p 281

Shackley S, Deanwood R (2002) Stakeholder perceptions of climate change impacts at the regional scale: implications for the effectiveness of regional and local responses. J Environ Plann Manage 05/2002; 45(3):381–402.

Sharma D, Tomar S (2010) Mainstreaming climate change adaptation in Indian cities. Environ Urban 22(2):451–465. doi:10.1177/0956247810377390

The Sustainable Development Strategy of the Baia Mare Town 2020 (http://www.baiamare.ro/Baiamare/Strategia%20de%20dezvoltare/Strategia%20de%20Dezvoltare%20Baia%20Mare%202020.1.pdf. Accessed 22 Jul 2015

The World Bank (2011) Guide to climate adaptation in cities. The International Bank for Reconstruction and Development, 106 p

UNEP (2012) UNEP Year Book: Emerging issues in our global environment. UNEP Division of Early Warning and Assessment, United Nations Environment Programme, 80 p

Chapter 17
Trends and Issues of Climate Change Education in Japan

Keiko Takahashi, Masahisa Sato, and Yasuaki Hijioka

Abstract This paper focuses on Climate Change Education for Sustainable Development, a flagship initiative of the UN Decade of Education for Sustainable Development. We attempt to identify trends and challenges for climate change education (CCE) programs currently underway in Japan and to provide recommendations for improving such programs by comparing a number of specific programs. With regard to the skills expected to be acquired through CCE programs, the paper relies on the literature review and classification of "key competencies in sustainability" employed by Wiek et al. (Key Competencies in Sustainability: a Reference Framework for Academic Program Development. Springer, 2011). In addition, regarding the content and methods of CCE programs, this paper organizes the connections between issues identified in the IPCC's Fifth Assessment Report (AR5) and key competencies in sustainability, and provides recommendations regarding the framework and educational methods required for capacity development programs based on previous research by two of the authors (Sato and Takahashi, Capacity Development Programme for Climate Change Education (CCE) in the context of ESD, based on the International Discussion on "Key Competencies in Sustainability" and the IPCC Fifth Assessment Report. J Energy Environ Educ 2015).

Among the issues revealed by our investigation are the lack of a common definition of CCE or established content or methods for CCE programs in Japan and a weighting of instruction towards knowledge which does not employ methods intended to develop key competencies in sustainability, with general suggestions to improve on current circumstances in Japanese CCE also provided.

Keywords Climate change education • Education for sustainable development (ESD) • Key competencies in sustainability • Comparative studies of selected cases

---

K. Takahashi (✉) • Y. Hijioka
Center for Social and Environmental Systems Research, National Institute for Environmental Studies, 16-2 Onogawa, Tsukuba, Ibaraki 305-8506, Japan
e-mail: takahashi.keiko@nies.go.jp

M. Sato
Tokyo City University, Tokyo, Japan

Introduction: Definition of Climate Change Education

In 2009, in the second half of the UN Decade of Education for Sustainable Development (DESD), the United Nations Educational, Scientific and Cultural Organization (UNESCO), the lead agency for DESD, proposed CCE as a flagship initiative, and climate change is specified as a key area of ESD in the subsequent decade, and is implementing the Climate Change Education for Sustainable Development initiative. The initiative defines adaptation and mitigation within CCE in the context of ESD not as two discrete activities but as essential elements to be linked, and as processes through which learners acquire the necessary knowledge, skills, and confidence (UNESCO 2013) to address climate change.

The *Kankyō Kyōiku Jiten* [Encyclopaedia of Environmental Education] (Japanese Society of Environmental Education [JSOEE] 2013), classifies environmental education in Japan into 18 categories, including conservation education; energy environmental education; and development education. However, no reference is made to CCE—indicating that CCE is not established as a field of environmental education.

Trends of Climate Change Education (CCE) in Japan

The publication of *Kankyō Kyōiku Shidō Shiryō* [Teacher's Guide for Environmental Education] (Ministry of Education 1991) can be said to be one of the initial catalysts that led to inclusion of global warming as a theme in Japanese environmental education. The development of the teacher's guide fostered a shared awareness regarding global environmental problems among all countries of the world, most prominently of global warming (Sajima 2007). This in turn led to the incorporation of instruction related to the limited nature of resources, energy conservation, and recycling in school education (JSOEE 2013).

The legal underpinnings for implementation of education on the prevention of global warming were subsequently established in October 1998 by the "Act on Promotion of Global Warming Countermeasures" (hereinafter, "Act on Global Warming"), which stipulated the development of a cadre of Climate Change Action Officers (hereinafter "action officers") to disseminate and implement knowledge regarding the current status and countermeasures to global warming at the local level; the creation of Prefectural Promotion Centers for Climate Change Action (hereinafter "prefectural centers") to support such activities; and the establishment of a national organization (Japan Center for Climate Change Actions, hereinafter "JCCCA") to support the activities of action officers and prefectural centers. As a result of this Act on Global Warming, there are presently approximately 7000 action officers throughout Japan, supported by 55 prefectural centers.

With the entry into force of the Kyoto Protocol in 2005, Japan initiated numerous citizen PR and outreach initiatives on climate change countermeasures. These continue today, as part of, for example, the climate change action campaign known as "Fun to Share". Based on this campaign, the government launched the IPCC Report Communicator initiative to heighten public awareness of anti-global warming actions in 2013 and communicate the status of climate change, ultimately encouraging adoption of low-carbon practices including "Cool Biz".

As described above, action officers, prefectural centers, and the JCCCA, established under the direction of the national and local governments, have played a central role in implementing CCE in Japan. In addition, environmental NGOs and NPOs, which ramped up their global warming-related activities on the occasion of the 1997 Kyoto Protocol Climate Conference, COP3, began to offer programs targeting the general public, in addition to instruction in formal educational settings.

Current Situation of CCE in Japan

The Act on Global Warming states that the role of action officers is to "[promote activities] for regional dissemination of knowledge regarding the current situation of global warming and related countermeasures" and to "[deepen] understanding of the general public concerning the current situation of global warming and the importance of global warming countermeasures." It further stipulates that prefectural centers and the JCCCA are to perform public awareness and public relations campaigns to increase public awareness (Ministry of Internal Affairs and Communications 2015). These provisions situate the significance of CCE largely in increasing public awareness of the current status—and knowledge regarding—global warming over its educational value.

Purpose and Target of This Study

The purpose of this study is to examine CCE programs currently being carried out in formal and non-formal educational settings to identify trends and issues of such programs by comparing the programs (1) educational characteristics, (2) evaluation status, (3) climate change-related content, and (4) implementing scheme. We identified and organized skills to be acquired through CCE programs based on the literature review and classification employed by Wiek et al. (2011). In addition, we compared the educational content and methods of CCE programs based on previous research by the authors (Sato and Takahashi 2015), which references work by Hoffmann (2014a), that organizes the connections between issues identified in the IPCC's Fifth Assessment Report and key competencies in sustainability and provides recommendations regarding the framework and educational methods required for capacity development programs.

Method

We conducted semi-structured interviews with the directors of major CCE programs currently being offered in Japan, selecting formal and non-formal education programs that have climate change or global warming as a main theme and are jointly planned and implemented by various organizations from programs mentioned in the proceedings of the 25th Meeting of the JSOEE (2014) in Tokyo, which presents examples of various environmental education or ESD programs; the Ministry of the Environment (MOE) 2013 ESD Environmental Education Program Guidebook; and the 2014 Local ESD Environmental Education Program Guidebook (MOE 2014).

The JSOEE meeting proceedings cover environmental education programs on diverse topics, currently offered in Japan. Of the programs mentioned in 165 oral presentations and 41 poster presentations, five programs from oral presentations and two from poster presentations were found to meet the above-mentioned selection criteria. Excluding programs appearing in both oral and poster presentations, the pool of potential programs was narrowed to five. Of these five programs, three were excluded as being either insufficiently complete as a program; or focused on the development of instructional material or the reporting of tools being used. Ultimately, two programs were selected as interview targets: ① a JCCCA program, and ② a program in Iida City, Nagano Prefecture on local impacts of climate change and mitigation/adaptation.

The 2013 ESD Environmental Education Program Guidebook (MOE 2013) contains examples of model environmental education programs, 20 of which were included in the pool of potential programs. Of these one was found to meet the selection criteria: ③ the Fifty/Fifty Project. The 2014 Local ESD Environmental Education Program Guidebook (MOE) summarizes local ESD programs tailored to the unique features of the natural environment, history, and culture of each region in each of the 47 prefectures, one of which was found to meet the selection criteria: ③′ The Fifty/Fifty Project implemented in Hokkaido, with the overlap in ③ and ③′, meaning that, in practical terms, three programs were selected for comparison. Two other programs (④ Children's Eco-life Challenge and ⑤ IPCC Report Communicator) were included in the sample based on recommendations from informants. The JCCCA offers 24 different programs, and we included the longest running and most popular program, "Can you carry a bag of Energy?"

Interviews were conducted from December 2014 to January 2015. Responses related to (1) CCE program overview, (2) educational characteristics, (3) aspects of climate change dealt with in the program, (4) status of program evaluation, and (5) program implementing structure were recorded (Table 17.1). For the ESD-related elements and key constructs for sustainable societies included in the target skills and attitudes and the educational content of the programs, we referred to the framework deemed necessary for creating and executing ESD courses of study by the National Institute for Educational Policy Research (2014).

Table 17.1 Interview items

		Details
1	CCE program overview	Program name
2		CCE program administrative body
3		Target audience, relationship with ministry's curriculum guidelines
4		Duration
5		Venue
6	Educational characteristics	Overall program goal
7		Individual program objectives
8		Skills and attitudes involved as program objectives 1. Critical: competence of critical thinking 2. Future: competence of future image anticipation, and planning 3. Multi-dimensional: competence of multi-dimensional and holistic thinking 4. Transmission: competence of communication 5. Cooperation: behavior of cooperation with others 6. Relation: behavior of respecting connections 7. Participation: behavior of active participation 8. Others
9		ESD elements included in learning contents (Framework concepts for society of sustainable development) 1. Diversity 2. Interdependence 3. Finiteness 4. Equity 5. Cooperation 6. Liability
10		Educational method (Methods and effect)
11		Structure of the program (Devised points for the structure, and reasoning)
12		Teaching materials (Teaching material content for program, and effect)
13		Required skills and knowledge level of trainers
14	Learning contents	1. How did you select the learning contents? 2. What were the foremost criteria when deciding on learning contents? 3. Presence or absence of persons involved in selecting learning content, and their role. 4. What were the foremost criteria when deciding the learning level? 5. Information resources on climate change, and ensured accuracy of the information
15	Evaluation	Type of evaluation (What were the foremost criteria in ensuring the effective implementation of your program?)
16	Implementation structure	Presence or absence of collaboration among municipalities, other organizations etc. and its content
17		Presence or absence of sharing contents of the program with other organization, and nature of shared content
18		Presence or absence of issues in implementing the program, and their content

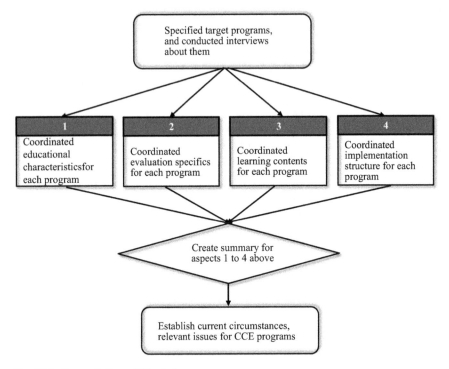

Fig. 17.1 Research flow of this study

The flow of this study is as shown in Fig. 17.1. The educational content of each program was compared based on the authors' previous research (Sato and Takahashi 2015) on the framework and learning methods required for CCE programs (Fig. 17.2).

With regard to the educational content of CCE, we believe that the causes, impacts, projections, and countermeasures to climate change comprehensively described in AR5 should be taught within an ESD framework. In addition to (1) identifying and organizing the educationally-relevant issues identified in AR5 based on the Summary for Policy Makers (IPCC 2013, 2014a, b), which outlines the main points of the AR5, and clarifying the areas of overlap between CCE in the context of ESD and the five key competencies in sustainability (Wiek et al. 2011), the authors (2) identified and organized the items to take into consideration when developing CCE programs (examples of competencies to be acquired and learning methods) in Sato and Takahashi (2015).

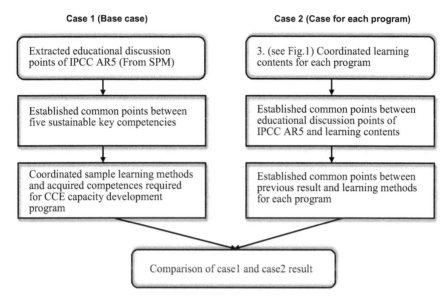

Fig. 17.2 Learning content comparison flow

Key Competencies in Sustainability: A Reference Framework for Academic Program Development (Wiek et al. 2011)

The literature review and classification proposed by Wiek et al. (2011) (hereinafter, the "aforementioned research") (1) identified research relevant to key competencies in sustainability, (2) created a coherent framework for synthesizing the identified competencies, and (3) attempted to address the greatest gaps in the conceptualization of key competencies. The aforementioned research defined "key competencies in sustainability" and developed a "coherent framework for sustainability research and problem solving" based on international discussion on sustainability and a review of previous research. Based on an analysis of these concepts, five key competencies were proposed: ① systems-thinking competence (ability to comprehensively analyze complex systems across different domains, such as society, environment, economy, etc. and different scales, from local to global), ② anticipatory competence (ability to comprehensively analyze, evaluate, and formulate rich "pictures" of the future in connection with problem-solving frameworks and sustainability-related issues), ③ normative competence (ability to comprehensively map, specify, apply, reconcile, and negotiate the values, principles, goals, and objectives of sustainability), ④ strategic competence (ability to comprehensively design and implement governance strategies to encourage intervention, transition, and transformative governance strategies towards sustainability), and ⑤ interpersonal competence (ability to motivate, enable, and facilitate collaborative and participatory sustainability research and problem solving).

Wiek et al. (2011) described problem solving as a stepwise process and proposed key competencies as stepwise processes interlinked within a coherent framework. We believe this framework is necessary for acquiring climate change-related problem solving skills in stepwise fashion and adopted it as a fundamental analytical concept. As the aforementioned paper does not propose learning methods specific to climate change as a means of acquiring each competence, we referenced the learning methods implemented by Hoffmann (2014b) to evaluate each CCE program.

Result of Interview Survey

Below are summaries and unique features of each program.

Case 1: The Fifty/Fifty Project (Implementing Organization: FoE Japan)
This project, based on the principles underlying the energy conservation Fifty/Fifty Project introduced in schools in Germany, enables participants to learn while experiencing hands-on the range of energy conservation-related processes, including the development of concrete methods for energy conservation; discovery of areas to be improved; development of tools for increasing awareness; implementation of energy conservation means; evaluation of impacts and challenges; and reporting of results. In this project, teachers serve as instructors, while NPOs take on the role of advisor. The 11-h program geared towards junior high school students is designed to promote hands-on learning on energy and environmental issues.

Case 2: Can You Carry a Bag of Energy? (Implementing Organization: JCCCA)
This program geared towards students in grade 4 and higher, is a one-time (approximately 30 min) participatory program to develop citizens capable of engaging in anti-global warming activities. The instructional materials comprise bags of different weights designed to get participants to experience differences in energy quantity as differences in weight, to encourage participants to grasp (realize) the reasons behind such differences and think about their relationships to energy. The program encourages proactive participant learning by combining (1) hands-on experience, (2) diverse movement, and (3) a developing of the participants' own themes. The program places value on "understanding (realization)" and adopts a quiz format. Although multiple stakeholders are involved in program development, the involvement of teachers and others directly familiar with the field is low.

Case 3: Children's Eco-Life Challenge (Implementing Organization: Kiko Network)
This program, geared toward 4th through 6th grade students in Kyoto City, deals with global warming prevention and was jointly designed and launched in 2005 in partnership between environmental NPOs, Kyoto City, and elementary schools. The program is conducted by NPO members and consists of three stages

emphasizing, respectively, learning, practice, and reflection. Specifically, these entail pre study (2 h); implementation (during summer or winter vacation); and reflective study. Following the program the application, continuation and development of things learned during the program is emphasized (focus on continued implementation). The program was planned and designed in consultation with the Board of Education and teachers; its content is tailored to the students' developmental stage with the overall objectives of not only facilitating learning about climate change related problems, but also encouraging students' continuous behavior to prevent global warming. The program employs a quiz format and group work that involves stimulating thinking by means of discussions.

Case 4: Mitigation and Adaptation Program About Local Impacts by Climate Change in Iida City, Nagano Prefecture (Implementing Organization: Hosei University Center for Regional Research)
The objective of this program is to develop human resources capable of contributing to solving climate change-related problems. The goal is to get environmental leaders in the local community to help residents investigate and share information on local impacts of climate change and thereby deepen understanding of adaptation and mitigation. Another emphasis is understanding of the impacts of climate change on socio-economic factors. The learning methods consist of ① a pre-program survey, ② the creation of a timeline and maps of climate change impacts, and ③ a workshop to think about adaptation and mitigation strategies. However, since ② is actually carried out by the program implementers, ① and ③ represent the actual learning opportunities for program participants. The program is jointly developed by a community center, local NPOs, businesses, and the municipal office.

Case 5: IPCC Report Communicator (Implementing Organization: MOE)
This is a one-off 8-h program designed to train communicators who are able to widely disseminate accurate information about AR5 to the public. The program also tries to get trainees to connect climate change with related problems and impacts in their own communities and to thereby contribute to efforts to mitigate global warming impacts in their respective regions. The program is geared towards individuals with previous knowledge of climate change or those in a position to teach about climate change. In addition to lectures in which experts present the latest information from climate change research, trainees engage in role-playing while learning from communication professionals about communication methods used by weather forecasters, etc.

As can be seen in Table 17.2, Stage I (systems-thinking competence) is included in the educational content of all five programs. Cases 1 and 2 do not cover any of the five main components of Stage II (anticipatory competence), while Cases 3 and 4 cover one component, and Case 5 covers four components. Cases 1, 2, and 3 do not cover any of the four main components of Stage III (normative competence), while Case 4 covers one component, and Case 5 covers two components. Cases 1, 2, and 3 do not cover any of the five main components of Stage IV (strategic competence), while Case 4 covers one component, and Case 5 covers four components. Cases 1, 2, 3, and 4 do not cover any of the three main components of the

Table 17.2 Organized connections between learning contents of each program and key competencies in sustainability

Learning development stage and competency	Educational discussion point derived from SPM (Sato and Takahashi 2015)	Case 1	Case 2	Case 3	Case 4	Case 5
Stage I: [A] Complex problem constellations in the current situation and their history Systems-thinking competence	1. Observed changes, impacts, vulnerability, and exposure in the climate system	+	+	+	NA	+
	2. Detection and attribution of climate change	NA	NA	+	+	+
	3. Future global and regional climate change (understanding of new set of scenarios, and impacts on nature systems)	NA	NA	NA	NA	+
	4. Climate stabilization, climate change commitment and irreversibility (relationship between total emissions of CO_2 and global mean surface temperature, understanding geoengineering means, and impact assessment)	NA	NA	NA	NA	+
	5. Key risks across sectors and regions and potential for adaptation (understanding and consideration of new risks for natural and human systems)	NA	NA	NA	NA	+
	6. Principles for effective adaptation, managing future risks and building resilience (understanding and consideration of dynamics of vulnerability and exposure and their linkages with socioeconomic processes, sustainable development, and climate change)	NA	NA	NA	NA	NA
	7. Climate-resilient pathways and transformation (Understanding of risk management included iterative processes)	NA	NA	NA	NA	+

(continued)

17 Trends and Issues of Climate Change Education in Japan

Table 17.2 (continued)

Learning development stage and competency	Educational discussion point derived from SPM (Sato and Takahashi 2015)	Case 1	Case 2	Case 3	Case 4	Case 5
	8. Long-term mitigation pathways, sectoral and cross-sectoral mitigation pathways and measures (understanding of mitigation scenarios and CO_2 concentrations, GHG concentrations; relationship with aggregate economic cost of mitigation; relationship between various kinds of energy technology and low carbon)	NA	NA	NA	NA	+
Stage II: [B] Non-intervention future scenarios Anticipatory competence	1. Understanding the evaluation of climate models and climate systems responses (understanding magnitude of global warming in response to past and future forcing)	NA	NA	NA	NA	NA
	2. Future global and regional climate change (projections of changes based on a set of scenarios)	NA	NA	+	NA	+
	3. Climate stabilization, climate change commitment and irreversibility	NA	NA	NA	NA	+
	4. Sectoral and regional risks and potential adaptation (risks and potential adaptation within and beyond the twenty-first century)	NA	NA	NA	+	+
	5. Long-term mitigation pathways, sectoral and cross-sectoral mitigation pathways and measures (understanding of mitigation scenarios such as future energy systems, land-use changes etc.)	NA	NA	NA	NA	+

(continued)

Table 17.2 (continued)

Learning development stage and competency	Educational discussion point derived from SPM (Sato and Takahashi 2015)	Case 1	Case 2	Case 3	Case 4	Case 5
Stage III: [C] Sustainability visions Normative competence	1. Adaptation experience and principles for effective adaptation (recognition of interests, circumstances, social and cultural context required for implementing adaptation plans at all levels of government)	NA	NA	NA	NA	NA
	2. Sectoral risks and potential adaptation (increased capacity, voice, and influence of low-income groups in urban areas and vulnerable communities and need for their partnerships with local governments)	NA	NA	NA	+	NA
	3. Responding to climate-related risks involves decision making	NA	NA	NA	NA	+
	4. Sectoral and cross-sectoral mitigation pathways and measures (understanding of a key mitigation strategy)	NA	NA	NA	NA	+
Stage IV: [D] Sustainability transition strategies Strategic competence	1. Responding to climate-related risks involves decision making	NA	NA	NA	NA	+
	2. Sectoral risks and potential for adaptation (scenario planning needed for building resilience and flexible and regret-less solutions)	NA	NA	NA	+	NA
	3. Adaptation experience and principles for effective adaptation (understanding of effective risk reduction, and planning and implementation of adaptation plan)	NA	NA	NA	NA	+
	4. Climate-resilient pathways and transformation (realization of climate-resilient pathways by transformation in economic, social, technological, and political decisions and actions)	NA	NA	NA	NA	+

(continued)

17 Trends and Issues of Climate Change Education in Japan

Table 17.2 (continued)

Learning development stage and competency	Educational discussion point derived from SPM (Sato and Takahashi 2015)	Case 1	Case 2	Case 3	Case 4	Case 5
	5. Mitigation policies and institutions (understanding of international cooperation for mitigation policies, technology policy, and integration of multiple policy objectives)	NA	NA	NA	NA	+
Cross-sectoral approach: [A][B][C][D] Interpersonal competence	1. Responding to climate-related risks involves decision making	NA	NA	NA	NA	+
	2. Sectoral risk and possibility of adaptation (effective multi-level urban risk governance, alignment of policies and incentives, strengthened local government and community adaptation capacity, synergies with the private sector)	NA	NA	NA	NA	+
	3. Policy linkages among regional, national, and sub-national climate policies	NA	NA	NA	NA	+

Case 1: The Fifty/Fifty Project; Case 2: Can you carry a bag of energy; Case 3: Children's Eco-Life Challenge; Case 4: Mitigation and adaptation program about local impacts by climate change in Iida city, Nagano prefecture; Case 5: IPCC Report Communicator
+: Confirmed by publication, documents of seminar etc.
NA Not applicable
Source: authors

Cross-sectoral approach (interpersonal competence), while Case 5 covers all three components. It can be seen that Case 5, whose goal is the dissemination of information contained in AR5, comprehensively covers all Stages I through IV, as well as the Cross-sectoral approach.

Current Situation of Program Evaluation

Current evaluation circumstances of each program are shown in Table 17.3.
As indicated, no program evaluation is conducted for Case 1, and other cases are evaluated using questionnaires. In Case 3, in addition to a questionnaire, the

Table 17.3 Current evaluation circumstances of each program

Evaluation		Case 1	Case 2	Case 3	Case 4	Case 5
Evaluation method	Questionnaire	NA	+++	++	++	++
	Others	NA	NA	++	NA	NA
Evaluator	Self evaluation (learner)	NA	NA	NA	NA	++
	Internal evaluation	NA	NA	++	++	++
	External evaluation	NA	++	++	NA	NA
Evaluation target	Learners' awareness	NA	NA	+	+++	NA
	Learners' knowledge	NA	NA	NA	+++	NA
	Learners' attitude	NA	NA	+	NA	NA
	Learners' behaviour	NA	NA	+	NA	NA
	Program	NA	NA	+	NA	+++
	Others	NA	+	+	NA	+++
Time of evaluation	Before the program	NA	NA	+++	++	NA
	During the program	NA	NA	+++	NA	NA
	Soon after the program	NA	NA	+++	++	++
	After the program	NA	++	+++	NA	NA

Case 1: The Fifty/Fifty Project; Case 2: Can you carry a bag of energy; Case 3: Children's Eco-Life Challenge; Case 4: Mitigation and adaptation program about local impacts by climate change in Iida city, Nagano prefecture; Case 5: IPCC Report Communicator

+: Confirmed by publication, documents of seminar etc.; ++: Confirmed by interviews; +++: Confirmed by interviews and publications

NA Not applicable

Source: authors

program is evaluated based on a self-evaluation filled out by participants regarding the degree to which they have adopted eco-friendly behavior. In three cases (Cases 3, 4, and 5), an internal evaluation is performed, with the program implementers evaluating the questionnaires filled in by the program participants. In two cases (Cases 2 and 3), an external evaluation is performed by a teacher who is not an instructor or program implementer. In Case 5, the program participants themselves evaluate program content using a questionnaire that also includes an item for self-evaluation.

In Cases 3 and 4, the areas of evaluation include the participants' learning and behavior, etc. In Case 3, children's attitudes and behaviors are evaluated based on observations by teachers; however, it is difficult for a teacher to keep track of changes in all children. In Case 2, the instructional materials are evaluated by users and, in Cases 3 and 5, the programs as a whole are evaluated by a teacher (third party) or by program participants.

In Case 3, evaluation was conducted at various time points from pre- to post-program. In the other cases, evaluation was conducted immediately after program conclusion, indicating that the evaluations are used to make future improvements in the program. In Cases 3 and 4, it is clear that evaluations are conducted both before and after the program because the evaluations focus on changes in the participants.

With regard to issues for such evaluations, comments have been received such as "We don't know what can be considered a change in behavior, so we can't measure change in behavior" (Case 2) and "the challenge is verifying what kind of affect [the program] has had on children" (Case 3).

Discussion

Issues of CCE Programs in Japan

Issues for each program are indicated as follows.

(a) Evaluation situation of the CCE program

As indicated by the comments above, one issue for the evaluation of programs was that, despite most of the programs having change in participant behavior as a goal, the programs are conducted without specifying what constitutes "concrete behavior". Thus, it is difficult to evaluate changes in the behavior of participants. Another issue was that only one case conducted evaluations throughout the program cycle from pre- to post-program and that evaluated both the program and learning by participants. This suggests that effective program improvements are not being made in most cases.

(b) Learning contents of climate change

Case 5 was the only program with educational content that covered all stages of key competencies in sustainability. Although systems-thinking competence was included in the educational content of all programs, we believe that the acquisition of anticipatory competence (Stage II) and higher competences, which in the scheme proposed by Wiek et al. (2011) are geared towards higher education students, would be difficult for elementary and junior high school students who are the targets of Cases 1 to 3. The only case in which learning method was varied based on the details of the competence was Case 4, however, in all cases learning methods based on the characteristics of each competence were not adopted as in the case of Hoffmann (2014a). In terms of methods for communicating information, participatory methods such as quizzes and group work were mixed in with lecture-style instruction.

Although the need to consider adaptation in relation to mitigation has been pointed out (UNESCO 2010), adaptation is not dealt with in Cases 1 to 3. In Case 4, the dissemination of information on adaptation measures is one of the program objectives. Many opinions opposing the introduction of adaptation in CCE programs have been raised, including "I am reluctant because it [talking about adaptation] feels like we are admitting defeat" and "adaptation is not a familiar concept for the general public, so it is difficult to talk about." Thus, it seems that it would be difficult to introduce elements of adaptation into Japanese CCE programs under the current circumstances.

(c) Implementation structure

In terms of program implementing structure, program development and execution suffer from a shortage of personnel (Cases 1 and 2). Related to this shortage of personnel, the fact that teachers and others with long-term experience with the target participants are not involved in program development means that the CCE programs are inadequately matched to the developmental stage of the learners. Against this backdrop, another reason for the shortage of personnel is the low demand for CCE programs. In the context of school education, given the limited number of instruction hours available for the various components of ESD, including environmental education, development education, education for disaster management, etc., it is difficult to implement programs strictly limited to CCE. In addition, for the planning and implementation of CCE programs geared towards the general public, attracting participants is an issue.

Another problem is the shortage of personnel with the requisite knowledge to conduct CCE programs. Although the organizations interviewed in this study comprise individuals with expertise in climate change, they are not education experts. As such, the reality is that each organization relies on their own methods of instruction, and there are few opportunities for organizations to exchange experiences and knowledge. Moreover, there are no systematic training programs geared towards action officers. Similarly, the IPCC Report Communicator initiative's limited target participants and narrow goal of information dissemination make it difficult to cultivate key competencies in sustainability among instructors.

Conclusion

The following points were revealed as regards the current status and issues of Japanese CCE:

(1) CCE is not established as a field of environmental education in Japan; there is no common definition of CCE.
(2) Each instructor conducts programs based on their own knowledge and experience; there are few opportunities for organizations to exchange methods and contents of CCE programs.
(3) In the majority of cases, the instruction was weighted heavily towards knowledge; methods intended to develop key competencies in sustainability were not employed.
(4) Since the programs are conducted without specifying concrete behavior, it is difficult to evaluate changes in the behavior of learners.
(5) It is difficult to introduce elements of adaptation to CCE programs under the current circumstances.

(6) There is a shortage of qualified personnel to conduct CCE programs; there are no systematic training programs geared towards trainers.
(7) CCE programs are inadequately matched to the development stages of the learners.
(8) Since there is a low demand for CCE programs, it is difficult to develop and implement systematic programs.

CCE has three different stepwise stages: 1. information dissemination, 2. environmental education, 3. Capacity development. The majority of Japanese CCE programs are situated in stage 1. However, in terms of methods for communicating information, such methods are mostly well established and developed. Furthermore, since CCE programs are implemented not only at government but also grassroots levels, the impacts of CCE are significant.

In order to overcome the above challenges and improve on current circumstances, it is necessary to address the following fundamental issues: (1) establish standard concepts, methods, and definitions for CCE programs in Japan, (2) develop and implement instructor training programs to nurture key competencies in sustainability through collaboration among teachers, experts, and educators, and (3) expand information exchange and share experiences to develop networking amongst organizations conducting CCE programs, by identifying the function and role of different stakeholders.

Specifically, we have to consider the following points: (1) evaluation: as an evaluation target, we should set proposed key competencies and introduce an evaluation method for acquiring it, (2) educational content: comprehensive components appropriate for learning development stage should be included in educational programs, (3) learning method: we should adopt learning methods based on the characteristics of each competence, and coordinate skills to acquire these, and then discuss the linkage between learning contents and methods, (4) target audience: we have to coordinate and develop a structure for learning content which gives consideration to development stage.

Finally, it is imperative that the process of developing CCE capacity building programs focuses on acquiring sustainable key competencies. Therefore, in our forthcoming research we intend to address this challenge by means of investigations of successful/advanced CCE programs that have been implemented overseas, and to analyze and compare these with Japanese programs. Aspects such as the dissemination of skills; networking with stakeholders; and program and learner evaluation and subsequent improvements to the program, as they relate to sustainable key competencies, should be considered as criteria for a program for trainers. To ensure optimized penetration across the scope of trainers, cooperation and use of existing implementation structures should be considered—in particular with action officers (JCCCA; 7000 people), IPCC report communicators (MOE; 300 people), etc.

References

Hoffmann T (2014a) Is there a specific ESD methodology? Schools for Sustainability a resource Toolkit for Teacher Training, pp 1–8

Hoffmann T (2014b) Klimawandel in Baden-Wuerttemberg Unterrichtseinheit als Beitrag zur Bildung fuer nachhaltige Entwicklung (in German)

IPCC (2013) Summary for policymakers. In: Stocker TF, Qin D, Plattner G-K, Tignor M, Allen SK, Boschung J, Nauels A, Xia Y, Bex V, Midgley PM (eds) Climate change 2013: the physical science basis. Contribution of Working Group I to the Fifth Assessment Report of the Intergovernmental Panel on Climate Change. Cambridge University Press, Cambridge

IPCC (2014a) Summary for policymakers. In: Field CB, Barros VR, Dokken DJ, Mach KJ, Mastrandrea MD, Bilir TE, Chatterjee M, Ebi KL, Estrada YO, Genova RC, Girma B, Kissel ES, Levy AN, MacCracken S, Mastrandrea PR, White LL (eds) Climate change 2014: impacts, adaptation, and vulnerability. Part A: global and sectoral aspects. Contribution of Working Group II to the Fifth Assessment Report of the Intergovernmental Panel on Climate Change. Cambridge University Press, Cambridge, pp 1–32

IPCC (2014b) Summary for policymakers. In: Edenhofer O, Pichs-Madruga R, Sokona Y, Farahani E, Kadner S, Seyboth K, Adler A, Baum I, Brunner S, Eickemeier P, Kriemann B, Savolainen J, Schlömer S, von Stechow C, Zwickel T, Minx JC (eds) Climate change 2014: mitigation of climate change. Contribution of Working Group III to the Fifth Assessment Report of the Intergovernmental Panel on Climate Change. Cambridge University Press, Cambridge

Ministry of Education (ed) (1991) Teacher's guide for environmental education (in Japanese)

Ministry of the Environment (ed) (2013) Guide book of ESD environmental education model program (in Japanese)

Ministry of the Environment (ed) (2014) A version of local Guide book of ESD environmental education model program (in Japanese)

Ministry of Internal Affairs and Communications (2015) http://law.e-gov.go.jp/htmldata/H10/H10HO117.html. Accessed 2 Mar 2015 (in Japanese)

National Institute for Educational Policy Research (2014) Final research report on education for sustainable development at school (in Japanese)

Sajima T (2007) Verification of the role of the "teacher's reference materials for environmental education". Environ Educ 17(2):13–18 (in Japanese)

Sato M, Takahashi K (2015) Capacity Development Programme for Climate Change Education (CCE) in the context of ESD, based on the International Discussion on "Key Competencies in Sustainability" and the IPCC Fifth Assessment Report. J Energy Environ Educ (accepted, in Japanese)

The Japanese Society of Environmental Education (ed) (2013) Kankyō Kyōiku Jiten, Kyoiku Shuppan (in Japanese)

The Japanese Society of Environmental Education (ed) (2014) The 25th annual meeting of the Japanese Society of Environmental Education, Tokyo (in Japanese)

UNESCO (2013) Education sector technical notes: climate change education

Wiek A, Withycombe L, Redman CL (2011) Key competencies in sustainability: a reference framework for academic program development, Integrated Research System for Sustainability Science, United Nations University, Springer

Chapter 18
Changes in Attitude Towards Climate Change and Transformative Learning Theory

Gherardo Girardi

Abstract The author conducts in-depth interviews with individuals who describe themselves as having undergone a change in attitude with regards to climate change. Determinants of attitudinal change, and characteristics of the process, are identified and compared with those found in the literature on attitudes towards climate change. Given that changes in attitude are often linked with transformation in perspectives, the author proposes transformative learning theory as a framework with which to interpret and explain the experiences of the respondents.

Keywords Climate change • Change in attitude • Transformative learning theory

Introduction

Recent studies of the public's attitudes towards climate change indicate that there has been an increase in scepticism towards the idea that anthropogenic climate change is actually taking place (Pidgeon 2012a; Brulle et al. 2012; Capstick et al. 2015), in spite of a broad consensus among scientists that it is indeed happening and that its consequences could be very damaging. What might explain this scepticism[1,2]? One possible reason could be that media coverage of the issue has fallen since 2006 (Pidgeon 2012a), as, it would seem, has political concern, with politicians giving priority to the economy, in particular the economic slowdown which affected many countries following the 2008 recession. Another

---

[1] This scepticism may be part of a wider scepticism towards institutions (the scientific community, governments, etc.), as reflected in declining participation in democratic elections (Wolf and Moser 2011).
[2] Whilst the scientific community is right—indeed has an obligation—to point out the potentially catastrophic consequences of climate change, one should not exclude the possibility that the public may be right in its assessment of the dangers of climate change in relation to other dangers it faces.

G. Girardi (✉)
London Guildhall Faculty of Business and Law, London Metropolitan University, 84 Moorgate, London EC2M 6SQ, UK
e-mail: g.girardi@londonmet.ac.uk

possible reason that is often found in the literature is 'information fatigue', i.e. the prolonged exposure to the same message leading to reduced absorption of the message on the part of the public (Gifford 2011).

Pidgeon (2012b) comments that it is not clear if the public is aware of the magnitude of the change in lifestyle that is required to adequately address the problem of climate change. Even though calls about the importance of making lifestyle changes have been made by scientists, the media, and people in position of political and religious authority such as and President Obama[3] and Pope Francis,[4] it is not clear to what extent these calls have been heeded in practice. The public in certain parts of the world shows particularly low levels of concern with climate change, e.g. in China, Russia and the U.S. (Pidgeon 2012b). According to Pidgeon, the time gap between emissions and impacts is such that the public's perception of increased risk is likely to intensify only slowly. Some optimism may be warranted, however, as a BBC poll carried out in 2007 in 21 countries indicates that people know they will have to change lifestyle and that they are in favour of taxation if tax revenues are used to reduce emissions or maintain constant the overall tax burden (BBC 2007, in Pidgeon 2012b).

The literature identifies a number of barriers which prevent the public from taking the message of the majority of the scientific community on board (see the references above). These include 'psychological distance'—the belief that anthropogenic climate change is happening in places far away from us and/or will anyways happen in the distant future; a belief that technology in the future will be developed so much that the problem will be solved ('technosalvation'); the fear that if one does something to address climate change, other people will do nothing (i.e. they will 'free-ride' on what the individual does), causing the individual to feel a sense of injustice which will induce him or her not to take any action to address climate change; and last but not least, uncertainty about the whole issue of climate change is used by some people as a (selfish) way to excuse themselves for not taking action.

This paper seeks to identify which factors are responsible for changes in attitudes towards climate change, in the hope of making (policy) suggestions which may foster positive attitudinal change, whilst being respectful of people's freedom to choose without being unduly pressurized by government or other

---

[3] In his speech on 21st January 2013 inaugurating his second term in office, President Obama appeared to place climate change high up his list of priorities when he stated "We will respond to the threat of climate change, knowing that failure to do so would betray our children and future generations" (Stevenson and Broder 2013).

[4] According to Pope Francis, protecting the environment is an integral part of a Christian's identity, not something a Christian can choose to opt out of (USA Today 2015). The Vatican has been using solar power energy since 2008 when it installed a new solar-powered roof for one of its auditoriums, and announced plans to increase its production of solar-powered energy, potentially making it an exporter of solar-power energy for distribution on the Italian national grid (Pulella 2008). The Dalai Lama is another well-known religious leader who has called on nations to collaborate to protect the global environment (Perry and Tait 2009).

institutions. The author does so by interviewing subjects who have undergone significant change of attitude towards climate change. Subsequently, the author seeks a framework which is able to accommodate and explain the broad range of experiences of the interviewees. Since two common threads across experiences are (a) a struggle to make sense of reports of climate change and to attribute a 'correct' meaning to these reports, and (b) the transformation in attitudes undergone by the interviewees, the author proposes that a plausible framework to interpret these experiences can be found in transformative learning theory, which incorporates both these aspects.

Methodology

In order to obtain information about the details of the process by which individuals undergo a change of attitude with regards climate change, the author adopted a qualitative empirical approach involving a number of one-to-one in-depth interviews. Such a focus is advocated by Wolf and Moser (2011) in their survey article of in-depth studies about individuals' perceptions of and engagement with climate change: "Future research should examine how and what could *stabilize or change* individuals' perceptions. [...] What internal processes and/or outside events can *change* them?" (italics not in the original).

An unstructured interview approach was adopted as this method is ideal for exploratory studies where new variables and relationships may emerge: the present study is exploratory in the sense that, in contrast with much of the literature, it focuses on individuals who have undergone a change of attitudes. The unstructured interview approach allows the interviewer much freedom to probe where he or she feels that doing so would be beneficial towards achieving the purpose of the study. In addition, the technique is ideal with a view to encouraging interviewees to "tell their story"[5]; furthermore, as attitudinal change can occur over the course of one's life, it is not unusual for such stories to be life histories,[6,7] Goodson and Sikes (2001) discuss the merits of the 'life history method', an important benefit in their view being that this method "disrupts the normal assumptions of what is 'known' by intellectuals [...]".

The author placed an advertisement on the news page of London Metropolitan University's internal website inviting students and staff who had undergone a change of attitude with respect to climate change to come forward for interview.

---

[5] Turk (2009) promotes the use of narrative data in the economics discipline by pointing to the fact that such data conveys "the internal conversations of key actors operational at the time under study as a real causal mechanism driving economic and social change".

[6] This was clearly the case for two interviewees.

[7] For a philosophical discussion of how narrative is ideally suited to bring together and understand the various aspects of one's life, see MacIntyre (1981).

The advertisement left unspecified the direction of change. Respondents were selected so that those who had not undergone a change of attitude were politely turned down—a number of respondents replied who simply had strong feelings about the importance of the environment.[8]

To the author's surprise, only 15 individuals out of a population of more than 30,000 responded, perhaps because some were still on their Christmas holidays and others were busy preparing their examination re-sits. Of these 15, 8 were selected, of which 7 were students and 1 was a member of staff working in administration. Not having any teaching member of staff in the sample, the author approached colleagues in the business school and found that two had undergone a change of attitude with respect to climate change, and so they too were invited for interview. Hence the sample consists of seven students, one administrator and two teaching member of staff, giving a total of 10 interviewees.

Of the 10 individuals, 4 were women. The ages of the 10 individuals varied between 20 and 58, with the average being 36. The relatively high average age reflects the fact that most of the students interviewed were mature or Master's students. The 10 individuals included 2 Americans (from the USA), 2 British, 1 Brazilian, 1 Estonian, 1 German, 1 Mauritian, 1 Tunisian and 1 Turkish. The sample reflects the multicultural population of London Metropolitan University, though it is under-representative of the African and Asian communities.

The setting adopted for the interviews was informal, usually the canteen, so as to make the interviewees feel at ease. Interviews did not last for a pre-specified length; instead, their length varied considerably, depending on the how rich the conversation was felt to be by the interviewee and interviewer. Interviews lasted between 45 min and 1.5 h. The aim was to generate a conversation in which the interviewee' beliefs were questioned but not challenged. It was left at the interviewer's discretion how and how much to intervene in the conversation.

The appendix reports 10 vignettes describing the experience of the respondents, who were asked to check that the vignettes accurately summarized their experience.[9] These vignettes constitute material which, if analysed by an interdisciplinary group of social scientists, may well yield insights that are far deeper and of greater value than those attainable by any one researcher rooted in a single discipline; indeed, the author, who is an economist, hopes that other social scientist will read and comment on the material in the vignettes. This approach is supported by Wolf and Moser (2011), who state: "Interdisciplinary research might yield interesting insights, for example on how stages in human/personal development (not just

---

[8] Not surprisingly, for most if not all of the individuals who responded to the advertisement (including those selected for interview), concern for climate change went hand in hand with generic concern with the environment, as has been found in other empirical studies (Pidgeon 2012b).

[9] The vignettes were adjusted to take into account the post-interview comments of the interviewees, with the exception of two vignettes, those of D.Y. and C.B., who did not respond to email invitations to comment.

chronological age), life experiences, level of maturity, social roles or personality types affect beliefs and perceptions of climate change".

Findings

Surprisingly, though a majority of interviewees had undergone a change of attitude in the direction of believing that anthropogenic change is taken place, one interviewee interpreted "change of attitude" to mean a strengthening of his stance that anthropogenic climate change is taking place, and one interviewee experienced a change in the opposite direction, i.e. away from believing that is taking place.[10]

The author's analysis of the respondents' experience is summarized in the table below:

The first columns reports the respondent's indicative initials, which, for the sake of safeguarding anonymity, do not correspond to the respondents' true initials, and serve merely to identify respondents in the analysis that follows. The second column of the table reports the word that most succinctly captures the experience of the respondent from the author's point of view—the choice of keyword is likely to become clear to the reader upon reading the vignettes in the appendix: the different keywords reflect the different experiences of the respondents. The table also reports the interviewee's age and gender. The subsequent columns capture key features observed in the interviews that are common to a number of respondents; the analysis which follows considers each of these factors in turn, by proceeding in the table from left to right.

Evolutionary Change Arguably, many if not all respondents can be understood to be on a life journey which, at some point and after a number of turns, included becoming aware of the importance of climate change, as part of a process of personal development. The table identifies only those interviewees (four in total) for whom the evolutionary nature of the process by which they came to this awareness is very clear. Furthermore, all four of the interviewees were aware that their change of attitude was evolutionary in nature. However, only three clearly saw this is as part of a life-long journey (F.B., D.O. and S.G.), whilst the fourth (C.B.) took a very rational approach of reading the news and reviewing his beliefs,[11] without displaying much emotion and apparently without connecting this activity to other important aspects of his life, such as his history and personal development.

---

[10] This last interviewee did so after reading some literature disputing the mainstream evidence about climate change. Yet another interviewee had undergone a radical change of attitude of a different sort, in the sense that she had lost confidence in humanity's ability to address climate change.

[11] In what appeared to be a Bayesian fashion.

Change Marked by Revelation The idea here is that the process by which a subject changed attitude towards climate change was due to some unexpected event which resulted in a sort of 'awakening'; such change is conceptually opposed to the idea of evolutionary change described above, though in practice an individual may well experience both at different times in his life. Among the interviewees, S.G. felt a calling to work as a professional in an organization aiming to address climate change when she attended a course on ecology during which she encountered the concept of 'eco-crime'; D.Y felt inspired to start recycling after coming across surprising information in Al Gore's documentary 'An Inconvenient Truth'; and C.T. was surprised to find that Nick Stern's Review on the Economics of Climate Change was convincing enough to be trusted, even though, prior to being obliged to read Stern's work as part of his assessment in Research Methods, C.T. had felt that he could not trust Nick Stern as Stern worked for the World Bank, an institution which, in C.T's view, might be in place to further the interests of the rich countries, and which was paying Nick Stern "to produce this kind of report".

The number of interviewees who had experienced a sort of revelation (three) was roughly the same as that of interviewees whose change in attitudes was evolutionary (four), with one interviewee being present in both camps (S.G).

Attitude Change Due to Simple Receipt of Information For three interviewees, a change of behaviour appeared to follow the mere receipt of fresh information about climate change, i.e. without any other major influence visible at the time of their change in behaviour.[12] The literature suggests that normally the mere receipt of information is insufficient to induce people to change behaviour in relation to climate change [see for example the discussion in Pidgeon (2012b)[13]]. Furthermore, even if behavioural change takes place, it is likely to be superficial. A manifestation of superficial change is the 'rebound effect' (Gifford 2011), where an individual feels that, because he is taking some action to address climate change, he is allowed some 'compensation' in the opposite direction, for example C.B. feels that that he is entitled to fly as a reward for his conservation efforts.[14]

Peer Pressure Two interviewees experienced peer pressure and did so in a negative way, in the sense that the people around them, who behaved in an

---

[12] Though one cannot exclude the possibility that the subject was experiencing deep change at a sub-conscious level which made him more responsive to the arrival of fresh data about climate change, a possibility not picked up during the interviews.

[13] Wolf and Moser (2011) paint a rather complex picture of how the public handles information about climate change: "Clearly, individuals enact climate-relevant behaviour without or with an incomplete and sometimes misguided understanding of climate change, while others understand the problem full well and do or do not act to reduce the emissions."

[14] Kellstedt et al. (2008) go further and in an empirical study of the US find that more informed respondents feel less responsible for global warming and show less concern for it than less informed respondents. The authors suggest that "[p]erhaps this simply reflects an abundance of confidence that scientists can engineer a set of solutions to mitigate any harmful effects of global warming".

ecologically irresponsible way or at least were perceived by the interviewees to be doing so, induced the interviewees to change their behaviour in the direction of being less eco-friendly. One interviewee (K.M.) was affected in this way as a result of observing her flat-mates not recycling; another interviewee (M.B.) was affected in this way partly as a result of observing that young people did not care about the environment, or at least not as much as she would have liked them to.

Friend's Input Two interviewees (C.B. and S.G.) underwent a change of attitude towards climate change during conversations with friends, whose opinion they valued and trusted and who nudged the interviewees in the direction of being more eco-friendly. The input of trusted ones was pivotal in convincing these interviewees to believe that climate change is occurring and to take action to address it, a phenomenon which may be important in the general population.

External Influence Other Than Peer Pressure and Friend's Input According to the interviewees, external influence of this kind came from two sources: the media, and having to perform certain tasks as part of an assignment in an educational context.

One interviewee (C.B.) reported following the media very closely, and even admitted candidly that if he were to live in a country where the media reports about climate change were substantially different from those in the UK, his views about climate change would change accordingly. This is consistent with the literature on attitudes towards climate change, which suggests that an individual's views tend to change significantly over time, and that these views tend to follow the latest representation of the media (Brulle et al. 2012). Given the clear influence of the media in forming the public's views (a fact which at least some members of the public are aware of[15]), the question arises as to what extent the media can be trusted, particularly in the light of the fact that many sources of news adopt a political stance, rather than simply reporting the news in as neutral a way as possible. Wolf and Moser (2011) ask: "How can selective information uptake in highly polarized contexts be circumvented?"; in response, two possible answers emerge from the interviews. Firstly, according to D.M., even though one cannot have perfect trust in the media, "there is no smoke without fire". Secondly, according to S.G., one can read news from sources with different political stances (e.g. liberal and conservative). However, doing so is a time consuming activity in a society where time is of the essence, and is unlikely to be practiced by a majority of people who, in all probability, simply wish to read information expressed through a particular set of lens which they approve of.

External pressure can also take the form of having to perform certain tasks within an educational context. So, for example, C.T. was 'obliged' to read the Stern Review as part of his assignment for a course on research methods; after having read the Review, he was glad that he had been obliged to read it, and changed his

---

[15] As C.B. remarked, "If you'll repeat the message over and over again, you'll believe it".

views about climate change in line with the approach and recommendations of the Review. In another example, J.E. warmly recollected the time when, as a school boy in Estonia, he was made to clean the local park on a Saturday; when Estonia switched from socialist to capitalistic principles, however, such an activity was considered too 'illiberal' and so it was stopped; as a consequence, the park become dirty. Partially compensating for the end of this compulsory activity is the fact that the market, including J.E.'s family's business, responded by taking advantage of new opportunities, in particular by offering services and equipment to dispose of/recycle waste material. External pressure in the form of compulsory eco-friendly educational activities was therefore seen as desirable by the interviewees who were made to perform these activities. In a similar vein, Fujii and Gärling report that temporarily forcing car drivers to adopt other means of transport resulted in long-term reduction in car use (Fujii and Gärling 2003), with important implications for policy makers wishing to reduce emissions.

Rational Approach One of the barriers preventing people from seriously engaging with climate change is that they feel that whatever action they took would have so little impact that there would be hardly any point in doing it, which in turn can result in feelings of hopelessness and futility [see Lorenzoni and Pidgeon (2006) in Roeser (2012)]. Roeser (2012) suggests that emotions can help an individual overcome such a "rational" approach. For example, if K.T., travelling on the underground, had acted in a purely "rational" way, she would have done nothing and simply left the discarded magazines lying around, but in fact she intuitively felt her problem to be so important that, morally speaking, she could not remain passive, and, after much interior struggle, decided to carry pick up *a few* magazines (i.e. some intermediate number between a lot and zero).[16] A generous disposition, too, can help an individual overcome a sense of futility causing her to take no conservation action, by increasing the benefits of such an action so that they do not just accrue to her, but also to others. As Corner et al. (2014) report in their review of the literature about the link between climate change and human values, individuals who display high levels of altruism and self-transcendence are more likely to show concern for environmental risk and to engage in activities like recycling.[17]

Gifford (2011) points out that one of the barriers which prevent people from being active in engaging with climate change is the fear that other people may be passive and simply "free-ride" on the actions which one has taken at some personal

---

[16] Curiously, in a different environment—her hall of residence—K.T. came up with a different strategy, namely not to recycle. This apparent contradiction between her behaviour in the hall and her behaviour on the underground may be explained by a number of factors, including the fact that she knew the people in the hall better than those in the underground. In other words, psychological factors may be at work which explain the apparently contradictory behaviours of some interviewees.

[17] Altruism and rationality need not be seen as at odds with each other, quite the contrary, they may be regarded as entirely consistent with each other, in particular if one considers that altruism is an end, whilst rationality is a means.

cost. K.T. may well have felt this way. The way she resolved her dilemma may was, as stated above, to pick up only a few magazines; such a solution seems quite balanced—it involves not foregoing one's principles and at the same time it doesn't involve taking on board the impossible burden of collecting all the newspapers left on the underground; it may well be that other members of the public who face the same dilemma would benefit from being aware that such a solution is possible. F.B. came up with a different, though related, solution: F.B. recalled a Brazilian folk tale and saw himself as a bird letting drops of water fall on a forest fire; the fire here represents climate change and the drops represent the small actions that an individual can take. F.B.'s approach involves regarding the means, i.e. letting the drops fall, as the goal itself; in this way the goal comes within one's reach and so becomes feasible; furthermore, the free-rider problem is entirely avoided in that what others do is of no relevance to the individual concerned.

Strong Sense of Identity Various respondents displayed a strong sense of identity, i.e. of having clear objectives in life with clearly identified strategies to go about achieving them; of these, four displayed a strong "ecological" self, i.e. a personality marked by a strong desire for ecological balance and engagement with the issue of climate change. Van der Werff et al. (2013) find that values and past behaviour both influence *perceived* self-identity, which in turn affects behaviour. The experience of two interviewees, D.Y. and S.G., is consistent with this finding, in that both claimed that once one starts recycling, it's hard to stop; seen through the lens of Van der Werff et al.'s framework, recycling could be interpreted as past behaviour, which could affect one's perception of one's ecological self, which could in turn result in continued recycling. Van der Werff et al.'s findings imply that policies which targets perceived self-identity may be effective. However, they a note of caution is provided by these authors: people with a strong sense of ecological identity still struggle to make changes in their lives that would reduce emissions effectively. More generally, the authors observe that "Even those [individuals] most motivated to act struggle with making their actions meaningful in terms of emission reductions". This may be due in part to the fact that many people in economically developed countries do not live in contact with nature, so they may find it more difficult to attribute value to ecologically useful actions.

Strong Sense of Justice Five interviewees exhibited a strong sense of justice; these, probably not incidentally, coincided with those who had a strong sense of identity.

Perception of Personal Experience of Climate Change Four interviewees explicitly connected their own experience of the weather with the belief that climate change is indeed happening, e.g. when M.B. spoke about the fruits in her garden not ripening at the right time. Although "the attribution of specific events of climate change is difficult" (Pidgeon 2012a), "it is probable that people's experience of the weather shapes their expectation about long-run climate processes" (Pidgeon 2012b). Recent research is emerging which backs this hypothesis, e.g. the study by Joireman et al. (2010); however, it is unclear if people directly affected by natural disasters are more willing to engage in adaptive or mitigative activities (Wolf and Moser 2011).

Artistic Element Two interviewees felt inspired by the arts in their conservation efforts, a factor which was harmoniously integrated in the interviewees' rational discourse. F.B. identified with a bird in Brazilian folk tale which let drops of water fall onto a forest fire, whilst D.O. clearly recalled the songs he had heard in his youth which were about the environment. Research in this area is scant and may well yield innovative insights, possibly with a view to understanding which communication strategies are most effective.

Environment as Part of a Package of Values As Wolf and Moser (2011) indicate and as common sense suggests, individuals' attitudes towards climate change are invariably bound up with attitudes toward all kinds of other issues, and reflect an individual's particular set of values (see Corner et al. 2014).[18] Four interviewees clearly displayed that their attitude towards climate change constitutes part of a wider attitude towards important questions and issues in life; for example, for S.G. climate change is primarily an issue of social justice, for F.B. it is an issue of spirituality as well as of social justice, for D.O. it is part of an investigation into business practices and of the camouflaged lack of financial health which corporations suffer from and which green accounting would expose.

Change in Attitude Towards the Environment Accompanied by Change in Attitude Towards People Two interviewees manifested a change of attitude towards the problem of climate change which was accompanied by a change in attitude towards people. M.B. reported reduced trust in people (especially in young adults) and at the same time reduced trust in the humanity's ability to successfully tackle the problem of climate change (see Gifford (2011) for a discussion of the important role of trust which the public places in scientists and government officials). S.G. reported that, as she was becoming more ecologically conscious, she was also becoming more aware of the needs of other people, whilst before, she said, "it had been all been about [her]self". This is consistent with the above mentioned finding in the literature that altruism and self-transcendence are associated with greater pro-environmental effort (Corner et al. 2014).

The discussion of change in attitude towards climate change which is accompanied by a change in attitude towards people can be understood as a part of the discussion of the environment as a package of values. An interesting and empirically testable hypothesis on the idea that the environment is part of a package of values, and specifically of values concerning one's relations with other people, is put forward by Benedict XVI: "In order to protect nature [...] not even an apposite education is sufficient. If there is a lack of respect for right to life and natural death [...] the conscience of society ends up losing concept of human ecology and, along

---

[18] One's set of values may determine and/or be determined by political affiliation; for a discussion, see Pidgeon (2012b).

with it, that of environmental ecology. [...] The book of nature is one and indivisible [...]." (Benedict 2009). Some support for this position is provided by one interviewee (M.B.), who stated that "people don't care about others in the present so how can we expect from them to care about the future or future generations".

Importance of Cultural and Religious Background Four interviewees reported that their cultural background helped them become more active in addressing the problem of climate change: two were Muslims (K.T. and C.T.), one was Christian (F.B.), and a 4th (D.O.) commented that his parents, who were Christians, raised him with a strong sense of fairness. In addition, K.T. spoke of the family education she had received while growing up in Tunisia, in particular of the values of sobriety and frugality which she had been taught to respect. These accounts indicate that one's cultural background and formation can be a very important determinant of attitudes towards climate change, and can support a move towards increased responsibility in this area, as in the previously reported case of F.B.

Spiritual Dimension Two interviewees clearly displayed a spiritual attitude. F.B. identified conservation action to address climate change as part of the idea of '*convivencia*' (harmonious co-existence) between living beings on Earth. For F.B., there is a sense in which the dignity of the Earth has been violated; perhaps his spirituality is best reflected in the statement "We have received freely, let us give freely".[19] S.G. too displayed a spiritual attitude when she said that "the attitude we have towards the Earth is the attitude we have towards ourselves". As these quotes suggest, the theme of spirituality, because of its holistic character, naturally overlaps with the discussion of cultural and religious background, of the environment as part of a package of values, and of change of attitude towards the environment accompanied by change of attitude towards people.

Simple, Positive Message for Humanity The experience of some interviewees suggests a number of simple lessons which may help foster pro-environmental attitudes and/or ring a note of optimism with regards to humanity's ability to successfully address climate change:

1. Starting to take some action to address climate change can be enjoyable and, if repeated, can result in habitual behaviour such that, if stopped, reduced satisfaction obtains; in the words of S.G., "Once you start, it's hard to go back". In addition, D.Y. described recycling as a sort of joyful sacrifice.
2. Individuals may take more action to address climate change if they focus not so much on addressing climate change directly, which may well be perceived as an enormous problem beyond anyone's reach, but simply on taking part in some limited way in the collective effort to make the problem better, even if one's contribution is very small. In other words, individuals may take more action to address climate if they treat the means (taking some action) as the objective, and

---

[19] A very similar statement can be found in the New Testament in Matthew 10:8.

take their eyes off the Leviathan of global climate change, as F.B. did when he identified with a mythical bird in the Brazilian folk tale described earlier.
3. Obliging people to become familiar with the climate change literature may be a possible way forward, as the experience of C.T. suggests.[20] In addition, it may be advantageous to a society if it relies on both the contribution of the free market (e.g. innovation in pollution abatement equipment), and on compulsory activities such as having children in secondary education spend a couple of hours a week picking up litter in local parks, as J.E.'s account suggests.
4. When the free-rider problem is clearly an issue, i.e. when others are not 'doing their fair share' for the environment and in do doing create a disincentive for an individual who is keen to get involved, a balanced solution for the individual concerned could be to take a moderate degree of action, for example by recycling on behalf of others, but only within certain limits with which he or she is comfortable with.

Table 18.1 does not include certain columns—i.e. themes—as they did not seem to apply to a sufficient number of interviewees. For example, Wolf and Moser (2011) suggest that if opinion leaders (such as political leaders, civic leaders, business leaders and religious leaders) led the way in addressing climate change, people would follow; however, there is no obvious evidence of this attitude on the part of interviewees. Furthermore, in contrast with what the literature suggests (e.g. Pidgeon 2012b), there is no evidence that interviewees blame the government for not taking action; in fact on interviewee, D.Y., thinks that consumers, rather than government or firms, are responsible for pollution, since demand generates supply. In addition, the interviews contain evidence of the importance of 'structural issues' (other than government/opinion leaders). For example, when K.T. found out that the underground was not recycling leftover newspapers, she began to recycle herself; and D.M. recycled partly because recycling facilities were conveniently placed within easy reach.

A final theme commonly found in the literature which does not appear in Table 18.1 concerns the type of message that the media and/or the authorities should adopt to inform members of the public about climate change and to 'nudge' them in the direction of being more active in addressing this issue. D.M. says "maybe [...] messages [of fear] are needed to get to people", in contrast with the literature, which on the whole asserts that the message should be a positive one (Roeser 2012; Wolf and Moser 2011) and should avoid catastrophic messages (Pidgeon 2012b), not least because terror management theory suggests people don't like to be reminded about their death (Gifford 2011[21]).

---

[20] In Germany, climate change is a compulsory part of the syllabus in many secondary schools.

[21] This in turn raises the question of whether what the public wants necessarily coincides with what is good for them.

Table 18.1 Characteristics of respondents and of their change in attitude

Respondent	Keyword(s)	Age	Gender	Evolutionary change	Change marked by revelation	Attitude change due to simple receipt of information	Peer pressure	Friend's input	External influence other than peer pressure and friend's input	Rational approach	Strong sense of identity	Strong sense of justice	Perception of personal experience of climate change	Artistic element	Environment as part of a package of values	Change towards the environment accompanied by change towards people	Importance of cultural and religious background	Spiritual dimension	Simple, positive message for humanity
D.Y.	Exploration	27	M			Y				Y	Y		Y						Y
K.T.	Struggle	25	F				Y			Y	Y								Y
M.B.	Disillusionment	45	F				Y												
C.B.	Rationality	49	M	Y				Y		Y									
F.B.	Meaning	27	M	Y												Y			
D.O.	Mulling over	58	M	Y						Y	Y			Y			Y	Y	
S.G.	Justice	46	F		Y			Y			Y			Y		Y	Y		
D.M.	Periphery	38	F			Y			Y		Y	Y		Y					
C.T.	Coercion	25	M		Y	Y			Y			Y					Y	Y	
J.E.	Paradox	20	M						Y			Y						Y	Y

where 'Y' stands for 'Yes'

Interpreting the Results Through the Lens of Transformative Learning Theory

As Wolf and Moser (2011) argue and as common sense suggests, "information about climate change is always and inevitably filtered through the lens of pre-existing cultural worldviews". Research on identifying cultural factors at work has been up to now limited (Pidgeon 2012b), and the discussion below around transformative learning theory can help fill this gap in so far as transformation can take place away from a particular set of culturally received norms, and towards a new one. Because of the interdisciplinary nature of the role of culture in causing, but also in remedying, the problem of climate change, a fruitful discussion of the cultural issue surrounding the problem of climate change needs to be interdisciplinary, involving, among other groups, representatives of both the physical and social sciences. An innovative example of such interdisciplinary study is that of Karlsson et al. (2004), who are respectively a psychologist, an economist and a writer. Whilst these authors do not explicitly focus on either culture or climate change, they focus on the process by which people give meaning to events and ideas, a process that is an essential component of culture: culture offers a lens through which one can put climate change into context, assess its importance in relation to other problems, and decide if and how to respond to it—in other words, a lens through which one can give meaning to the problem of climate change.[22,23]

Transformative learning theory is closely connected with the question of meaning-making: it is a branch of educational studies which, at heart, identifies the purpose of education as one of promoting transformation of perspectives in students, and therefore of the of the way in which students give new meaning to facts and ideas, so as to benefit the students. The founding figure in transformative learning theory is Jack Mezirow, who argued that, through a clear use of logic made possible through unprejudiced group discussion in class, an individual can come to

---

[22] On the subject of meaning-making, Karlsson et al. (2004) write: "many psychologists (e.g. Bruner 2002; Kegan 1982), see meaning-making as the fundamental activity of human existence. [...] Without meaning, psychologist and philosophers argue, even the most prosperous existence isn't worth living. [...] [T]he capacity to find meaning can attenuate even the most severe hardships (Taylor 1983). Victor Frankl (1963) is especially associated with meaning as a result of his popular book *Man's Search for Meaning*. [...] To Frankl it is people's innate will to find meaning, and not their striving for pleasure, power, or wealth, that is the strongest motivation for living. To the extent that economics is the science of promoting well-being with constrained resources, then, meaning should be part of the equation."

[23] It might be possible to accommodate a discussion of transformation of perspectives within the economics discipline by interpreting transformation of perspectives as a conscious attempt to allow second order preferences to dominate first order preferences; see George (1998) for a discussion of first and second order preferences.

know his or her inaccurate assumptions about reality, often picked up in childhood when the individual was not able to assess the accuracy of the education he or she was exposed to, reject the distorted views based on these inaccurate assumptions, embrace new, more realistic assumptions and ultimately live a happier and more fulfilling life [see Taylor's (1998) review of transformative learning theory]. If the UN scientists are correct in their assessment of the seriousness of the problem of climate change, and the public in general is not, then transformative learning theory can, if applied correctly in the classroom, be indeed helpful for students in the way envisaged by Mezirow. Subsequently to Mezirow's contribution, Boyd focused on the sub-conscious and those factors which escape easy observation on the part of one's conscious self, and so limit the effectiveness of the kind of logical discussion which Mezirow had argued strongly in favour of Boyd (1991) in Taylor (1998).

Researchers working with transformative learning theory adopt different definitions of the theory, which arguably reflect different degrees of ambition as to what the theory is supposed to achieve. At one end, Cranton states that "Broadly, transformative learning occurs when people critically examine their habitual expectations, revise them, and act on the revised point of view" (Cranton 2006, in Thomas 2009). At the other end, O'Sullivan argues that "Transformative learning involves experiencing a deep, structural shift in the basic premises of thought, feelings, and actions. It is a shift of consciousness that dramatically and irreversibly alters our way of being in the world. Such a shift involves our understanding of ourselves and our self-locations; our relationships with other humans and with the natural world; our understanding of relations of power in interlocking structures of class, race and gender; our body awareness, our visions of alternative approaches to living; and our sense of possibilities for social justice and peace and personal joy" (O'Sullivan 2003).[24] In between these two positions lie a range of views involving different understandings of the degree of depth of transformation of perspectives which transformative learning is supposed to achieve.

The focus of this paper is not on the idea that that transformative learning can take place in the classroom, which is the usual setting for the application of the theory,[25] but that it takes place 'naturally' as a result of climate change, both directly, as a result of exposure to unusual weather conditions, and indirectly, though exposure to the media and government policies (not to mention other agents of change such as local associations). As Homer-Dixon states, "surprise, instability and extraordinary events will be regular features in our lives" and these events will "transform our

---

[24] One could add to O'Sullivan's definition a sense of becoming more compassionate towards others and of being part of a single human family, consistently with the approach taken by researchers studying spirituality such as Zohar and Marshall (2000).

[25] As Thomas (2009) points out, the UK's Higher Education Academy is keen to develop in students "problem-solving skills in a non-reductionist manner for highly complex real-life problems": climate change is clearly an example of such a problem.

outlook forever" (Homer-Dixon 2006, in Sterling 2011). Sterling (2011) adds: "Transformative social learning..., whether precipitated by energy price shocks...or global warning ...[,] is already with us, shaking public assumptions".

Indeed, the sort of change envisaged by the literature on climate change and climate change policy bears strong similarities with the sort of change which transformative learning theorists grapple with. According to Brody et al. (2012), "Continued habitual anti-environmental behaviour is difficult to change without a transformation of underlying values and preferences (Dahlstrand and Biel 1997; Monroe 2003; de Groot and Steg 2009), or a disruption of context (Verplanken and Wood 2006)": the concept of disruption of context is arguably very similar to the notion of disorientating dilemmas found in transformative learning theory, these being events which—as described in greater detail below—in a sense oblige one to reconsider one's set of values and beliefs. A transformation of the system of values promoted in industrialized (and industrializing) societies is advocated by some commentators; for example, in the fast growing field of behavioural economics, Gowdy (2008) writes: "The current crisis of sustainability cannot be resolved within the confines of the system that generated it. For economic analysis, this means stepping outside the Walrasian system with its emphasis on one part of human nature (greed and egoism) to the neglect of the other facets of human nature (cooperation and altruism)". In a similar vein, Jackson (2009) argues that "prosperity without growth" is both needed and possible, his argument finding support in recent findings from the well-being literature in psychology and economics that enjoying good relationships (e.g. with friends and family) is a robust determinant of life satisfaction, whilst income does not appear to be as robust (see, for example, Bruni 2010).

Transformative learning theory was very recently been proposed by Chen and Akilah (2015) as a valid theoretical underpinning for environmental education, in particular for promoting sustainable behaviour. They see transformative learning theory as a useful ally in challenging the "consumerist and capitalistic values that are ingrained into the fabric of culture and government" in a "significant, if not dominant, portion of the world". The authors go on to propose role-play simulations—involving, for example, the mimicking of international climate change negotiations—as a way of achieving transformative learning. Their pioneering work is substantially different from that of this paper in that at the core of this paper is the recognition that reality itself, through climate change, initiates and promotes transformative learning.

Table 18.2 below summarizes the experience of the interviewees (rows) organized into themes (columns) that are typically found in the transformative learning theory literature:

Table 18.2 Evidence of transformative learning in the interviewees' experience

Respondent	Keyword(s)	Age	Gender	Evolutionary change	Change marked by revelation	Attitude change due to simple receipt of information	Peer pressure	Friend's input	External influence other than peer pressure and friend's input	Rational approach	Strong sense of identity	Strong sense of justice	Perception of personal experience of climate change	Artistic element	Environment as part of a package of values	Change towards the environment accompanied by change towards people	Importance of cultural and religious background	Spiritual dimension	Simple, positive message for humanity
D.Y	Exploration	27	M		Y	Y				Y	Y	Y	Y						Y
K.T	Struggle	25	F				Y				Y	Y				Y	Y		Y
M.B	Disillusionment	45	F				Y									Y			
C.B	Rationality	49	M	Y		Y		Y		Y									
F.B	Meaning	27	M	Y							Y	Y		Y	Y		Y	Y	
D.O	Mulling over	58	M	Y					Y	Y	Y	Y		Y	Y		Y		Y
S.G	Justice	46	F	Y	Y			Y			Y	Y			Y	Y		Y	
D.M	Periphery	38	F			Y				Y			Y						
C.T	Coercion	25	M		Y	Y			Y	Y			Y				Y		Y
J.E	Paradox	20	M						Y				Y						Y

Where 'Y' stands for 'Yes'

Disorientating Dilemmas In transformative learning theory, disorienting dilemmas are unexpected events which are inconsistent with an individual's beliefs and so oblige the individual to re-assess these beliefs—the alternative being rather unhealthy, such as denying that an event took place at all, or adopting an interpretation inconsistent with reality, both strategies having as aim the avoidance of the discomfort that often comes with abandoning one's point of view. The role of disorientating dilemmas is therefore fundamentally very positive.[26] Wolf and Moser (2011) ask: "what internal processes and/or outside events can change one's beliefs?"; these 'outside events' can be understood to be disorientating dilemmas. Six interviewees have undergone such dilemmas. One example is that of D.Y., who saw the documentary "An inconvenient truth", was struck by it and reflected on it, believed its message and began to recycle.

Willingness to Have One's Attitudes Challenged Wolf and Moser (2011) go on to ask: "What psychological capacities are needed for people to confront the 'inconvenient truths' of climate change?" An important psychological capacity to be able to successfully undergo transformative learning, and one which unfortunately is often underdeveloped in people, is the willingness to have one's attitude challenged. Four interviewees display this characteristic, for example D.Y., who, having seen the documentary "An inconvenient Truth" and having believed its message to the extent that he began to recycle, then read some scientific literature which raised doubts about the existence of climate change, and found its arguments to be convincing. D.Y. therefore repeatedly manifested a willingness to have his attitudes challenged.

Willingness to Re-evaluate Beliefs and Assumptions Derived from Others This is another fundamental attitude which is needed for transformative learning to take place successfully. It is part of one's willingness to have one's attitudes challenged, and indeed the individuals who displayed it mostly coincided with those individuals who were willing to have their views challenged, indicating an openness of mind to new ideas and perspectives.[27]

Receiving Symbolic Content and Analyzing It As Dirkx explains (Dirkx 2012), Jungian or post-Jungian psychoanalysis seeks to foster a process of holistic devel-

---

[26] Disorientating dilemmas are often painful, but the pain can be proportional to the gain in terms of increased depth of meaning-making. Loewenstein (2009) describes his father's experience of being "interned in a French prisoner of war camp, hungry to the point where he dug up worms for food and chewed on shoe leather. But he once reported to me that being in the camp was the peak experience of his life". Lowenstein then goes on conclude that "[...] happiness is only one of many things that make life worthwhile, and many of the other things, such as meaning, wisdom, values and capabilities often come at the expense of happiness."

[27] C.T. would be included in this category in the table if his prejudices about the World Bank derived from comments made by other people, but the author does not know if this is the case.

opment which envisages "people as naturally moving toward wholeness through recognition of and relationships with the unconscious and consciousness". In particular, "as the ego gains strength and dominance in our adult lives, the unconscious compensates by manifesting itself in dream images and in other states in which the ego's control is reduced, such as emotional arousal". In this process, a dual relationship can be observed in which "the inner worlds of the individual come to be a voice for the inner worlds of the collective (Sardello 2004), and [...] the inner worlds of the collective shape the meanings we construct at an individual level". The "gentle knocking of the soul on the doors of consciousness", as Dirkx metaphorically put its (Dirkx 2012), can manifest itself as "imaginative engagement and elaboration of the inner stories, the private myths (Stevens 1995) that seem to implicitly guide and inform our lives". The process of development is also one of healing, whereby subconscious thoughts and feelings are allowed to emerge at the level of consciousness; "deepening [one's] understanding of our relationship with the unconscious requires [one's] ability to name, describe, elaborate, and give life to the stories that reflect the voice of the soul" (Dirkx 2012).

The experiences of two interviewees are well described by the Jungian school of thought: F.B. and D.O. both developed strong pro-conservation attitudes which were supported by "collective myths". F.B. in particular, identified with the bird of a Brazilian folk-tale which let drops of water fall on forest fire, and D.O. acknowledged that the lyrics of the songs which he had listened to as a youth, and which he remembered fondly, had had a clear impact on the values which he held dear.

The kind of learning and change entertained in the Jungian perspective "underscores the complexity of the human condition and how our sense of self and ways of coming to know are intimately bound up with our deep relationships with ourselves, as well as one another, our social contexts, and the broader world" (Dirkx 2012). In other words learning about our attitudes towards the environment, ourselves, and others come as a single "package"; in this way Jungian thought provides a theoretical basis for the term "the environment as a package of values" which appears in one of the columns in Table 18.1.

Action Fear, Rosaen, Bawden and Foster-Fishman (2006), in Sterling (2011) emphasize that transformative learning requires not only critical thinking and reflection, but also action. The majority of interviewees took some kind of action, most commonly recycling, as a consequence of undergoing a change in attitude towards climate change.

Grieving This is the process by which one suffers as a result of a loss of a dearly held view such as a particular perspective on life which one no longer sees as valid, usually as a consequence of experiencing a disorienting dilemma. Three interviewees display grieving, e.g. C.B. for knowing that his views are very much based on reports in the media and feeling that the media's message could well change. A number of interviewees underwent discomfort of some kind, e.g. K.T. for being one of the very few people to recycle newspapers in the underground, but her and others' experience of discomfort are not as a result of undergoing a change of views, so it cannot be classified as 'grieving' as defined in transformative learning theory.[28]

Depth of Transformative Learning According to Sterling (2011), transformative learning can take place at various levels of depth. Following Sterling's approach, an approximate indicator of the level of depth of transformative learning experienced by the interviewees has been included in Table 18.2, whereby '0' represents no transformative learning, a '1' represents a significant degree of transformative learning, whist a '2' represents a very significant degree of transformative learning.[29]

If no transformative learning has taken place, the subject has received some information and has simply adjusted his or her behaviour in response to it, but has not undergone a change in perspective; such a process is described as 'change within changeless' by Clark (Clark 1989, in Sterling 2011) or as 'doing things better' by Sterling (2011). None of the interviewees fit neatly this description of undergoing no transformative learning at all, though one could argue that in the case of three interviewees (D.Y., K.T and D.M) the re-assessment of own beliefs had not been deep enough to be described as transformative.

If a significant degree of transformative learning has taken place, an individual has put in a considerable effort in reconsidering his or her beliefs, and has devoted much thought to 'doing better things' (Sterling 2011). A number of interviewees display this behaviour, for example D.O., who, in his late fifties, describes himself as having moved away from a "it's plain right" view, to one which is "more complex" and sees himself "moving through arguments, and still moving through arguments", but with a "passion behind it". Finally, if very a significant degree of transformative learning has taken place, the individual has undergone a paradigmatic change (for O'Sullivan (2003), a dramatic shift in consciousness), is aware of such a change and, in the eyes of some commentators,[30] has come to a 'transper-

---

[28] Grieving in an educational context can be a necessary and unavoidable part of coming to a better (healthier, more realistic, etc.) perspective on life, and contradicts the cheap slogan that 'learning should always be fun' (Sterling 2011). Learning can be most uncomfortable, even traumatic, but, if successful, leads to greater personality integration.

[29] In Sterling's terminology, no transformative learning is labelled 'first-order change', a significant degree of transformative learning is labelled as 'second-order change', and a very significant degree of transformative learning is labelled as 'third-order change'.

[30] See for example Zohar and Marshall (2000).

sonal ethical and participative sensibility' (Sterling 2011).[31] A good illustration of this description is given by S.G., who in previous days had been, in her words, "a child of the throwaway society", but had subsequently moved away from a perspective on life where "it had all been about [her]self", towards one whereby she "noticed things more" in relation to people and the environment.

Conclusion

This article's contribution to the literature on attitudes towards climate change has been twofold: it has focused on individuals who have claimed to have undergone a change of attitudes with regards to climate change, and it has argued that transformative learning theory can be a useful framework with which to study the changes in attitudes of the individuals in question.

A number of results emerge from the interviews. Firstly, a conservation activity such as recycling can be an enjoyable exercise, a "joyful sort of sacrifice" in the words of an interviewee, so much so "once you start it's hard to stop", according to another interviewee. Climate change communication experts may well wish to incorporate this message in their communication strategy, consistently with the consensus in the literature that the message should be in essence a positive one, rather one primarily of fear (Wolf and Moser 2011).[32] Recycling that is so integrated into one's lifestyle suggests that transformative learning has taken place: indeed, according to O'Sullivan (2003) definition, transformative learning involves a "shift in consciousness that dramatically and *irreversibly* alters our way of being in the world. Such a shift involves [...] our visions of alternative approaches to living; and our sense of possibilities for social justice and peace and *personal joy*" (italics added).

Another interesting result that emerged from the interviews relates to the potentially positive role of compulsory activities. For example, one interviewee was obliged as part of his university assignment to read Nick Stern's Review, an activity which *ex post* he was very pleased to have carried out, and which enabled him to overcome prejudices about the World Bank which had prevented him from reading Stern's work. Having read it, he was convinced of its validity. Compulsory activities of this kind fit very well the definition of 'disorientating dilemma' in

---

[31] Some commentators believe that 'higher order learning' comparable to in-depth transformative learning is necessary to generate a change of worldview necessary for (ecological) sustainability (see Lyle 1994, in Sterling 2011).

[32] As Hulme suggests (Hulme 2009), climate change can be seen as an opportunity for doing something creative and imaginative. Addressing climate change in this way can promote personal development.

transformative learning theory, these being (normally painful) events which make it possible for people to re-assess what they believe to be true. Compulsory activities would ideally be introduced in a balanced fashion; in higher education, for example, this could be done by providing students with information by both scientists who believe that anthropogenic climate change is happening and those who don't, and encouraging students to make their minds as to which side is more likely to be correct.[33]

Another important result which emerges is that successful transformation of attitudes towards climate change is sometimes accompanied by images, songs and other various artistic factors which steer the individual in the right direction, or support him in his conservation activities by providing meaning and inner strength. This is consistent with a Jungian approach to transformative learning, in so far as an individual is able to consciously access these images and songs which are considered manifestations of the subconscious.

Focusing on songs and images can help overcome the free-rider problem that is commonly found in the economics literature, the argument being that "rational" (and self-interested) people will wait for others to incur the cost of addressing climate change, and take no action themselves. Images and songs seem to help the idealistic, healthy part of us to emerge and to prevail over the "rational" one understood in a very narrow sense (that is, in a sense that does not allow for altruism). Images and songs also help people to focus on intermediate goals, as seen in the experience of the interviewee who identified with a bird in a Brazilian folk tale, and whose objective was not to so much to put out a forest fire, but to drop tiny drops of water onto the fire, so that the instrument (dropping tiny drops, representing the limited contribution that each one of use can make to the problem of climate change) becomes the objective. In addition, if, as mentioned above, recycling is enjoyable or at least becomes habitual, this too can help to overcome the free-rider problem and more generally one's barriers against taking conservation actions.

Perhaps not surprisingly, interviewees who underwent a deep change in attitudes in general were also the ones who displayed a high degree of transformative learning.[34] They displayed greater personality integration in so far as they had reflected deeply on the problem of climate change and had been able to give it personal meaning, and found an orderly place for it within their own, new perspective on life. These individuals are clearly intrinsically motivated to take action to address climate change, rather than being primarily motivated by, say, financial incentives, though such incentives clearly help. Intrinsic motivations help individ-

---

[33] An interviewee who had taken a course in research methods had benefited from being exposed to both sides of the debate.

[34] These are the individuals with a '2' under 'depth of transformative learning' in Table 18.2.

uals overcome any free-rider consideration described earlier. As Pidgeon argues, "Framing climate change to resonate with people's (moral) norms and values (e.g. Crompton 2011), rather than as a simple issue of economic choice [...] is part of the challenge" (Pidgeon 2012b).

The fact that all interviewees who underwent a change of attitude with regards to climate change exhibited some degree of transformative learning makes transformative learning theory a plausible framework with which to understand people's changes in attitudes in relation to climate change. The depth of transformative learning experienced, however, varied significantly from individual to individual, with some exhibiting only a limited degree of re-structuring of perspectives and grappling with meaning-making, whilst others went much further, exhibiting an in-depth transformation of views, a change in paradigm and new sense of connectedness with others and with the world around them.

Whilst this study is merely an exploration of the application of transformative learning theory to the public's attitude towards climate change, the author hopes to have shown that there is merit in pursuing this approach. Future research may consider particular aspects of the theory, such as the role of disorientating dilemmas, and delve more deeply on ways in which policy makers may use transformative learning theory to enable people to have positive and meaningful (though probably not pain-free) experiences of dealing with climate change.

Acknowledgements The author would like to thank Craig Duckworth and Kmar Makni for conducting some of the interviews. In addition, the author is grateful to London Metropolitan University for financially supporting the project.

About the Author

Gherardo Girardi is a senior lecturer in economics and business strategy at London Metropolitan University, UK. His research covers, in addition to matters related to climate change, teaching and learning in higher education, and industrial economics.

Appendix: The Participants' Narratives

D. Y. is an American student in his late 20s studying business economics at undergraduate level. D. Y. saw Al Gore's documentary "An Inconvenient Truth", which "made him think" and decided to start recycling. Then sometime later he read "The State of Fear" by Michael Crighton (the author of Jurassic Park, D. Y. explained), a "fictional book", "but it has a very good bibliography of scientific articles". D. Y. "checked the articles", which challenged Al Gore's extremely worrying view. For example, said D. Y., Crighton's book claims 90 % of global ice is in Antartica, and that the ice there is getting thicker. Now, when D. Y. hears about global warming, his attitude is one of scepticism. Nonetheless, D. Y. is still recycling, as "once you get into the habit, you see it's not difficult". He now feels bad about throwing in the bin what he knows he can recycle. He recycles out of a "greater sense" of the common good, not out of peer pressure, and not out of fear that the humanity will "disappear". He regards recycling as a joyful sort of sacrifice and not very costly. He thinks that consumers (rather than governments or firms) are responsible for pollution, since demand generates supply. D. Y. thinks that there is a lot of "inequality in the world, pollution, war…we can't fix them now, but you want to make things easier for future generations". He thinks that "every generation wants the next one to do better than them", and also thinks that the majority of people think this, though this is "the silent majority". D. Y. was intending to go to a war torn country to build homes for victims of landmines.

K. T. is a Tunisian student studying International Business at Master's level. K. T. thinks that most people in Tunisia, independently of their social level and of the place they live in, "don't really care about streets' cleanliness and are not really aware about the fact that throwing away food is a bad thing". K. T. said that where she lives in Tunisia there is a lady who works as a cleaner, collects plastic bottles from the house where she works and from her friends, and takes the bottles to a factory in return for some money. This lady pays her customers about 750 millims per kg of plastic bottles, with which one can buy about four baguettes or just under half a packet of "decent" cigarettes. K. T. said that when she went as a tourist to Germany, she was quite impressed by the extent of recycling there. Then she came to London to study for her International Business master degree, and felt indignant about the tabloids left lying on the tube. Initially she used to simply leave the papers on the tube because she believed that the underground would recycle them. When she realized that the underground was not doing so, she began collecting the papers herself. However, she felt very discouraged that she was the only one doing so; she was expecting more from people living in a

(continued)

"developed country" and thought the whole thing "quiet depressing if you think about a better future". The indifference of her fellow passengers contrasted vividly with the way she had been brought up: "At home we would never throw away a jar of Nutella, we would wash it and reuse it as a recipient for rice, pasta, or any kind of food that could be kept in a glass; we would never throw away clothes, we would either give them to people in need or use the ones that couldn't be worn as cleaning towels, for dusting or for use in the garden; we would never throw away bottles made of glass, we would rather use them to store flower essences or olive oil...and all this not because we couldn't buy more but because it was the logical and most natural thing to do". Ultimately she decided to stop collecting papers, otherwise, she said humorously, she would have ended up seeing a psychotherapist. However, she did not stop recycling entirely: she still did a small amount of collecting. To make things worse, the "expensive student hall" in which she lived did not separate the garbage. She recycled and told others to do the same, but in the end she "essentially gave up" recycling in her hall as she saw no point in doing it if other people did not do it, since on her own she was not going to able to "change the world". Instead, she says, recycling is an area for government intervention. If she was the government, she'd forbid paper envelopes and fine people for littering.

M. B. is British woman in her mid 40s and works as a manager at university. She has three children and has a background in natural sciences. M. B. used to work in the private sector and then moved to university. She accepts climate change is happening not only because of what she reads, but also because personal observations such as the fruit in her garden not maturing at the right time. M. B. was hopeful and confident about the government's ability to tackle climate change and pollution in general when she was working in the private sector. However, since she started working in the public sector, i.e. at university, she underwent a change in perspective. This happened for two reasons. Firstly, she describes how she was shocked to discover that students were selfish and indifferent, which led her to lose faith in young people and their willingness to care for future generations (she says het her generation was more politically active). Secondly, and at the same time, a friend of hers, working in another publically funded university, explained to her that colleagues in the friend's university doing research on the environment produced a fake report. Hence, she lost her trust in universities' ability to perform their public duties. She became very cynical and now does not believe that things will change for the better in the future. She says that "people don't even care about others in the present so how can we expect from them to care about the future or future generations". The only way for things to change is if the government "hit [...] people where it hurts, i.e. in their pockets". She

(continued)

believes that there is no point of informing people of the harmful causes of climate change since they won't do anything to change anyway; she argued that this phenomenon is like smoking, you know it is bad but you do it anyways, even if there are warnings on the packet itself; or like crossing the street when the traffic light is red: people know it is dangerous but they do it anyways. Even though she recycles the garbage (though not the newspapers due to there not being adequate bins), she thinks that recycling is not a solution as it is very expensive and not really effective; instead, we should curb excessive consumption and waste, which are consequences of people's greed. Finally, she suggested that if people want to maintain the same "life style" they have today, nuclear energy could be the only solution.

C. B. is a white British man aged 49, studying Japanese part time in the evening and unemployed for approximately a year. He is not married and has no children. He says he went from being entirely sceptical of the idea that climate change is anthropogenic, to being convinced that it's actually true. His conviction stems from having heeded the message in the media: "I don't like saying this, but if you repeat the same message over and over again, you'll believe it; that's what happened to me." Nonetheless, C. B.'s careful use of language suggests a rational and calculating approach to climate change (though his demeanour displays a certain reserved warmth). C. B. worries about the validity of the information he reads in the media. Having said that, he says he is more comfortable believing the British media than the UN's IPCC (Intergovernmental Panel on Climate Change); when asked why, he says "maybe it feels more like home", and "perhaps just because the logic fits my mind". He wonders how qualified he is to interpret the data he reads in the press. Then he says he does not trust the British media, instead he trusts a friend who "has an alternative way of thinking". When asked whether his conviction that climate change is anthropogenic would lessen as a result of moving to a country where the media is less supportive of this view, he said yes, though not completely. He then manifested detachment from what the media says by saying that he has always cycled and recycled, and that if more reports of worsening climate were to reach him, his behaviour would not change (this includes minimizing waste and plastic bags usage) as "we've already reached the stage where things are very hard or impossible to change". He is at ease with his level of environmental conservation, in fact he's happy to fly by plane as he sees that as a reward for his conservation activities. He thinks that, although new technologies may be found which mean that we don't have to cut consumption now to leave a clean planet for future generations, nonetheless we should be responsible and cut out excesses, and is in favour of government regulation like compulsory recycling. It is unlikely that we care about future generations more than

(continued)

about ourselves, as "everyone has a need to preserve himself", and we all seek some level of comfort, except for rare exceptions like nuns.

F. B. is a Brazilian 27 year old man studying for the MA in International Sustainable Tourism. He is not married and has no children, describes himself as a practising Catholic and has lived the in UK for 6 months prior to the interview. In 2005 he began reflecting on his life choices, in particular looking for ways to have a "worthwhile existence": he began to think about "not only [...] my career but choices, things that could make me feel accomplished and happy". He wanted to influence policy, but also to make a personal contribution of sorts. At the same time, he was exposed to news on TV and in the press on environmental degradation. He responded to this message by taking practical steps such as not buying a second computer (so as to reduce industrial wastage), and by spending less time in the shower (though he also did this to reduce his energy bill). While he was effecting this transition in behaviour, a Brazilian folkloristic tale came powerfully to his mind, that of a bird which tries to put out a forest fire by dropping tiny drops of water on it, uncaring of whether he succeeds in putting out the fire or not. In the story, we don't know whether the fire is ultimately extinguished (i.e. we don't know the story's ending, in a sense), instead, we appreciate that the bird's action is important and valuable in itself. F. B. feels at a deep level that we do not own the Earth; instead, his views of life are strongly captured by the notion of "convivencia", a word which, he says, has no exact translation in English, the closest being "co-existence". He is passionate about the Earth's beauty and feels a need to "apologize to the planet" for the way we have treated it. F. B. is very family rooted and maintains strong links with his relatives. He feels that he knows he's right about his vision and purpose of life. When asked why we should care about future generations, his answer is "We are a future generation—somebody cared about us—or are we just lucky?". Furthermore, "We have received freely, let us give freely". Finally, he thinks it is important to affect change in teenagers, who don't care about the future, and so we need to work on their "sensitivities".

D. O. is an American in his late 50s; he lectures in business and has a teaching interest in sustainability issues. When he was fifteen, he memorized the lyrics of "There goes the mountain", a 1960s song about strip mining. He says strip mining was very important in this period. Also he remembers that Pete Seeger sang about the need to get the Hudson river cleaned up. D. O. recollects this period as one in which he was searching for identity, which he sees this as a typical teenage thing. His parents, who had a stronger religious approach to life than him, raised him with a strong sense of fairness. At this time, D. O. also sympathized with the civil rights movement and the anti-

(continued)

Vietnam protesters. "It was civil rights, Vietnam and the environment. It seemed plain right to be concerned about these issues". Gradually he began to appreciate economic activity as a way of improving human welfare. As he became more involved in business issues, he began to "see good business leaders", and his sympathy for business increased. In 2004 a major event occurred which left a mark in D. O.: he read Lomborg's "The sceptical environmentalist". Lomborg and others, said D. O., argued that we should select priorities based on evidence, and suggested that fighting malaria be given priority rather than climate change. This encouraged D. O. to move away from a "it's plain right" view, to one which is "more complex" (i.e. it is not obvious that addressing climate change is a priority). This attitude has continued to this day and he sees himself "moving through arguments, and still moving through arguments" (but he is "emotional nonetheless", with a "passion behind it"). For example on the one hand he believes that, given time, technologies will be developed to address environmental problems; on the other hand he also believes in responding quickly to problems as they arise, for example consumers should "stop buying 4by4s", and firms should "stop making cars that go at 120mph". His view of nature is that that there are natural "healing and adjustment processes" and so we need to allow nature to heal itself; however, "we could get out of equilibrium and not be able to get back" (i.e. there may not be a natural healing process in some circumstances), for example if the Gulf stream comes to an end (he backed this last point with reference to systems theory). Currently his professional view is that accounting systems should take into account externalities, and that genuinely green accounts would show companies making a loss. He also sympathizes with shareholder activism, whose founding figure is Bob Monks, where activists buy shares to affect companies' actions.

S.G. is a German citizen who has lived in the UK since late 1989. Although recycling was "big" in Germany, she took little interest in ecological matters when she lived there, describing herself as a "child of the throw-away society". About 12 years before the interview, two friends of hers got her interested in alternative holistic medicine (such as herbal medicine), in the environment and in the idea that everything is interconnected, though she never actually adopted any particular "belief system" (she does however reject an anthropocentric view of the universe). She joined a vegetable and fruit delivery scheme and began to read widely on the environment, in a process which she describes as a "self-feeding system: once you start, it's very hard to go back". She said that up until this time, "it had been all for myself", after which she "started to notice things more" in relation to other people and the environment. S.G. began to see inequality in many aspects of life, for example in terms of unequal access to freedom of expression and

(continued)

economic access. Since moving to the UK, she has been working in the voluntary sector, including a human rights organization which protects indigenous people in economically less developed countries. Then about a year before the interview she attended a course in criminology during which the instructor spoke about "eco-crime". It suddenly dawned on her that this is the specific area which in the future she would like to work in. She feels that she had "needed time to come through things", i.e. to make this discovery. When asked whether she thought she had discovered her "true preferences", S.G. gave a resounding yes. She says that the attitude we have towards the Earth reflects the attitude we have towards ourselves. When asked whether she thought that pro-environment attitudes had to be formed or awakened, she replied "formed". To do this, she said we should appeal to people's survival instinct. She said that in our times it is possible to change one's belief system in a lifetime, "unlike in the distant past". She felt that even though "the media is not fair", as "it is governed by who owns it", one could overcome the problem of bias by reading both "liberal and conservative papers". She felt that the one could trust scientists (who claim that climate change is taking place) as there have been "thirty years of long debate and controversy" before reaching that conclusion.

D. M. is a member of staff from Mauritius and is in her 30s. She read a lot about the environment in the UK through the media, whilst in Mauritius "you never talk about these things". She says she knows that climate change is taking place because this is confirmed by her experience in Mauritius, where "it is becoming hotter". Her family in Mauritius share her view about climate change in Mauritius, as do other family members living in South Africa. D. M. recycles "maybe a little out of guilt for being a vibrant consumer", but most importantly because she feels she "must do so as a responsible citizen"; also to an extent because recycling is "not too bothersome" for her, as the council "makes it easy" by placing recycling facilities within reach. However, she often does not make the connection between recycling and climate change. She says she does not feel any peer pressure to recycle, it is her conscience that puts pressure on her. She puts pressure on her husband to recycle. She thinks that a lot of the media coverage about climate change is "overblown" and "sensationalist", opinion is often presented as fact, and fear is created. Nonetheless, she thinks there is an element of truth in it: "if it is not there, is it is usually contested" and "there's no smoke without fire". Also "maybe these messages [of fear] are needed to get to people". She has become very conscious of her carbon footprint, for example she travels by air a lot less, taking the train instead. She says, with a dash of humour, that she cares more about her carbon footprint than about the environment. She thinks one could motivate people to become eco-friendly by encouraging them to make

(continued)

significant but inexpensive changes, and suggested a book entitled "Change the World for a Fiver".

C. T. is a Turkish student studying International Business at Master's level. He is 25, single and has no children. Up until taking the course on Research Methods, C. T. had refused to read the work of Nicholas Stern and in particular he had not read the Stern Review on the Economics of Climate Change. This is because C. T. was sceptical of the work of the World Bank, the institution for which Nicholas Stern used to work. C. T. wondered whether the World Bank "was there to help rich or poor countries" and thought that Stern "was getting paid to produce this kind of report". When taking Research Methods, C. T. had to read the Stern Review as part of the module's compulsory reading. He also read some of the work of William Nordhaus, whose message is in many ways opposite to Nick Stern's (Nordhaus' message does not inspire the same sense of urgency as Stern's). As a result of reading this material, C. T. changed his mind from being sceptical about climate change to believing that it is really happening and that it is anthropogenic. His own experience of climate change in Turkey, he says, matches the conclusion of these authors. This new perspective however has not affected his behaviour since he was "recycling even before reading the Review". C. T. says that being a Muslim makes him socially responsible as in Islam "everybody should clean up their mess" and "must be totally environmentally friendly". C. T. regards himself as someone who bases his views on what he reads and believes he does not blindly follow the crowd. His view is that people, if informed about climate change, will change their behaviour. Indeed he regards the diffusion of "information as the number one priority for both developed and developing countries".

J. E. is second year undergraduate studying finance and economics. He has lived all his life in Estonia except for the last one and a half years, during which he has been living in London. Up until 1991, Estonia had been part of the Soviet Union, after which it became an independent state. At the time when J. E. was in secondary school, every Saturday the school sent its pupils "in groups to clean the forests", and "the monuments of soldiers who died" in the Second World War. The pupils did this for free, as part of their school duties. However, says J. E., "When we became capitalist, we stopped doing this", so that the forests and other places are now dirty. He also says "Now everybody screams about democracy and freedom, so you cannot make them do it [the compulsory cleaning of the forests by children]. We were taught to do it, didn't think about it, and when the area became dirty you could see the benefit of the Saturday". Today secondary students in Estonia are not obliged to clean up, rather they are taken to landfills "to see where the garbage goes

(continued)

and [...] speak with specialists". J. E. went to secondary school when the education system went through this transition process, so he did the cleaning of the forests in his first year and was taken to landfills later. He approves both of compulsory cleaning (even if not counted in GDP, he says) and of visiting the landfills. At the same time, the arrival of capitalism in Estonia also presented new commercial opportunities, so J. E.'s family realized there was profit to be made by making container skips and rubbish compactors, which are used extensively in shopping malls to dispose of packaging. Now his family is making and selling such products. The sector is very competitive, and the firms in the sector (including J. E.'s family business) are putting pressure on the Estonian government to introduce policies that will make Estonian industry greener, directly through measures like taxes on the use of landfills, and indirectly by persuading Estonian manufacturers to adopt a more eco-friendly attitude (and equipment). The EU is applying a similar pressure on the Estonian government. Firms like that of J. E.'s family set the prices of their products to be just below the cost of disposing of rubbish via landfills.

References

Benedict XVI (2009) "Caritas in veritate" (charity in truth). CTS, London

Brody S, Grover H, Vedlitz A (2012) Examining the willingness of Americans to alter behaviour to mitigate climate change. Clim Pol 12(1):1–22

Brulle RJ, Carmichael J, Jenkins JC (2012) Shifting public opinion on climate change: an empirical assessment of factors influencing concern over climate change in the U.S., 2002–2010. Clim Change 114(2):169–188

Bruni L (2010) The happiness of sociality. Economics and eudaimonia: a necessary encounter. Ration Soc 22(4):383–406

Capstick S, Whitmarsh L, Poortinga W, Pidgeon N, Upham P (2015) International trends in public perceptions of climate change over the past quarter century. Clim Change 6(1):35–61

Chen JC, Akilah RM (2015) Role-play simulations as a transformative methodology in environmental education. J Transform Educ 13(1):85–102

Corner A, Markowitz E, Pidgeon N (2014) Public engagement with climate change: the role of human values. WIREs Clim Change 5(3):411–422

Dirkx JM (2012) Nurturing soul work: a Jungian approach to transformative learning. In: Taylor EW, Cranton P, and associates (eds) The handbook of transformative learning: theory, research, and practice. Wiley, San Francisco, CA

Fujii S, Gärling T (2003) Development of script-based travel mode choice after forced change. Transport Res F: Traffic Psychol Behav 6(2):117–124

George D (1998) Coping rationally with unpreferred preferences. East Econ J 24(2):181–194

Gifford R (2011) The dragons of inaction. Am Psychol 66(4):290–302

Goodson IF, Sikes P (2001) Life history research in educational settings. Open University Press, Buckingham

Gowdy JM (2008) Behavioral economics and climate change policy. J Econ Behav Organ 68(3–4):632–644

Hulme M (2009) Why we disagree about climate change. Cambridge University Press, Cambridge

Jackson T (2009) Prosperity without growth. Earthscan, London

Joireman J, Barnes Truelove H, Duell B (2010) Effect of outdoor temperature, heat primes and anchoring on belief in global warming. J Environ Psychol 30(4):358–367

Karlsson N, Loewenstein G, McCafferty J (2004) The economics of meaning. Nordic J Polit Econ 30(1):61–75

Kellstedt PM, Zahran S, Vedlitz A (2008) Personal efficacy, the information environment, and attitudes toward global warming and climate change in the United States. Risk Anal 28(1):113–126

Loewenstein G (2009) That which makes life worthwhile. In: Krueger AB (ed) Measuring the subjective well-being of nations: national accounts of time use and well-being, NBER. University of Chicago Press, Chicago, IL

MacIntyre A (1981) After virtue. Gerald Duckworth, London

O'Sullivan E (2003) Bringing a perspective of transformative learning to globalized consumption. Int J Consum Stud 27(4):326–330

Perry M, Tait P (2009) Dalai Lama says climate change needs global action. Reuters, 30 November

Pidgeon N (2012a) Climate change risk perception and communication: addressing a critical moment? Risk Anal 32(6):951–956

Pidgeon N (2012b) Public understanding of, and attitudes to, climate change: UK and international perspectives and policy. Clim Pol 12(1):S85–S106

Pulella P (2008) Vatican unveils ambitious solar energy plans. Reuters, 27 November

Roeser S (2012) Risk communication, public engagement, and climate change: a role for emotions. Risk Anal 32(6):1033–1040

Sterling S (2011) Transformative learning and sustainability: sketching the conceptual ground. Learn Teach High Educ 5:17–33

Stern N (2008) The economics of climate change. Am Econ Rev 98(2):1–37

Stevenson RW, Broder JM (2013) Speech gives climate goals center stage. New York Times, January 21

Taylor EW (1998) The theory and practice of transformative learning: a critical review. Center on Education and Training for Employment, Information series no. 374

Thomas I (2009) Critical thinking, transformative learning, sustainable education, and problem-based learning in universities. J Transform Educ 7(3):245–264

USA Today (2015) It's Christian to protect the environment, 9 February

Turk JD (2009) Traction in the world: economics and narrative interviews. Int J Green Econ 3(1):77–92

Van der Werff E, Steg L, Keizer K (2013) The value of environmental self-identity: the relationship between biospheric values, environmental self-identity and environmental preferences, intentions and behaviour. J Environ Psychol 34(June):55–63

Wolf J, Moser SC (2011) Individual understandings, perceptions, and engagement with climate change: insights from in-depth studies across the world. Clim Change 2(4):547–569

Zohar D, Marshall I (2000) Spiritual intelligence: the ultimate intelligence". London Bloomsbury, London

Chapter 19
Societal Transformation, Buzzy Perspectives Towards Successful Climate Change Adaptation: An Appeal to Caution

Sabine Trõger

Abstract Societal Transformation (ST) is taken as the promise towards successful Climate Change Adaptation (CCA) in recent discourses. With the example of the Nyangatom, one of the World Cultural Heritage (UNESCO) pastoralist ethnic groups in South Omo/South Ethiopia, any unanimous belief in the power of ST is challenged! In interpretation of data from ethnographic research, perspectives of actors on the ground highlight imminent dangers of well-recognized and accepted adaptive measures. The 'committee', central requisite of Enclosed Rangeland Management schemes, translates into a ST without return! The paper is meant to warn against a too ready euphuism towards ST as well as to give a voice to pastoralists in their needs and concerns, a voice, which is generally refused to them in given political environments in Ethiopia and in Africa in general.

Keywords Transformative adaptation • Societal transformation • Token of change

Setting the Stage: 'Transformative Adaptation', the Promise of Hail in Climate Change Adaptation

The present and future relevance of climate change in its various articulations showing up on the African continent and impacting its peoples is generally recognized. Climate change is a serious threat to human health and well being, particularly to vulnerable populations (e.g. IPCC reports 2007, 2014). Recently and with special and explicit high-ranking reference to the worldwide recognized IPPC-report published 2014, the formula of 'transformative adaptation' has emerged to be taken as the ultimate promise of 'hail' and success towards climate change resilience.

S. Trõger (✉)
Department of Geography, University of Bonn, Meckenheimer Allee 166, 53115 Bonn, Germany
e-mail: troeger@geographie.uni-bonn.de

Already in 2012 and way ahead of the '14-IPCC-Report an explicit distinction between 'incremental' against 'transformational' adaptation had been suggested with emphasis on transformational adaptation in cases of highly severe vulnerabilities and risk. Transformational adaptation is here qualified as a 'large scale adoption' of measures, truly 'new to a particular region or resource system' and with the consequence to 'transform places and shift locations' (Kates et al. 2012: 7156). In the same year 2012, 'transformation' had as well already turned to be the cornerstone in international development cooperation as, for example, to be understood by the GIZ (Deutsche Gesellschaft für internationale Zusammenarbeit) stating: " ... transformation should be a central paradigm of international cooperation (GIZ 2012: 19).

When in March 2014 the 5th IPCC-Assessment Report is released, 'transformative adaptation' is re-stated and put in place in its decisive role towards a systemic status of 'resilience': "Transformations in economic, social, technological, and political decisions and actions can enable climate-resilient pathways (high confidence). Strategies and actions can be pursued now that will move towards climate-resilient pathways for sustainable development, while at the same time helping to improve livelihoods, social and economic well-being, and responsible environmental management" (903). This trust in 'transformation' taken as the key concept for sustainable development was further mirrored in 'The Bonn Conference for Global Transformation' which was staged in May 2015 as one of the numerous activities in preparation of the SDGs (Sustainable Development Goals), to be signed in September 2015 by the UN General Assembly (www.bonn-conference.net/de/).

The paper challenges a firm belief in the power of transformation and better draws attention on a closer reflection on what 'transformation' and explicitly 'societal transformation' can mean in local contexts. Already a closer look at the above quoted IPCC-Report 2014 warns against any unambiguous belief in transformation as the ultimate goal of successful and sustainable development. This report stresses an appeal to caution: "As a fundamental change in a system transformation may involve changes in actors' objectives and associated values. Therefore transformational adaptation is not without risks and costs ... the greater level of investment and/or shift in fundamental values and expectations required for transformational change may create greater resistance" (IPCC 2014: 922). Transformative adaptation, as can be concluded, may be accompanied by threats and endangers those to be supported in their search for livelihood security and wellbeing in the face of climatic and altogether environmental change.

Focused on the example of one of the pastoralist communities in south Ethiopia, the Nyangatom in South Omo, and with additional reference to a further pastoralist group, the Borena situated to the east from South Omo, the hail promising formula of 'transformative adaptation' will be reflected in its comprehensive localized meaning.

Using village-level ethnographic data the paper supports the following thesis: The formula of 'transformative adaptation' does not necessarily mean altogether a change towards more resilience and sustainable development, i.e. towards the positive. Climate change adaptation in many observed cases translates into innovative measures and techniques in e.g. the fields of agriculture, land use systems, and energy generating schemes. In consequence the underlying reasoning in favor of these innovations is ecologically motivated to be then paired with the ideal of 'governance' as deducted from the 'northern' concept of democracy. Propagated measures in improved land use techniques in the broader sense go hand in hand with a management ideal of equitized representation.

In favor of a better and more substantial understanding of processes initially kicked off by changes in terms of 'transformation', the paper turns against a too euphoric interpretation of these. Climate change, its causes, consequences and possible response options remain contested by different groups in society and are associated with different perspectives in power and dominance. Against notions, which continue to primarily discuss social effects of climate change in terms of tolerable risks for societies as a whole (Adger et al. 2009), the more substantive aspects of social justice need attention:

- What distributional effects will societal transformations have within societies?
- How are questions of societal justice, participation, and the distributive effects of climate change addressed in societal debates?
- What changes in modes of governance could be essential for transformative processes (see Driessen et al. 2013: 3f)

The paper outlines the argument in four steps: In order to set the analytical context the argument refers to theory discourses, which conceptualize agency of local actors in reflection on their social environment. In correspondence the Nyangatom landscape of risk is depicted in global to local scales. In answer to this setting of risks the scheme of 'enclosed rangeland management' is subsequently addressed in its ecological reasoning as well as in its impacts in terms of societal transformation. The argument then moves on to the Nyangatom societal environment with reference to theory discourses on the idea of 'travelling model' as the analytical representation of particular aspects of reality. Though processes of societal transformation are still ongoing and cannot be evaluated in their ultimate outcomes, comments of actors 'on the ground' finally draw a picture of societal sceneries characterized by alienated governance and external determination.

Nyangatom Pastoralist Landscapes of Risk

> Pastoralism is the finely-honed symbiotic relationship between local ecology, domesticated livestock and people in resource-scarce, climatically marginal and highly variable conditions. It represents a complex form of natural resource management, involving a continuous ecological balance between pastures, livestock and people ... (Nori and Davis 2007: 7).

Individual and social adaptation to climate is nothing new, neither as an empirical nor theoretical construct. Within the range of arguments, pastoralism is taken as highly adaptive to harsh and at the same time fragile environments. Would not these flexible and environmentally alert societal systems be predestined to likewise adapt to environmental transformations in the course of climate change—especially, as climate variability always has been some accompanying feature of a pastoralist's life?

Evidence seems to talk another language: Recent studies on pastoralist lives and livelihood system express concern about altogether the future persistence of this livelihood system. Connotations like 'weakening' (Pavanello and Levine 2011: 1), 'rangeland degradation and a rise in internal disputes' (Beyene 2009: 487), and 'alienation of pastoralists from productive lands' (Abbink et al. 2014: 1) predominate perceptions and analyses in academic discourses on pastoralism in eastern Africa.

On questioning pastoralist adaptation to climate change and its implicit meaning for sustaining given socio-ecological systems the paper focuses on the perspective of 'threshold' in adaptation in the understanding of Adger et al. (2009: 337). In accordance with this line of argument the question of limits to adaptation is conceptualized as a contingent question in dependency on socially guided constructs responding to values and the respective interpretations of 'signposts' when navigating through given landscapes of risks and opportunities, an image adopted from Schatzki, which conceptualizes various interpretations by human actors in confrontation with the very 'same objective special expanse of the world' (2010, quoted in Mueller-Mahn and Everts 2013: 26).

Correspondingly, the argument refers agency in the face of the risk of climate change induced droughts to the complexity of further motives of agency. It relates interpretations and subsequent agency of people to their own overarching idea of a landscape of their social environment, "for these landscapes are eventually navigated by agents who both experience and constitute larger formations, in part from their own sense of what these landscapes offer" (Appadurai 2006: 628). The argument recognizes landscape as lived and storied, thereby incorporating an emotional and socially valued geography. These imagined landscapes exhibit a semiotic, which will be shifting in the face of new environmental concerns and priorities, which implies that the attention has to go beyond a consideration of the ordinary object defined by its function. The object is rather to be considered in how it is experienced, by which processes and practices people relate to it, and in which ways social meaning accumulates around it. This enables us to study present-day and future questions of contested imaginaries, scientific interpretations and physical manifestations of climate change in relation to apparently unremarkable, every-day objects that are rendered remarkable for the individual subject. By this obtained perspective a closer insight into the transformation of a formerly culturally firmly established and resilient livelihood system, as of now fluid and subject to new interpretations and definitions in confrontation with adverse forces and negative impacts, climate change being only one of these, can be achieved.

The above referred to rather skeptic and pessimistic view on pastoralist livelihood perspectives in eastern Africa and Ethiopia corresponds to forces, which are located on the various levels from the global to the local, which are perceived of and reacted to by the social actors on the ground. This is their 'landscape of risks and opportunities' (Fig. 19.1): First of all the climate change imperative is to be seen as

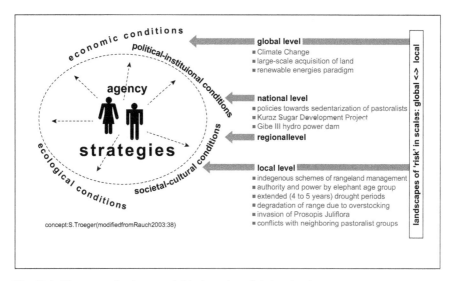

Fig. 19.1 Nyangatom landscapes of risks in scales: global ⇔ local

one of the global forces, which defines its corresponding regional impact as described below. Additionally, the large-scale acquisition of land by states and investors is to be considered as a global movement, which in most cases forcibly settles or displaces pastoralists. Sedentarization has long been the standard formula for 'development' of the pastoral sector of the world. In relation to this globalized threat to pastoralism, the urge for renewable energies in the shape of partly huge dimensioned hydro-power dams claims especially those lands, which are taken to be 'under-utilized'.

In correspondence to these global trends, provisions in the existing federal policy and law of Ethiopia reaffirm policies on the national level to expropriate land in pastoral areas for development (Mulatu and Bekure 2013: 193), which in the South Omo and Nyangatom case means the 'Kuraz Sugar Development Project', which claims 250,000 ha of pastoralist land in combination with Gibe III, a hydropower dam up-stream river Omo. This latter deprives the pastoralists, who are actually agro-pastoralists, of the chance for river-retreat agriculture on the riverbanks of the Omo in the retreat of the seasonal river floods (for further details see Abbink et al. 2014).

The local scenery, to complete the picture, is defined by the specific articulation of the pastoralist livelihood system as practised by the ethnic group in focus, the Nyangatom: Pastoralists that manage their livestock by managing their rangelands and their mobility across those rangelands. There are two sets of logical principles underpinning livestock migration. The first determines movement between wet and dry season grazing areas. The second is linked to the management of rangeland quality. Rangeland management means that pastoralists have to get the balance right, using seasonal combinations of intensive grazing of high quality with movements to zones with higher levels of vegetation, but usually of lower quality, and often with greater disease burden. Altogether, by management of the range the pastoralists create a resource for themselves, a resource out of what nature provides. This management is, or was guided by the 'elephant' age-group of the elders, which being the political and decisive group in the respective communities will be initiated to be in power for some years and subsequently be replaced by the next elephant-group in succession. These elders are respected and listened to for their life-experience and wisdom (source: oral narratives).

Further threats and influences on the local scale are the specific face of the climate change imperative in South Omo, the invasive plant Prosopis Juliflora, which has some 6–8 years ago turned the former cultural center Kibish inaccessible, and, the threat by ever increasing conflicts with neighboring pastoralist groups, especially the Turkana from North Kenya invading the grazing territory of the Nyangatom.

The further argument basically follows those academic discourses, which conceptualize agency in risky environments as the aggregated reflection of risk-interpretations by the actors bound in their specific realm of values and norms, which again produces and re-produces given landscapes of risk (see Mueller-Mahn and Everts 2013). But, in spite of this recognition of altogether complex interpretative frameworks, the argument singles out one line of perceptions and subsequent interpretations by the pastoralist actors. This is the line in the course of the interplay between the perceived impact of climate change, and, interpretations of a new technology for climate change adaptation, the enclosed rangeland management.

The Lay of the Land: Defined Fragility in Climate Change Prone Environments

In order to shed some light onto the interface between pastoralists' livelihood concerns in south Ethiopia and adaptive agency in reflection of the severity of given climate change impact patterns, a short outlook on the latter perspective, i.e. the pattern of climate change in the southern low- and dry-lands, is given.

Meteorological data on annual rainfall patterns partly suggest doubts about the severity, even altogether the reality of climate change impacts with relation to precipitation patters in Ethiopia. While there is secure data on the increase in temperature by 1.3 °C between 1960 and 2006 and estimates reckon with a further increase between 1.1° and 3.1° by 2060, annual rainfall data summarize no clear perspective and trend. In contradiction the World Bank ranked Ethiopia in February 2009 among the worldwide 12 most impacted countries with explicit focus on drought, ranking second position (IRINNEWS 2009).

In recent climate change related research increasing attention has been given to individual's and communities' knowledge and experience of past and recent climate, and the way these shape the actors' perceptions of future climate. Attention has turned towards people on the ground, the farmers and pastoralists, who experience 'their' climate pattern from day to day and season to season. In 13 months (2/2009 to 3/2010) climate change and adaptation research was realized at altogether 13 sites in 4 regions—Tigray, Amhara, Oromia, SNNPR—by intercultural teams of 8 German MA-students together with at total 18 woreda (district) experts. Their fieldwork offers a sound basis for a comprehensive insight into the climate change imperative as experienced by the people in Ethiopia in regional differentiation.

The methodological approach of this research made use of various assessment tools from approaches broadly addressed by the term 'ethnographic research', namely tools known as PRA, participatory observation, open narratives, and open, i.e. semi-structured interviews. Research activities along with the design of the methodological approach were adjusted to the definition of target groups. The arrangements were clearly distinguished with respect to gender and generational aspects. The validity of the data was safeguarded by methodological triangulation, which was especially relevant, because we were well aware of the fact that of course, farmers, pastoralists, meteorologists and climate scientists 'measure' different things when they talk about 'climate change'. The amount of rain measured is not taken as the arithmetic mean in isolation by farmers and pastoralists, but in relation to what it is supposed to do, that is, in relation to the water requirements of certain crops or of livestock.

The analysis of the accumulated data emphasizes a twofold pattern of climate change impacts in Ethiopia as depicted in Fig. 19.2.

According to these localized assessments, doubts about the severity of the climate change imperative must be rejected. Climate change takes place everywhere in the country, and as well at places, which might register annual rainfall quantities as before. Climate change just shows up with different faces: while the western highlands region of Ethiopia is characterized by more irregularities and weather extremes, the eastern-southern lowland of the country is clearly impacted by accelerated frequencies of droughts. The Nyangatom territory right in the southern lowlands is impacted by this climate change pattern, which can be

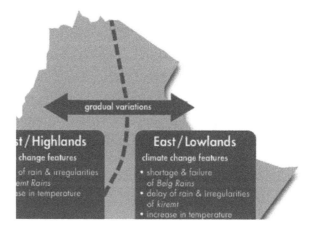

Fig. 19.2 Map: climate change patterns in Ethiopia

especially characterized by considerably shortened Belg rains, these rains being the most important rains in the eastern hemisphere at the same time. The Belg rains due end of February until end of May are part of the customary Ethiopian rainfall pattern. Eastern and southern regions of Ethiopia demarcated by the Rift Valley are highly dependent on this specific rainfall in their production in the course of the year.

The Nyangatom at present discuss spatial and quantitative changes of temperature, precipitation and flood. To their perception all deviations take their start off in the year 1989 and show a gradual decline in precipitation and floods. More severe changes are perceived since 1998, and encompass the increase of temperature, as well as a stronger variability of precipitation. Comparing the two rain periods, the Belg rains (i.e. Akoporo rains) are the most affected (Fig. 19.3).

In the year 2011, disaster scenarios projected by IPCC (2007) showed up earlier than foreseen. Large parts of the Horn of Africa were struck by one of the worst droughts in 60 years, taking shape in a severe food crisis. In July 2011, the United Nations declared a famine, but already some months before, in January, considerable quantities of the Nyangatom livestock had died. Severe droughts had all times done harm to the people and had left some pastoralists with all animals dead. But this drought, be it caused by climate change imperatives in combination with the effect of a very harsh La Niña event as suggested, took place in the succession of 4–5 years with more or less failed Belg rains.

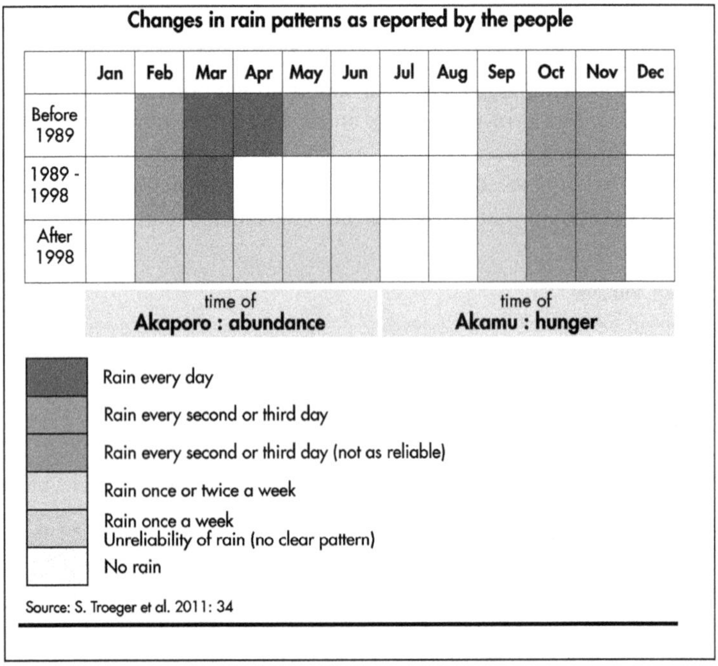

Fig. 19.3 Changes in rain patterns as reported by the people

'Enclosing the Range', a Means of Climate Change Adaptation in Pastoralist Drylands

In the face of the above outlined drought in 2011, which kept by the beginning of 2012 still 4.6 million people in Ethiopia, among these especially the pastoralists in the south and south-east of the country, in immediate need of food aid (GIZ 2012: 7), the initiative was started to establish the system of enclosed rangeland management with the Nyangatom pastoralist people. Since May 10, 2014 agreed upon community rangeland management groups, the 'committees', are legally by virtue of the district government (woreda) in place to oversee the sustainable management of the enclosed resources of their respective community land. Before focusing on this instrument of rangeland management the history of rangeland enclosure is shortly reflected on.

The technology of rangeland enclosures is a worldwide highly acknowledged measure to combat land degradation in many of the world's semi-arid rangelands, and in recent years it has become highly acknowledged with perspectives of climate change adaptation. Some decent surveys have investigated the ecological potential of these enclosures, which offer nature some time to recover and rehabilitate. From this perspective there is plausible and good reasoning in favor of the rangeland enclosures.

Enclosure does not represent formal private property rights, but it fulfills a four-fold bundle of rights comprising of 'access', 'withdrawal', 'exclusion' and 'management'. However, enclosing the formerly open and non-fragmented communal pastoralist grazing territory can mean, and in the Ethiopian pastoralist cases always means, a transformation away from undoubted 'truths' and livelihood-frameworks, which people knew by inherited experience. Against this view, the argument follows three steps: it summarizes documented experiences with the introduction of rangeland enclosures to the neighboring pastoralist community of the Borena, it outlines theoretical perspectives for further reflection, and, it reflects on perspectives of future developments in the course of the Nyangatom turn towards a climate change adaptation via the newly introduced scheme of rangeland enclosures.

The scheme of enclosed rangeland was already introduced to pastoral areas and livelihoods in Ethiopia ahead of today's emphasized attention on the climate change imperative and its implications on vulnerabilities of human livelihoods. In the 1960s the first fenced enclosures were introduced to the Borena pastoral livelihood systems partly promoted by the World Bank. An increasing number of pastoralists turned towards farming on those plots, which had been enclosed before. Cropland encroached into traditional grazing areas and enclosed rangeland was turned into privately owned farmland.

Looking at this process of enclosing rangelands more closely it is to be acknowledged that enclosing is a strategic action to isolate a resource unit from the commons, serving the purpose of 'legitimizing' individual action. As a consequence, this formerly common property cannot be used by all members of the community as before, but they must be admitted to enter the enclosure. In the

Borena case this very obvious 'logic' turned against the poorer and less empowered members of the community: in the course of these establishments some powerful individuals among the pastoralist groups appropriated larger areas of land, while the Borena customary institutions were not capable to prevent this privatization. All in all a recent study concludes that about one third of the pastoralist communities' members have "moved out" to undocumented destinies, whereas the former wealthier and more powerful members of these communities have "moved up" to become wealthier than before (Napier and Desta 2011). In a similar tone another study states: "… land use policies that favor private use of the rangeland, using land enclosure as an entry point, can potentially harm pastoral livelihoods rather than supporting sustainable pastoral development" (Beyene 2009: 480). As a conclusion the same study states in general and in reflection of the over all experiences with range enclosures in Ethiopia:" Though the econometric model predicts new forms of benefits are associated with the expansion with enclosure, the social relationships distorted by private land uses in fragile ecology leads to the conclusion that the practice of enclosure does not represent a transition from a less to a more efficient land use system" (ibid. 487).

Managing the Range by Enclosures, a Token of Societal Transformation

Moving on to the Nyangatom case of rangeland enclosures, the interpretative framework of 'travelling model' is introduced. This model was set by Rottenburg in 1996 and further elaborated on by Behrends et al. in 2014. The perspective of the 'model' basically contends that, first of all, it is to be understood essentially as an analytical representation of particular aspects of reality. Models are not diffused by themselves, but are transferred by translation, which is bound to interpretations by the actor in charge of the process of translation (Behrends et al. 2014: 1 pp). The authors refer to this process of translation with the term 'token' taken as a sign or symbol to be filled with societal meaning in the process of translation. It is how this translation with respect to the rangeland enclosures occurs which this paper is interested in.

From the very start of the establishment of the system of rangeland enclosures high attention was paid to the question of how to govern this societal innovation and avoid the analyzed above process of societal segregation. In the beginning the author herself guided this process by the approach of participatory action research. In the meantime this process has been appropriated by the regional government and the woreda (district) administration and guided by an 'expert' in formulating rules and regulations for managing the enclosures from Addis Ababa. A 'committee' has been put in place composed of at first glance 'democratically' selected members of the Nyangatom community in order to govern the process of rangeland management.

This process and governing structure is attentively reflected by the community, which is illustrated by the by the comment box (source: interview-assessment in March 2015, unpublished; Fig. 19.4).

Pastoralist woman, elected women's representative, approx. 50 years

- *"The people come from the Woreda or the Government in Addis Ababa and say the enclosures have to be ruled by the committee and not by the elders, and everything of the government is related to Food-for-Work. So you will be forced to accept and act according to their will, as otherwise no one will give you food and some necessary things".*
- *When we come to the decision of the elders, sometimes this was not fair. It did not take the situation of the people into consideration. It was not made for the weak and poor people. If they only had few family members, they could not send someone on migration with the cattle. Then they were severely punished. Some even committed suicide, because they were so afraid of the punishment".*

Pastoralist from the age group of the 'elephants', the politically ruling age group according to traditional regulations, approx.. 65 years

- *"Right now, things have changed. Even the decisive age group has become so weak and cannot pass its decisions to the community any more. It has been said the rangeland would be managed by the committee. The leaders of the committee are from different age groups, and they are now in power of administering the community".*
- *"With the by-laws of the new arrangement I have a problem. If these were ruled by the elders, I could accept them. But these people of the committee cannot be respected. Who would listen to those un-experienced youngsters?"*

Pastoralist from the group of herders, approx. 45 years

- *"The government and some other people, who used to come and pay us for fencing the rangeland, said there were some left activities to be done. So we just sat down without working and visiting these places which needed additional construction work. Then the wind came and destroyed the fence".*
- *"The Woreda and Addis Ababa people told us which rules for which punishment were the right ones".*
- *"I have no idea of what can be the future of my life here in Nyangatom. When I think of migrating long distances with my animals in search of grass , I have a lot of conflicts with other pastoralists of other tribes on the little grazing potential of our region. I think I have to safe my life by shifting my life style. Maybe I could become a merchant?"*

Fig. 19.4 Comment box

From the cited comments it becomes obvious that the 'committees' and their composition will outdate existing and long established conventions of governance and societal dominance. Seen from a northern perspective, these processes of seemingly democratic governance might be appreciated. But the question remains, whether these participatory processes did not ascertain power and social control "not only by certain individuals and groups, but also of particular bodies of knowledge ... power is everywhere, and can be particularly analyzed through the creation of social norms or customs that are practiced throughout society" (Kothari 2004: 142f). In line with the above reflected concept of 'travelling model', the 'committee' turns into a token to be filled with societal meaning in translation. The pastoralist woman appreciates this change, as the former class in power, the elders, had not considered specific livelihood situations of households. People had been punished for disobedience, which means punishment in case households did not have the younger men and sons to follow the rules of migration set by the elephant group. She states 'un-fairness'. The present situation is qualified by dependency on food-for-work schemes, which were introduced by the Ethiopian Government. Power has been and still is taken out of the hands of the community, fringed by the outside controlled committee. The elder of the elephant group translates the 'model' into a threat to power along with the inherited societal hierarchy. To him the enclosures have turned a synonym of disorder and lack of governance. From the perspective of the younger herder terms of obedience and dis-obedience have shifted away from arguments deducted from insights into ecological frameworks, i.e. the 'law of nature', towards the 'law of money'. He and his fellow Nyangatom immediately stopped the construction of the fences when no money was paid for the work any more and deliberately risked the destruction of the just before built fence structures. In correspondence, perspectives for the future life have faded away and are seen in the vague and not materialized perspective of a 'merchant'.

Conclusion: The Terms of Pastoralist Live Ahead

On reflecting the initial question of 'transformative adaptation' a societal transformation is to be stated in the case of the people of the Nyangatom. The framework of climate change and climate change adaptation has turned into an issue of power and has staged a struggle for power. Even if from a western democratic understanding the turn towards governance in the hands of stakeholders in balanced age and gender representation seems to be wanted, the question of modes of governance has after all been taken from the hands of the community. People perceive themselves as externally governed and determined. By the model of a 'committee—"Whenever new things are initiated by the government, a 'committee' is established" (statement by woreda official)—people in tendency resign from a constructive and self-confident handling of future perspectives, as in the case of the younger herder. Processes are at a preliminary stage and cannot be evaluated in their final outcome. Whether the given situation will in the end safe the community from those processes of societal segregation observed in neighboring pastoralist communities like the Borena, remains to be waited for. Enclosures, the token of climate change adaptation, this can already be concluded, will not be much in favor of a societal transformation towards a more resilient climate change adaptation.

References

Abbink J et al (2014) Lands of the future: transforming pastoral lands and livelihoods in eastern Africa. Working Paper No. 154. Max Planck Institute for Social Anthropology, Halle/Saale

Adger N et al (2009) Are there social limits to adaptation to climate change? Clim Chan 93:335–354

Appadurai A (2006) Disjuncture and difference in the global cultural economy. In: Durham MG, Kellner DM (eds) Media and cultural studies. Blackwell, London

Behrends A, Park S-J, Rottenburg R (2014) Travelling models, introducing an analytical concept of globalisation studies. www.booksandjournals.brillonline.com/content/books/69789004274099s002

Beyene F (2009) Locating the adverse effects of rangeland enclosure among herders in eastern Ethiopia. Land Use Policy 27(2):480–488

Dessai S, Hulme M (2004) Does climate adaptation policy need probabilities? Clim Pol 4:107–128

Driessen PPJ et al (2013) Societal transformations in the face of climate change. JPI, Brussels

GIZ (2012) Transformation: shaping complex change processes. akzente, the GIZ magazine, vol. 2. Bonn & Eschborn

IPCC (2007) Climate change 2014: impacts, adaptation, and vulnerability. 4th Assessment Report. World Meterological Organization, Geneva

IPCC (2014) Climate change 2014: impacts, adaptation, and vulnerability. 5th Assessment Report. World Meterological Organization, Geneva

IRINNEWS (2009) Global: twelve countries on climate change list. www.irinnews.org/

Kates RW, Travis WR, Wilbanks TJ (2012) Transformational adaptation when incremental adaptations to climate change are insufficient. Proc Natl Acad Sci USA 109(19):7156–7161

Kothari U (2004) Power, knowledge and social control in participatory development. In: Cooke B, Kothari U (eds) Participation, the new tyranny. Zed Books, London, pp 139–152

Mueller-Mahn D, Everts J (2013) Riskscapes. In: Mueller-Mahn D (ed) The special dimension of risk. Routledge, London, pp 22–36

Mulatu A, Bekure S (2013) The need to strengthen land laws in Ethiopia to protect pastoral rights. In: Catley A, Lind J, Scoones I (eds) Pastoralism and development in Africa: dynamic change at the margins. Routledge, Abingdon, pp 186–194

Napier A, Desta S (2011) Review of pastoral rangeland enclosures in Ethiopia. PLI Policy Project funded by US Aid, by support of Feinstein International Center and Tufts University, un published.

Nori M, Davis J (2007) Change of wind or wind of change? Climate change, adaptation and pastoralism. Working paper prepared fort the World Initiative for Sustainable Pastoralism. http://cmsdata.iucn.org/downloads

Pavanello S, Levine S (2011) Rules of the range: natural resources management in Kenya-Ethiopia. Humanitarian Policy Group (HPG) working paper, London

Rauch T (2003) Bessere Rahmenbedingungen allein beseitigen Armut nicht! Geogr Helv 58(1):35–46

Rottenburg R (1996) When organization travels. In: Czarniawska-Joerges B, Sevon G (eds) Translating organizational change. De Gruyter, New York, pp 191–240

Troeger S et al (2011) Failing seasons, ailing societies: climate change and the meaning of adaptation in Ethiopia. Heinrich Boell Foundation, Addis Ababa

Walshe R (2014) Concept paper: a new systems approach to resilience. www.bonn-conference.net/de/

Chapter 20
Analysis of Climate Change Adaptation Measures Used by Rural Dwellers in the Southeast and Southsouth Zone of Nigeria

C.C. Ifeanyi-Obi

Abstract The renewed quest for specific climate change adaptation measures that will address the needs of rural dwellers in their specified location prompted this study. This study analysed climate change adaptation measures used by rural dwellers in southeast and Southsouth Nigeria. Structured questionnaire, interview schedule, focused group discussion and direct observation were used to collect data from 320 respondents selected using a multi-stage sampling method. Data obtained were analysed using descriptive, multivariate and inferential statistics namely mean, percentages, frequency count, Varimax rotated factor analysis, analysis of variance and Turkey HSD post hoc multiple comparison test. The major occupation of the rural dwellers in the study area was farming (70.9 %) and 95 % have a farm size of less than 1 ha of land. The majority (83.7 %) had at least 10 years farming experience and earned an average monthly income of less than fifty thousand naira (77.2 %). The majority (60.3 %) maintained a household size of between 6 and 10 persons. Cassava and yam were the major crops cultivated while poultry, goat and fish were the major animals reared. The results further showed that 97 % of the respondents were already aware of the change in climate mainly through personal experience (78 %). Variation in rainfall patterns (46.9 %) was found to be the most prominent climate change variable in the study area. The Varimax rotated factor analysis identified eight major climate change adaptation measures used by farmers in the study area namely improved livestock management practices and mixed farming, soil fertility improvement measures and use of information from Agricultural Extension agent, use of improved crop species and diversification, change in planting/harvesting dates, application of agricultural chemicals, irrigation facilities and insurance, increase farm size and reliance on family labour, use of cover crops and avoidance of sale of remaining seedling after planting. The result of the ANOVA revealed that significant differences were found to exist in the adaptation

---

C.C. Ifeanyi-Obi (✉)
Faculty of Agriculture, Department of Agricultural Economics and Extension, University of Port Harcourt, Port Harcourt, Rivers State, Nigeria
e-mail: clara.ifeanyi-obi@uniport.edu.ng

measures use amongst the rural dwellers in the study area. Based on the findings of this study, it was recommended that there is a need for expansion in the role of agricultural extension agents in the nation. In addition, more studies addressing the differences in climate change adaptation measures are encouraged as this will equip the extension agencies and other stakeholders in climate change issues to develop relevant location-specific climate change adaptation measures that will effectively address the needs of farmers.

Keywords Climate change • Adaptation measures • Rural dwellers • Differences

Introduction

Climate change has over the years become a topical issue in the agricultural sector as well as other sectors of the Nigerian Economy. This could be due to the devastating effects it has on agricultural production. The term climate change according to the report of Intergovernmental Panel on Climate Change (IPCC 2007) is statistically significant variations that persist for an extended period typically decades or longer. It includes shifts in the frequency and magnitude of sporadic weather events as well as the slow continuous rise in global mean surface temperature. This change in climate manifest itself in numerous ways which includes; change in earth's surface temperature, change in sea water level, change in duration and intensity of rainfall, increased drought event etc. Climate change has been noted to have negative effects on agricultural activities which culminates to decrease crop and animal production (Agoumi 2003; Issa et al. 2008; Eboh 2009; Anyadike 2009; Okhimamhe 2009; Ifeanyi-Obi et al. 2011, 2012; Ifeanyi-Obi and Asiabaka 2014).

Nigeria is highly Vulnerable to climate change and ranked tenth among countries that are most vulnerable to the effects of climate change (Maplecroft 2013). Similarly, the increasing flooding in the coastal and non-coastal regions of the country is indicative of the climate change and its expected adverse effect (Vanguard Newspaper, 2012). According to the report, there are two extremes of the expected challenges of the climate change in Nigeria, namely, increase in rainy and dry seasons with each lasting approximately six months on the average, that is, April to October and October to March, respectively. This is already taking place in Nigeria as longer rainy season is experienced in the country. Heavy rains with thunderstorms which are usual with peak of rainy season (June and July) are now experienced in February. Climate change has become a global issue in recent times manifesting in variations of different climate parameters including cloud cover, precipitation, temperature ranges, sea levels and vapour pressure (Ministry of Environment of the Federal Republic of Nigeria (MoEFRN) 2003: 43). According to the report, the variations in climate parameters affect different sectors of the economy such as agriculture, health, water resources, energy etc while the main cause of climate change has been attributed to anthropogenic (human) activities.

Mendelsohn et al. (2000) estimated that by 2100, Nigeria and other West African countries are likely to have agricultural losses up to 4 % of GDP due to climate change. Parts of the country that experience soil erosion and operate rain fed agriculture could have decline in agriculture yield of up to 50 % between 2000 and 2020 due to increasing impact of climate change (Agoumi 2003; IPCC 2007). Desertification, sea level rise and flooding are gradually effecting many states in Nigeria including the wealthiest areas of Victoria Island in Lagos and oil rich Niger Delta (Oneworld.net 2010). Similarly, Agwu (2008) stated that by 1985, deforestation claimed 1544 sq miles of the nation's forest land. Between 1983 and 1993 alone, Nigeria lost 20 % of its forest and woodland areas; in northern Nigeria, especially the Sudan Sahelian region, where desertification is a key environmental problem, droughts have been re-occurring for the past three decades thereby affecting food security and increasing cardio-respiratory health issues resulting from the increasing dust pollution.

Specifically in the Southeast and Southsouth Nigeria, climate change is already exerting its diverse effects on the people. The major effects of climate change in this area comprises increased gully erosion, soil fertility loss, decreased agricultural yield, deforestation, fishery resource decline, flooding, coastal/marine erosion, health decline, natural disasters, reduction in the water availability, food insecurity, heat waves, air pollution, social dislocation, infectious diseases and erosion. (Akuegwu et al. 2012; Nnadi et al. 2012; Ozor 2009; Chinweze and Abiola-Oloke 2009; Etuonovbe 2007).

These devastating effects of climate change are most likely to be felt by the rural dwellers whose major livelihoods are agricultural activities which are done at subsistence level as a result of low income capacity of the farmers as well as poor infrastructural development (Akpabio 2005; Ekong 2010; Mgbada 2010). Farmers are already trying to adapt to climate change using experiences from their farming activities as well as few recommendations at their disposal.

Onyeneke and Madukwe (2010) identified six major adaptation measures used by crop farmers in Southeast rainforest zone of Nigeria to be Portfolio diversification, soil conservation, change of planting dates, planting trees, changing tillage operation and use of irrigation. Similarly, Adesina and Odekunle (2011) listed adaptation measures to address crop production in Nigeria to be provision of accurate and timely weather forecasting, enhancing agricultural extension services, extending and optimizing existing irrigation infrastructures, adoption of drought tolerant and early maturing crop varieties, diversifying livelihood to improve income, increasing and upgrading crop storage facilities, control of pests, stabilizing gullies and erosion sites, helping farmers to secure agricultural insurance, growing more cover crops like potatoes and melon to protect soils from erosion, improving the monitoring and evaluation of agricultural activities with realistic and measurable indicators. Though various adaptation measures have been identified by researchers, farmers still face some challenges in using these adaptation measures. One of these challenges is the problem of climate variation in close locations. In some instances, zones who share common boundaries have been observed to differ in their climatic conditions and as such require different climate change adaptation measures. Areghero (2005) noted that Nigeria is a country of marked ecological diversity and climatic contrasts. Therefore, it is necessary that adaptation measures become more location-specific for it to be effective since adaptation measures that are effective in a zone may not be effective for others around it. In view of this, this research work identified the adaptation measures use by rural dwellers in southeast and Southsouth zone of Nigeria and went further to examine if differences exist in the adaptation measures used by rural dwellers in the two zones studied.

Objectives of the Study

The general objective of this study is to analyse the adaptation measures used by rural dwellers in the southeast and southsouth Nigeria. The specific objectives were to:

- describe the socio-economic characteristics of the rural dwellers in the study area;
- identify the rate and level of climate change awareness
- identify the most prominent area of climate change
- identify the sources of climate change information
- assess the climate change adaptation measures use by rural dwellers in southeast and Southsouth Nigeria, and
- ascertain whether differences exist in the adaptation measures used by rural dwellers in southeast and Southsouth Nigeria.

Hypothesis of the Study

Choice of adaptation measures used by rural dwellers in Southeast and Southsouth Nigeria are not significantly different.

Methodology

Study Area

This study was conducted in Southeast and Southsouth geopolitical zones of Nigeria. Nigeria as a nation lies wholly within the tropical zone with wide variability in climate across different parts of the country. Near the coast, the seasons are not sharply defined. Temperatures rarely exceed 32 °C (90 °F), the humidity is very high and the nights are hot. In the inland, there are two distinct seasons: a wet season from April to October, with generally lower temperature, and a dry season

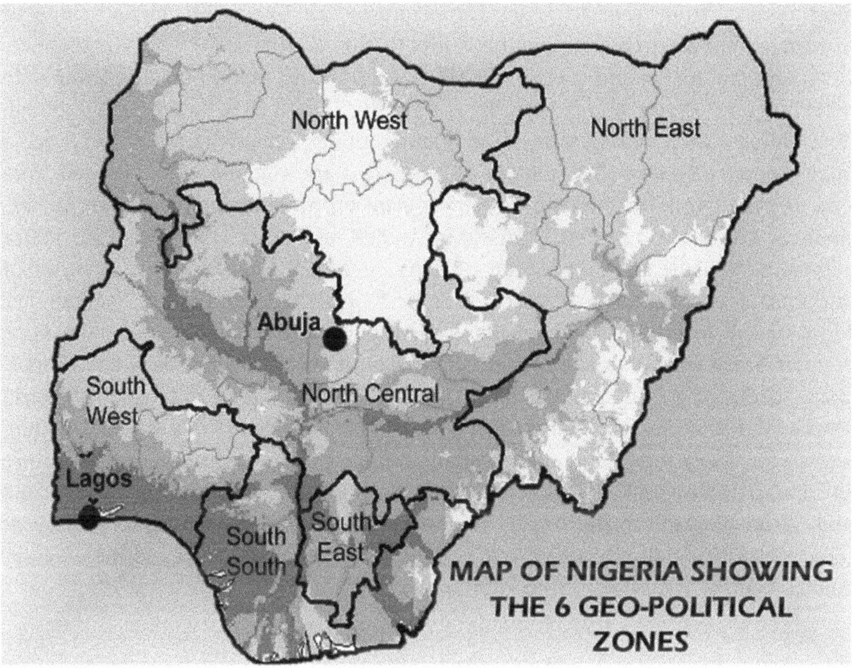

Fig. 20.1 Map of Nigeria showing the six geopolitical zones of Nigeria. *Source*: Nigerian budget states analyses

from November to March, with midday temperatures that surpass 38 °C (100 °F) and relatively cool nights. As in most of West Africa, Nigeria's climate is characterized by strong latitudinal zones, becoming progressively drier as one moves north from the coast. Rainfall is the key climatic variable, and there is a marked alternation of wet and dry seasons in most areas (Fig. 20.1)

The Southeast zone of Nigeria is located between latitude 04° 15 Min and 70 N and longitude 05° 50 Min and 09° 30 Min (Chukwu and Mbanaso 1999). Its lowland rain forest lies between the rainforest—savanna ecotone and the salt and fresh water swamp forest along the coast (Igbozurike 1975). It shares boundary in the North with North West zone and in the south with south west zone. Rainfall is the key climatic variable and there is a marked alternation of wet and dry seasons in most areas. The rainy season which usually commences from late February or early March and continues into late October or sometimes early November is bi-modal with the major peak occurring July and September. There is usually a short dry spell in August this so-called 'August break' or 'little dry season' may begin as early as the last week in June and last several weeks in succession, or occur in alternate weeks. The rainy season may in some years linger into the middle of November and has an average length of 241 days. From November, the weather clears up rapidly with the wind system shifting to become the North-easterly dusty 'harmattan' current which ushers in the dry season, which ends in late February. While the mean annual rainfall of the area is very high, there is considerable variability in the total rainfall from year to year. Temperatures are high with annual range of maximum and minimum temperatures less than 6 °C.

According to Unamma et al. (2004), Soil type is predominantly loam with percentages of sandy, loamy and clay soils as 26, 24 and 29 % in the zone although distribution varied among states. The main cropping systems in the zone is mixed inter cropping system. The number of crops that are mixed-cropped ranged from two to over five with the overall predominant crop combination as yam, maize, cassava, vegetable and egusi. Diseases, weeds and pests constituted the major constraint to increased production. A ranking of the pests/diseases in order of importance was weed > diseases > insects > rodents > birds > nematodes. Most of the farmers owned mostly locally manufactured hand tools; the most popular being knives (machetes and cutlasses), hoes of various sizes and shapes, axes, spades, shovels, diggers, forks and hand diggers. The Southsouth Nigeria shares similar climatic condition with the Southeast zone, it is a tropical region, known for heavy rainfall. Major occupation of the people in the zone is agriculture and the extractions of raw minerals, such as limestone, oil etc. It has one of the highest oil deposits in the world. The vegetation of the zone consists mainly of forest swamps. Rainfall in the zone is heavy due to the closeness to the equator. Annual rainfall totals vary from 2400 to over 4000 mm.

Sample Selection

The population of this study comprises rural dwellers that live in the Southeast and Southsouth zone of Nigeria. The Southeast zone is made up of five states namely; Anambra, Imo, Abia, Enugu and Ebonyi State while the Southsouth zone comprises of six states namely AkwaIbom, Bayelsa, CrossRivers, Rivers, Edo and Delta state. The sampling procedure used is Multi-stage sampling technique. The first stage was a random selection of two states from each of the zone used for the study; this gave a total of four states for the study. The second stage was the selection of two agricultural zones from each of the selected states. Out of all the agricultural zones in each of the selected State, two were purposively selected based on rurality (the dominance of rural communities as constituents). This gave a total of eight agricultural zones. Third, two blocks were randomly selected from each agricultural zone and finally two circles were randomly selected from each block giving a total of 32 circles. The sampling frame was the list of rural dwellers that have lived in the selected circles for a very long time and was compiled with the help of village heads, Chiefs and key informants. From this sampling frame, a sample size of 320 respondents was proportionately and randomly selected for the study.

Data Collection

The primary data for this study were information gathered from the field survey. The major instruments used for data collection were; structured questionnaire, interview schedule, focused group discussion and personal observation.

The questions in the questionnaire were structured to capture and elicit information from the respondents based on the objectives and hypotheses of the study. Interview schedule and focused group discussion were employed to complement the use of structured questionnaire for detailed and in-depth information from the respondents. Also, transect walk was taken for on-the-spot observation and assessment of practices. Secondary data were sourced from existing and related literature from journals, institutional publications, internet, textbooks and previous research reports.

Methods of Data Analyses

Data obtained were analyzed using descriptive and inferential statistics namely; frequency counts, percentages, mean, Varimax rotated factor analysis, and Analysis of Variance (ANOVA). Turkey HSD (post hoc test) was used to detect where exactly the identified difference lies.

Climate change adaptation measures were captured using a 25 statement 4 point likert type rating scale which was assigned weight of 4,3,2,1 respectively for positive items and 1,2,3,4 for negative items.

Model Specification

Null Hypothesis Choice of adaptation measures used by rural dwellers in Southeast and Southsouth Nigeria are not significantly different.

For the test of this hypothesis, one way Analysis of Variance model was used.
Generally, ANOVA formula (Udofia 2011) is given as:

$$F = \frac{\text{Between} - \text{Samples variance}}{\text{Within} - \text{Samples variance}}$$

$$= \frac{MSSB}{MSSW}$$

$$= \frac{SSB/k - 1}{SSW/n - k}$$

$$= \frac{SSB\,[n - k]}{SSW\,[k - 1]}$$

$$SST = SSW + SSB$$

$$SSW = \sum_{i=1} nj \sum_{j=1}^{k} (O_{ij} - \bar{O}_j)$$

$$SSB = \sum_{j=1}^{k} nj(O - \bar{O})$$

Where:
SST = the total sum of squares
SSB = the between sum of squares
SSW = the within sum of squares
nj = sample size from population
k = number of samples
$\bar{O}_j$ = mean of sample population
$\bar{O}$ = grand mean
n = number of observations
O_{ij} = ith observation from the population j
k − 1 = degrees of freedom for between samples
n − k = degrees of freedom within sample.

The result was further subjected to a post hoc multiple comparison test (Turkey HSD) to specifically find out where exactly the difference in adaptation measures lie.

Results and Discussion

Socio-Economic Characteristics of Respondents

Out of the 320 respondents that were interviewed in this study, 64 % were male indicating that majority of the respondents used for the study were male. As regards age of the respondents, it was shown that 79.9 % of the respondents were above 40 years of age. The mean age was found to be 50 years. A possible implication of this result is that respondents have lived long enough in the studied area to be able to give reliable responses to the questions asked. Result further showed majority (76.9 %) of the respondents were married. The mean number of years spent in school by the respondents was 12 years implying that majority of the respondents attended secondary school. It could be inferred that inhabitants of the south east and southsouth zone of Nigeria are literate enough to read and write. This could serve as an impetus in getting aware on issues of climate change as well as adopting new ideas in combating climate change. According to Obinne (1991), education is an important factor influencing adoption of farm innovations. Also Agwu and Anyanwu (1996) in the same line noted that increase in educational status of farmers positively influence adoption of improved technologies and practices. the majority of the respondents (77.2 %) earned a monthly income of less than N50,000, indicating that inhabitants of the study area are majorly subsistence farmers. Respondents had average of 22 years of farming experience. This indicates that majority of the respondents had long-term farming experience and could have over the years experienced the change in climate as well as the corresponding effects. As a result, they must have acquired wealth of knowledge and effective adaptation measures for cushioning the effects of climate change as well as identified obstacles militating against effective adaptation. Thus Mapuno et al. (2008) noted that farmers could be in a better position to identify challenges and opportunities on climate change. Average household size was seven persons with a minimum of one person and maximum of 20 persons. This result is in line with the findings of Ozor and Nnaji (2010) which found that farmers in southern Nigeria have an average household size of seven persons. In the same vein Adesope et al. (2012), Ozor and Nnaji (2010) and Akinnagbe and Ajayi (2010) in their findings reveal that majority of households in rural areas of Nigeria maintain household size of 6–10 persons. The implication of this is that more family labour will be readily available since relatively large household size is an obvious advantage in terms of farm labour supply. This could probably serve as an insurance against short falls in farm labour supply.

Farming (70.9 %) was the major occupation hence respondents were capable of giving information regarding the effects of climate change on agricultural activities and the adaptation options. This is not surprising as this study is targeted at rural dwellers whose major livelihood activity is known to be farming (Table 20.1). This agrees with the findings of Ifeanyi-Obi et al. (2011), Enete et al. (2010) and Nzeh and Eboh (2010) that majority of rural inhabitants have farming as their major occupation. Majority (95 %) of the rural dwellers in the study area have farm size of less than 1 ha. This is in line with the findings of Nzeh and Eboh (2010) which indicated that 70 % of the respondents in a study in the three agricultural zone of Enugu State had access to 3 ha of land or below. The implication of higher percentage of respondents having access to less than 1 ha of land for farming shows the increasing pressure on land as more people depend on fewer portion of land for farming. It was also observed that land in the study area was being used for other activities like building of commercial residential houses, markets, etc. The pressure on land use could add to deforestation which is one of the likely causes of climate change. Mixed farming (45.6 %) was found to be the major farming activity. The dominance of mixed farming is not surprising as it is a popular farming practice in the country particularly in the southern part of the country.

Table 20.1 Distribution of rural dwellers in the study area by their socio-economic characteristics

Variable	Frequency	Percentage
Gender		
Male	204	64
Female	116	36
Age		
21–40	17	19.1
41–60	171	53.4
61–80	82	25.6
Above 80	6	1.9
Marital status		
Single	36	11.3
Married	284	88.7
Level of education		
0 (no formal education)	7	2.2
1–6	58	18.1
7–13	13.1	40.9
14–19	121	37.9
19 and above	3	0.9
Monthly income		
≤₦50,000	247	77.2
₦51,000–₦100,000	65	20.3
₦101,000–₦151,000	6	1.9
₦151,000–₦200,000	2	0.6
Farming experience		
1–10	52	16.3
11–20	118	36.9
21–30	99	30.9
Above 30	51	15.9
Household size		
1–5	108	33.8
6–10	193	60.3
11–15	17	5.3
16–20	2	0.6
Major occupation		
Farming	227	70.9
Non-farming	93	29.1
Farm size		
1 ha	304	95
1–2	11	3.4
3–4	3	0.94
Above 4	2	0.66
Major farming activity		
Mixed farming	146	45.6
Livestock	22	6.9
Crop production	133	1.6
Fishery	19	5.9

(continued)

Table 20.1 (continued)

Variable	Frequency	Percentage
Major aim of production		
Sale	41	12.8
Consumption	30	9.4
Both	245	76.6
Others	4	1.3

Source: Field Survey (2012)

According to most of the farmers, mixed farming is always preferred in order to guard against crop or animal failure. Also, they indicated that in mixed farming the manure from animals is usually used as cheap organic fertilizers for the crops while the crop residues after harvest serve as animal feed. Similarly, Ozor and Nnaji (2010) found that mixed farming was the predominant farming practice for farmers in Southern Nigeria. This also further implies that farmers actually will be vulnerable to climate change impacts as crop and livestock form the main categories affected by the change in climate. Both sales and consumption (76.6 %) was found to be their major aim of production. A possible explanation of this is that majority of the respondents feed their households from their farm produce thereby reducing the market share of their produce.

Crops Cultivated by the Respondents

Result shows that out of the ten crops investigated only five are greatly cultivated by the respondents with cassava being the most cultivated with a mean of 2.8. Cassava was followed by yam with a mean of 2.6, maize (2.5), vegetables (2.2), plantain/banana (2.1), cocoyam (1.9), cowpeas/beans (1.9), rice (1.2), cocoa (1.2) and cashew (1.2).

Cassava is not only a major staple food but also a major source of farm income for the Nigerian farmers (Nweke 1996). And compared to other crops, cassava is the most resistant crop to extreme weather events. Adebayo (2006) noted that its high resilience and adaptability to a wide range of ecological conditions has sustained its production through many generations in Sub-Saharan Africa since it was introduced into the region in sixteenth century. It is therefore most often described as a hardy crop and may in this sense be the most adaptable crop to climate variations (Enete 2003). Benhin (2006) reported that one of the strategies which served as an important form of insurance against rainfall variability is increasing diversification by planting crops that are drought tolerant and/or resistant to temperature stresses. Also cassava as a crop could be processed into many products namely garri, fufu, starch, cassava flour, tapioca, animal feed and industrial starch. This peculiar characteristic has been observed to be one of the reasons why most farmers go into cassava production.

Table 20.2 Crops cultivated and animals reared by the study respondents

Crops	Cultivated	Moderately cultivated	Not cultivated	Mean
Cassava	270 (84.4)	38 (11.9)	12 (3.8)	2.8
Yam	231 (72.2)	60 (18.8)	28 (8.8)	2.6
Maize	197 (61.6)	85 (26.6)	37 (11.6)	2.5
Cocoyam	94 (29.4)	98 (26.6)	127 (39.7)	1.9
Rice	22 (6.9)	21 (6.6)	277 (86.6)	1.2
Cashew	16 (5.0)	31 (9.7)	272 (85.0)	1.2
Plantain/banana	134 (41.9)	103 (32.2)	81 (25.3)	2.1
Cocoa	18 (5.6)	22 (6.9)	280 (87.5)	1.2
Vegetables	144 (45.0)	95 (29.7)	79 (24.7)	2.2
Cow peas/beans	19 (5.9)	57 (17.8)	244 (76.3)	1.9
Animals reared				
Cattle	38 (11.9)	38 (11.9)	244 (76.3)	1.9
Goat	89 (27.6)	64 (20.0)	167 (52.2)	2.4
Poultry	111 (34.7)	68 (21.3)	141 (44.1)	2.5
Fish	66 (20.6)	35 (10.9)	219 (68.4)	2.4
Sheep	12 (4.1)	13 (4.1)	249 (91.9)	2.3
Rabbit	9 (2.9)	23 (7.2)	288 (90.0)	2.3
Pig	13 (4.1)	19 (5.9)	288 (9.0)	2.3

Source: Field Survey (2012)
Mean score ≥2 is significant
Figures in parentheses are percentages

Also, yam being the second most cultivated crop is not surprising as yam is known to be a cherished and respected staple food in the south east Nigeria where there is generally an annual celebration in honour of the crop in most part of the zone.

As regards animals reared in the study area, it was shown that poultry was the most important animal production enterprise in the study area as it ranked first with a mean value of 2.5. This was followed by goat and fish with same mean values of 2.4, sheep, rabbit and pig all have the mean values of 2.3 while cattle has the least mean value of 1.9. This is not surprising as the study area is not a cattle rearing zone in the country (Table 20.2).

Rate and Level of Awareness of Climate Change

Rate of Awareness of Climate Change

Table 20.3 shows that majority (96.9 %) of the respondents were aware of the climate change while 3.1 % were not yet aware. Nzeadibe et al. (2010) found out that 94.8 % of farmers in Niger Delta have knowledge of climate change on farming practices.

Table 20.3 Distribution of respondents by rate of awareness of climate change

Are you aware of the climate change	Frequency	Percentage
Yes	310	96.9
No	10	3.1
Total	320	100

Source: Field Survey (2012)

Comparing the two results, one could imply that climate change has gained more awareness in the recent times. This shows that the different emphasis being laid on climate change by the different sectors of the economy is making appreciable impact. This is very important as it is expected that farmer's awareness of climate change will arouse their interest on adaptation and mitigation issues concerning climate change. Hassan and Nhemachena (2008) stated that the awareness of climate problems and potential benefits of taking action is an important determinant of adoption of agricultural technologies. In the same vein, Maddison (2006) stated that farmers' awareness of change in climate attributes (temperature and precipitation) is important to adaptation decision making.

Duration of Time Respondents have Noticed Climate Change

Table 20.4 shows that more than half of the respondents (53.1 %) have noticed climate change for relatively low number of years (5years and below). This may be responsible for the relatively low emphasis on effects of climate change by farmers as compared to the adverse effects climate change have on their production activities. Relatively high numbers of the respondents (28.1 %) have been aware of climate change for 6–10 years while only 18.8 % have noticed climate change for more than 10 years.

Level of Awareness

Table 20.5 shows that respondents were highly aware of climate change in the following areas; rainfall pattern (75.6 %), atmospheric temperature (76.6 %) and solar radiation i.e. sunshine (74.4 %). The implication of this is that these areas may be the most noticeable areas of climate change in the study area as most of the respondents indicated having high level of awareness in those areas. It was shown in the table that majority of respondents have moderate level of awareness of climate change on humidity and drought. Oyekale et al. (2009) noted that rainfall, temperature and sunlight are major climatic variables in the Southeast Nigeria.

Table 20.4 Distribution of respondents by duration of time respondents have noticed climate change

Number of years	Frequency	Percentage
5 years and below	170	54.9
6–10 years	90	29.0
Above 10 years	50	16.1
Total	310	100

Source: Field Survey (2012)

Table 20.5 Distribution of respondents by level of awareness

Areas of climate change	High	Moderate	Low	Mean
Aware of variation in rainfall pattern	237 (75.6)	56 (18.4)	17 (5.9)	2.7
Aware of variation in atmospheric temperature	240 (76.6)	58 (19.1)	12 (4.4)	2.7
Aware of variation in humidity	115 (31.5)	134 (42.8)	61 (19.7)	2.1
Aware of variation in drought	104 (34.1)	105 (33.8)	101 (32.2)	2.0
Aware of variation in solar radiation (sunshine)	233 (74.4)	55 (18.1)	22 (7.5)	2.7

Source: Field Survey (2012)

The Most Prominent Climate Change Variable

Table 20.6 revealed that rainfall pattern (46.9 %) is the most prominent climate change variable in the study area. This is followed by variation in atmospheric temperature (23.4 %), humidity (11.9 %), drought (5.6 %) and solar radiation (12.2 %).

Mbakwe et al. (2004) stated that rainfall is the key climatic variable of the south east agro-ecological zone. Also, Oyekale et al. (2009) found rainfall to be the most important climate variable in cocoa production in Nigeria. In the same vein, Ozor (2009) noted that farmers in Nigeria depend on rainfall signal for their agricultural activities. According to him the onset of rains signals the beginning of crop production activities. This goes a long way in explaining the significance of rainfall to farmers in the study area.

Sources of Climate Change Information

Table 20.7 revealed that the main source of information on climate change was through personal experience (78.8 %). This may be as a result of the fact that majority of the respondents have farmed for fairly long period of time and as such have noticed the change in climate. Radio ranked second (56.9 %) as the source of information on climate change. The radio is known to be an effective channel of communication through which rural population, largely non-literate who seldom has access to written forms of information can be reached (Olawoye 1996). He further explained that the strength of radio as an agricultural communication tool is widely regarded to lie in its ability to reach illiterate farmers and provide them with information relating to all aspects of agricultural production in a language they can understand (local language). This enhances understanding by the farmers.

Table 20.6 Distribution of respondents by most prominent climate change variable

Areas of climate change	Frequency	Percentage
Aware of variation in rainfall	150	46.9
Aware of variation in atmospheric temperature	75	23.4
Aware of variation in humidity	30	11.9
Aware of variation in drought	18	5.6
Aware of variation in solar radiation	47	12.2

Source: Field Survey (2012)

Table 20.7 Distribution of respondents by sources of information on climate change

Sources of information	Frequency[a]	Percentage
Newspaper	117	36.6
Television	133	41.6
Extension agents	141	44.1
Through fellow villagers	118	36.9
Radio	182	56.9
Personal experience	252	78.8
Mobile phones	108	33.8

[a]Multiple responses were recorded
Source: Field Survey (2012)

Extension agents ranked third (44.4 %) implying that extension service delivery in terms of climate change information in the area can be said to be relatively satisfactory. This was followed by television (41.6 %), through fellow villagers (36.9 %), newspaper (36.6 %) and lastly mobile phone (33.8 %). The result reveals that the television channels are not left out in the fight against the menace of climate change as farmers indicated that they telecast climate change issues. Also it can be said that the farmers in the study area are relatively literate since about 36.6 % could read newspaper to get climate change information. According to the respondents, the low response on mobile phone (33.8 %) as a source of climate change information was as a result of network fluctuation, cost of recharge cards, unavailability of electricity for charging the phones as well as the fact that some farmers in the study area have not yet realized that mobile phones could be an effective channel of transferring agricultural information.

Adaptation Measures to Climate Change Effects Used by the Rural Dwellers in the Study Area

The 34 statement of the adaptation measures used by the rural dwellers in the study area were subjected to varimax rotated factor analysis using SPSS version 16. The suitability of data for factor analysis could be assessed based on sample size, relationship among the variables, Bartlett's test of sphericity and Kaiser-Meyer-

Olkin (KMO) measure of sampling adequacy, (Pallant 2010). The sample size should be above 300 and correlation matrix should have majority of the coefficients above 0.3, (Tabachnick and Fidell 2007). The Bartlett's test of sphericity should be significant ($p < 0.05$) and Kaiser-Meyer-Olkin (KMO) should have a minimum value of 0.6, (Tabachnick and Fidell 2007). Prior to performing the varimax factor analysis, the suitability of data collected in this study for factor analysis was assessed. Inspection of the correlation matrix revealed the presence of many coefficients of 0.3 and above. The Kaiser-Meyer-Olkin value was 0.82, exceeding the recommended value of 0.6 and Bartlett's Test of sphericity reached statistical significance $p = 0.000$ supporting the factorability of the correlation matrix.

In determining the number of factors to extract, Pallant (2010) Eigen value rule and Scree plot analysis was used. This rule suggests that only factors with Eigen value of 1.0 or more are retained for further investigation. Also the Scree plot analysis involves plotting each of the Eigen values of the factors and inspecting the plot to find a point at which the shape of the curve changes direction and becomes horizontal. Principal component analysis revealed the presence of eight factors with Eigen values exceeding 1, explaining 17.4, 13.9, 10.0, 5.0, 4.7, 4.2, 3.8 and 3.2 % of the variance respectively. An inspection of the screeplot revealed a change after the eight components. It was then decided to retain eight factors for further investigation. The eight factors explained a total of 62.15 % of the variance.

Table 20.8 shows the factor analysis procedure with varimax rotation applied to the data yielded an eight-dimensional solution. The communalities, which can be regarded as indications of the importance of the variables in the analysis are generally high (above 0.5). This shows that the variables selected for this study are appropriate and relevant. The eight factors which altogether accounted for 62.15 % of the total variance in the 34 original variables may be regarded as composite indicators defining adaptation measures used by rural dwellers in the study area.

Based on items loadings of the factor analysis conducted eight factors were isolated and named. These eight factors therefore represent the major adaptation measures used by respondents in the study area to combat climate change. These factors are improved livestock management practices and mixed farming (factor 1), soil fertility improvement measures and use of information from Agricultural Extension agent (factor 2), use of improved crop species and diversification (factor 3), change in planting / harvesting dates (factor 4), application of agricultural chemicals (factor 5), irrigation facilities and insurance (factor 6), increase farm size and reliance on family labour (factor 7), use of cover crops and avoidance of sale of remaining seedling after planting (factor 8).

Factor 1 (improved livestock management practices and mixed farming) accounted for 17.4 % of the total variance and is without doubt the most important factor. The specific issues that loaded high under this factor include; rearing of heat tolerant breed of livestock (0.882), rearing of disease resistant breed of animals (0.873), feeding of livestock more frequently than before to improve their productivity (0.861), using of automated water sprinkling system during hot weather to sprinkle water on livestock to reduce heat stress (0.817), keeping of animals under

Table 20.8 Pattern rotated factor matrix for the distribution of adaptation measures used by rural dwellers in the study area

Loadings

S/N	Statement	F1	F2	F3	F4	F5	F6	F7	F8	Communalities
1.	Increase in farm size	0.109	0.272	−0.076	0.011	0.326	0.173	0.624	0.107	0.630
2.	I move to a better farm land	−0.083	0.453	0.216	0.041	0.346	−0.254	0.233	0.182	0.531
3.	Increase in the quantity of material use for mulching	−0.001	0.666	0.032	0.079	−0.210	0.069	0.207	0.148	0.564
4.	Planting of different varieties of crops	−0.054	0.117	0.591	0.393	0.071	−0.014	0.159	0.127	0.567
5.	Carrying out of early planting crops	−0.024	−0.028	0.238	0.743	0.164	0.010	0.018	0.000	0.638
6.	Use of different planting dates for the crops	−0.016	0.074	0.190	0.748	0.182	−0.040	−0.147	0.142	0.677
7.	Treating of seeds with fungicides before sowing	−0.060	0.216	0.121	0.362	0.692	−0.018	−0.040	0.027	0.678
8.	Application of pesticides to plants	−0.008	0.518	−0.097	0.272	0.411	0.261	0.019	−0.200	0.628
9.	Use of herbicides to reduce the high rate of weed infestation	0.012	0.020	0.334	0.110	0.746	0.126	0.062	0.011	0.702
10.	Increase in the use of farm yard manure	0.083	0.658	−0.044	0.038	0.183	−0.023	0.170	0.085	0.514
11.	Increased use of fertilizer	0.022	0.725	0.239	−0.078	0.186	−0.056	0.016	−0.010	0.628
12.	I avoid selling remaining seedlings after planting	0.052	0.325	0.066	−0.164	0.341	−0.143	0.139	0.534	0.581
13.	Use of the available irrigation facilities	0.129	0.452	0.032	−0.024	0.053	0.542	−0.023	0.232	0.574
14.	Use of more drought tolerant species of crops	0.030	−0.145	0.727	0.026	0.251	−0.015	−0.242	0.246	0.733
15.	I secure insurance for my farm enterprise	0.131	0.314	−0.362	0.178	0.005	0.612	−0.084	0.029	0.661
16.	Use of more pest and disease resistant species of crops	−0.110	−0.002	0.768	0.096	0.191	−0.019	0.010	0.077	0.654
17.	I harvest early when adverse dry weather is expected	0.057	0.517	−0.124	0.460	−0.058	0.267	0.115	0.130	0.602
18.	Undertaking of other non-farm income generating activities	0.033	0.125	0.693	0.123	−0.007	0.023	0.011	−0.097	0.525
19.	Use of information from agric extension agents to combat climate change effects	0.064	0.698	−0.124	0.074	0.008	0.056	−0.219	0.084	0.570

(continued)

Table 20.8 (continued)

Loadings

S/N	Statement	F1	F2	F3	F4	F5	F6	F7	F8	Communalities
20.	Use of available credit facilities to increase my production	0.078	0.712	0.068	−0.151	0.048	0.271	−0.176	−0.001	0.647
21.	Increase reliance on family labour	0.112	0.075	0.488	0.091	−0.035	0.119	0.562	−0.087	0.603
22.	More frequent weeding	0.073	−0.184	0.441	0.445	0.013	0.133	0.121	0.345	0.583
23.	Increase in the planting of cover crops	0.029	0.078	0.180	0.229	−0.107	0.189	−0.054	0.675	0.597
24.	Increase in the use of fallowing	0.039	0.398	−0.020	0.139	0.047	0.160	0.089	0.502	0.467
25.	Increased planting by the river side	−0.049	−0.014	0.315	−0.033	0.089	0.689	0.302	0.118	0.690
26.	Combination of crop production and livestock management	0.635	0.239	0.145	0.108	−0.044	−0.179	0.309	0.116	0.636
27.	Keep animals under shade to reduce the heat stress on them	0.789	0.206	−0.044	0.109	−0.210	−0.144	0.118	0.044	0.760
28.	Rearing of disease resistant breed of animals	0.873	0.082	−0.016	0.049	−0.143	−0.050	0.000	−0.049	0.797
29.	Rearing of heat tolerant breed of livestock	0.882	−0.049	−0.030	−0.008	0.017	0.051	−0.026	0.028	0.786
30.	Use of manual or automated water sprinkling system during hot weather to sprinkle water on livestock to reduce heat stress	0.817	−0.056	−0.050	−0.029	0.078	0.153	−0.149	0.090	0.733
31.	Feeding of livestock more frequently than before	0.861	−0.038	−0.010	−0.032	−0.012	0.092	−0.086	0.006	0.761
32.	Administer artificial feed supplements to livestock	0.740	0.012	−0.032	−0.129	0.200	0.105	−0.037	−0.045	0.620
33.	Join cooperative societies	0.361	0.268	0.208	−0.194	0.091	0.215	−0.123	−0.253	0.416
34.	Application of indigenous knowledge	0.262	0.164	0.069	0.108	0.084	0.042	−0.507	−0.050	0.380
	Eigen value	5.90	4.71	3.39	1.71	1.61	1.43	1.29	1.06	
	Percentage variance	17.35	13.86	9.97	5.05	4.74	4.20	3.79	3.19	
	Cumulative %	17.35	31.21	41.18	46.23	50.97	55.17	58.96	62.15	

shade to reduce the heat stress in them (0.789), administering of artificial feed supplements to livestock to enhance their productivity (0.635).

Factor 2 (soil fertility improvement measures and use of information from Agricultural Extension agent) accounted for 13.9 % of the total variance, the issues that amplified this factor include; increase use of fertilizer (0.725), use of available credit facilities to increase production (0.712), resort to information from agric extension agents to combat climate change effects (0.148), increase quantity of mulching material (0.660), increase in the use of farm yard manure to improve soil fertility (0.658).

Factor 3 (use of improved crop species and portfolio diversification) accounted for 10 % of the total variance. The variables that loaded high under this factor include; use of more pest and disease resistant species of crops (0.768), use of more drought tolerant species of crops (0.727), undertaking of non-farming income generating activities (0.695) and planting of different varieties of crops (0.591). Onyeneke and Madukwe (2010) found that crop farmer in the southeast rainforest zone of Nigeria resort to portfolio diversification as the major adaptation measure in combating climate change. This calls for urgent consideration by all stakeholders and immediate action. Farmers should be taught or exposed to adaptation measures that improve their crop and livestock production rather than abandoning it for other non-farming income generating activities. Climate change should not be allowed to extinct the agricultural sector rather the agricultural sector should evolved ways of adapting to the effects of climate change.

Factor 4 (change in planting/harvesting dates) accounted for 5 % of the total variation. Under this factor, the variables that loaded high were use of different planting dates for the crop and carrying out of early planting of crops.

Factor 5 (use of agricultural chemicals to combat weed and pest) accounted for 4.7 % of the total variance. Two variables loaded high under this factor and they include; use of herbicides to reduce the high rate of weed infestation (0.746) and treating of seeds with fungicides before sowing (0.692).

Factor 6 (use of irrigation facilities and insurance) accounted for 4.2 % of the total variance. The variables that loaded high under this factor include; securing of insurance for the enterprise (0.612) and using of available irrigation facilities (0.542).

Factor 7 (increase farm size and reliance on family labour) accounted for 3.8 % of the total variance. Two variables loaded high under this factor and they include; increase in farm size (0.624) and reliance on family labour to reduce cost of production (0.562).

Factor 8 (use of cover crops and avoidance of sell of remaining seedling after planting) accounted for 3.2 % of the total variance. Two variables loaded high under this factor and they include; avoidance of selling remaining seedlings after planting (0.534) and increase in the planting of cover crops to reduce heat stress on crops (0.675).

Test of Hypothesis

Null Hypothesis There is no significant difference in the adaptation measures used amongst the rural dwellers in the southsouth and southeast zone of Nigeria.

A one-way between-groups analysis of variance was conducted to check whether there are significant differences in the adaptation measures used by rural dwellers in the southsouth and southeast zone of Nigeria. There was a statistically significant difference at $p < 0.05$ level: $f (3, 314) = 10.794$, $p = 0.000$. Also, Table 20.9 shows f-cal value of 10.794 is greater than f-tab value of 2.65 at 0.05 confirming that significant differences exist in the adaptation measures used by rural dwellers in southsouth and southeast Nigeria.

The effect size, calculated using eta squared was 0.09 (medium effect size). This implies that the relative magnitude of the differences between means of the two zones is quite significant. To further investigate where exactly the differences in the adaptation measures use by rural dwellers in the two zones lies, Turkey ASD post hoc test was carried out. Post-hoc comparisons using Turkey ASD test indicated that the mean score for Anambra state (in southeast zone) ($M = 95.26$, $SD = 13.10$) was significantly different from Bayelsa state (in southsouth zone) ($M = 86.91$, $SD = 14.42$) and Rivers state (in southsouth zone) ($M = 88.46$, $SD = 13.31$). Also the mean score for Imo state (in southeast zone) (96.80, $SD = 12.33$) was significantly different from Bayelsa state (in southsouth zone) ($M = 88.91$, $SD = 14.42$) and Rivers state (in southsouth zone) ($M = 88.46$, $SD = 13.31$). It was also shown that the mean score of Rivers ($M = 88.46$, $SD = 13.31$) did not differ significantly from that of Bayelsa ($M = 88.91$, $SD = 14.42$) (Table 20.10).

Table 20.9 ANOVA

Sources of variation	Sum of squares	df	Mean	f-cal	f-tab	Sig
Between groups	5737.400	3	1912.462	10.794	2.65	0.000
Within groups	55632.060	314	177.172			
Total	61369.459	317				

Table 20.10 Turkey HSD post hoc multiple comparison test

States		Mean	Difference	Sig.
1		2	−1.53750	0.885
		3	8.35000*	0.001
		4	6.80096*	0.008
2		1	1.53750	0.885
		3	9.88750*	0.000
		4	8.33846*	0.001
3		1	−8.35050*	0.001
		2	−9.88750*	0.000
		4	−1.54904	0.884
4		1	−6.80096*	0.008
		2	−8.33846*	0.001
		3	1.54904	0.884

*Mean scores significant at 0.05 level
1—Anambra, 2—Imo, 3—Bayelsa and 4—Rivers

The variations in the adaptation measures used show that the inhabitants of the studied area adapt to climate change using measures that are effective in their domain which may not be effective in other geographical locations. This suggests further research to identify the location-specific adaptation measures.

Conclusion and Recommendations

The study identified eight major climate change adaptation measures used by the inhabitants of the two zones studied. These factors are improved livestock management practices and mixed farming, soil fertility improvement measures and use of information from Agricultural Extension agent, use of improved crop species and diversification, change in planting / harvesting dates, application of agricultural chemicals, irrigation facilities and insurance, increase farm size and reliance on family labour, use of cover crops and avoidance of sale of remaining seedling after planting. Also, the result of the analysis of variance showed that differences exist in the climate change adaptation measures used by rural dwellers in southsouth and southeast zone. This may not be surprise as the two zones differ in their different components for instance, the southsouth zone comprises mainly of riverine areas while the southeast is upland areas.

The study therefore recommends that measures to help rural dwellers secure improved varieties of crops and livestock should be put in place. For instance, the Federal government of Nigeria can incorporate distribution of improve crop and livestock varieties in the existing e-wallet of farmers which is mainly for fertilizer distribution. This will help to better distribute these inputs to farmers. In addition, information technology outfits in the country can be encouraged to develop a cost effective platform or application for information sharing among rural farmers. This can take advantage of the already existing mobile telephony in the country. This will improve the agricultural extension agent dissemination of information hence improving farmer's adaptive capacity since they depend on extension agent for climate change information. Furthermore, since the result of this study indicated a significant difference in the adaptation measures used in the southeast and southsouth zone of the country, further research that could identify the specific adaptation measures that differs the zones is encouraged. This will facilitate the effort towards developing location-specific adaptation measures.

References

Adebayo K (2006) Dynamics of the technology adoption process in rural-based cassava processing systems in southwest Nigeria. International Foundation for Science, Sweden

Adesina AF, Odekunle OT (2011) Climate change adaptation in Nigeria: some background to Nigeria's response III. In: 2011 International conference on Environmental and Agricultural Engineering IPCBEE, vol 15

Adesope OM, Ifeanyi-Obi CC, Ugwuja VC, Nwakwasi R (2012) Effect of malaria on the farm income of rural households in Ekeremor Local Government Area of Bayelsa State. Int J Agric Econ Manage 2(1):27–36

Agoumi Y (2003) Vulnerability of North African countries to climate changes, adaptation and implementation strategies for climate change. Issues and analysis from developing countries and countries with Economies in transition. International Institute for Sustainable Development (IISD)/Climate Change Knowledge Network, 14 pp. http://www.cckn.net//pdf/north_africa.pdf

Agwu J (2008) Climate change, its impacts and adaptation: gendered perspective from south-eastern Nigeria. Heinrich Boll Stiftung (HBS), Lagos

Agwu EA, Anyanwu AC (1996) Socio-cultural and environmental constraints in Implementing the NALDA programme in south eastern Nigeria. A case study of Abia and Enugu State. J Agric Educ 2:68–72

Akinnagbe OM, Ajayi AR (2010) Assessment of farmer's benefits derived from Olam Organization sustainable cocoa production extension activities in Ondo State. J Agric Ext 14(1):11–21

Akpabio (2005) Rural and agricultural sociology. In: Nwachukwu I (ed) Agricultural extension and rural sociology. Snaap, Enugu

Akuegwu BA, Nwi-ue, Nwikina CG (2012) Climate change effects and academic staff role performance in Universities in Cross River State, Nigeria. High Educ Soc Sci 3(1):53–59

Anyadike RNC (2009) Climate change and sustainable development in Nigeria; conceptual and empirical issues, Enugu forum policy paper 10. African Institute for Applied Economics, Enugu

Areghero EM (2005) Nigeria. http://www.fao.org/ag/AGP/AGPC/doc/counprof/nigeria/nigeria.htm. Accessed Mar 2015

Benhin JKA (2006) Climate change and South African agriculture: impacts and adaptation options, Discussion Paper No. 21. Centre for Environmental Economics and Policy for Africa (CEEPA), University of Pretoria, South Africa

Chinweze C, Abiola-Oloke C (2009) Women issues, poverty and social challenge of climate change in the Nigerian Niger Delta context. In: Paper presented at IHDP Open Meeting, the 7th International Conference on Human Dimension of Global Environmental Change, UN Campus, Bonn

Chukwu GO, Mbanaso EO (1999) Crop water requirements of radish in southeastern Nigeria. J Sustain Agric Environ 1(2):236–241

Climate change: any impact in Nigeria? 29 Oct 2012. http://www.vanguardngr.com/2012/10/climate-change-any-impact-on-nigeria/

Eboh E (2009) Implications of climate change for Economic growth and sustainable development in Nigeria, Enugu forum policy paper 10. African Institute for Applied Economics, Enugu

Ekong EE (2010) Rural sociology. Dove Educational Publishers, Uyo

Enete AA (2003) Resource use, marketing and diversification decision in Cassava producing household of Sub-Saharan Africa. A Ph.D. dissertation presented to the Department of Agricultural Economics, Catholic University of Lovain, Belguim

Enete AA, Madu II, Mojekwu JC, Onyeukwu AN, Onwubuya EA, Eze F (2010) Indigenous agricultural adaptation to climate change: study of Imo and Enugu state in southeast Nigeria. Africa Technology Policy Studies Network: Working Paper Series No. 53

Etuonovbe AK (2007) Coastal settlement and climate changes: the effects of climate change/sea level rise on the people of Awoye, Ondo State, Nigeria. Strategic Integration of Surveying Services. FIGR Working Week. Hong Kong SAR, 13–17 May 2007. http://www.fig.net/pub/fig2007/papers/ts-8/fts-03-etuonovbe

Hassan R, Nhemachena C (2008) Determinants of African farmers strategies for adapting to climate changes: multinomial analyses. AFJARE 2(1):83–104

Ifeanyi-Obi CC, Asiabaka CC (2014) Impacts of climate change on sustainable livelihood of rural dwellers in southeast Nigeria. Nigerian J Agric Food Environ 10(3):70–76

Ifeanyi-Obi CC, Etuk UR, Jike-wai O (2011) Climate change, effects and adaptation strategies: implication for agricultural extension system in Nigeria. Greener J Agric Sci 2(2):053–060

Ifeanyi-Obi CC, Asiabaka CC, Matthews-Njoku E, Nnadi FN, Agumagu AC, Adesope OM, Issa FO, Nwakwasi RN (2012) Effects of climate change on fluted pumpkin farmers in Rivers State. J Agric Ext 16(1):50–58

Igbozurike MU (1975) Vegetation types. In: Ofomate GEK (ed) Nigeria in maps: eastern states. Ethiope, Benin, pp 30–32

Intergovernmental panel on climate change (IPCC) (2007) Impact, adaptation and vulnerability. Contribution of Working Group 1 of the intergovernmental panel on climate change to the Third Assessment Report of IPCC. Cambridge University Press, London

Issa FO, Iyiola-Tunji AO, Arokoyo JO, Aregbe BE, Owolabi JO (2008) Evaluation of the challenges of climate change to commercial poultry egg production in Zaria, Kaduna State, Nigeria. Savannah J Agric 6(1):75–84, www.savannahjournal.com

Maddison D (2006) The perception of an adaptation to climate change in Africa. CEEPA Discussion paper No 10. CEEPA, University of Pretoria, Pretoria

Maplecroft (2013) Climate Change Vulnerability Index 2013. http://www.preventionweb.net/files/29649_maplecroftccvisubnationalmap.pdf. Accessed 30 Mar 2015

Mapuno P, Chiwo R, Mtambanengwe F, Adjei-Nsiah S, Baijukya F, Maria R, Mvula A, Gilla K (2008) Farmers perception leads experimentation and learning. In: Charese J, Hampson K, Salm M, Schoubroeck F, Roem W, Rooyakkers, Walsum E (eds) Dealing with climate change. LEISA, pp 28–29

Mbakwe R, Ukachukwu SN, Muoneke CO, Ekeleme F (2004) Problem identification and development of research base. In: Unamma RPA, Onwudike OC, Uwaegbute AC, Edeoga HO, Nwosu AC (eds) Farming systems research and development in Nigeria; principles and practice in Humid and derived savanna southeast zone. Michael Okpara University of Agriculture, Umudike

Mendelsohn R, Dinar A, Dalfelt A (2000) Climate change impacts on African agriculture. Preliminary analysis prepared for the World Bank, Washington, DC, 25 pp

Mgbada JU (2010) Agricultural extension: the human development perspective. Computer Edge, Enugu

Ministry of Environment of the Federal Republic of Nigeria (2003) Nigeria's first national communication under the United Nations. Framework convention on climate change, Abuja Nigerian states budget analysis. http://www.slideshare.net/statisense/nigerian-states-budget-analysis

Nnadi FN, Chikaire J, Nnadi CD, Okafor OE, Echetema JA, Utazi CO (2012) Sustainable land management practices for climate change adaptation in Imo State, Nigeria. J Emerg Trends Eng Appl Sci 3(5):801–805

Nweke FI (1996) Cassava: a cash crop in Africa. COSCA Working paper no. 14, COSCA, IITA, Ibadan

Nzeadibe TC, Egbule CL, Nnaemeka AC, Agu VC (2010) Climate change awareness and adaptation in the Niger Delta region of Nigeria. African Technology Policy Studies Working Paper. Series No. 57

Nzeh EC, Eboh OR (2010) Technological challenges of climate change adaptation in Nigeria: insights from Enugu state. African Technology Policy Studies Network Working Paper. Series No. 52

Obinne CP (1991) Adoption of improved cassava production technologies by small-scale farmers in Bendel state. J Agric Sci Technol 1(1):12–15

Okhimamhe AA (2009) Current vulnerabilities and latest adaptation strategies: Nigeria situation as it relates to women in climate change' (first lady initiative). A paper presented at the awareness workshop on the challenges of climate change adaptation and sustainable livelihood, organizes by the federal ministry of Environment (FMEHUD) in collaboration with the Heinrich Boll foundation on the 25–28 June 2008, Shukura Hotel, Sokoto

Olawoye JE (1996) Utilizing research findings to increase food production. What the mass media should do. In: Bableye T (ed) Media forum for agriculture. IITA, Ibadan

Oneworld.net (2010) Nigeria climate change. http://www.oneworld.net.com/articles/Nigeriaclimatechange. Accessed Mar 2010

Onyeneke RU, Madukwe DK (2010) Adaptation measures by crop farmers in the southeast rainforest zone of Nigeria to climate change. Sci World J 5(1):2010

Oyekale AS, Bolaji MB, Olowa OW (2009) The effects of climate change on cocoa production and vulnerability assessment in Nigeria. Medwell J (Agric J) 4(2):77–82

Ozor N (2009) Understanding climate change. Implications for Nigerian agriculture, policy and extension. Paper presented at the National conference on climate change and the Nigeria environment. Organized by the Department of Geography, University of Nigeria, Nsukka, 29 Jun–2 Jul

Ozor N, Nnaji CE (2010) Difficulties in adaptation to climate change by farmers in Enugu State, Nigeria. J Agric Ext 14(1):106–122

Pallant J (2010) SPSS survival, manual, 4th edn. McGraw-Hill Education, England

Tabachnick BG, Fidell LS (2007) Using multivariate statistics, 5th edn. Pearson Education, Boston

Udofia EP (2011) Applied statistics with multivariate methods. Enugu, Immaculate, pp 187–189

Unamma RPA, Onwudike OC, Uwaegbute AC, Edeoga HO, Nwosu AC (2004) Linkage strategy for sustainable agriculture in Nigeria. Research-Extension-Farmer-Input-Linkage System (REFILS). Micheal Okpara University of Agriculture, Umudike

Chapter 21
Science Field Shops: An Innovative Agricultural Extension Approach for Adaptation to Climate Change, Applied with Farmers in Indonesia

C. (Kees) J. Stigter, Yunita T. Winarto, and Muki Wicaksono

Abstract For implementation of climate change adaptation in communities on the islands of Java and Lombok, Indonesia, a new extension approach was designed and carefully further developed while establishing it. It is called "Science Field Shops (SFSs)". In these "Shops", farmers, scientists/scholars and (where possible, so ideally) local extension officers meet to discuss and solve vulnerabilities and actual local problems expressed by farmers. In such context, agricultural extension is defined best as: "bringing new knowledge to farmers" and this is coordinated at these SFSs.

We use SFSs to temporarily bridge the gap in availability and training of extension intermediaries by using farmer facilitators (FFs). Farmers are confused by consequences of climate change and want answers on questions that are related to local climate problems but also many other issues in growing their crops, among which rice is most important. Giving and discussing answers and predictions demand real dialogues in an agrometeorological learning approach to response farming.

Farmers start to believe in their attempts to understand and reduce yield differences with the past and between them, by actively learning about consequences of

---

C.(K).J. Stigter (✉)
Cluster Response Farming to Climate Change, Faculty of Social and Political Sciences, Department of Anthropology, Center for Anthropological Studies, University of Indonesia (UI), Fl.6/Building H—SeloSoemardjan Room, Kampus UI, 16424 Depok, Indonesia

Group Agrometeorology, Faculty of Agriculture, Department of Soil, Crop and Climate Sciences, University of the Free State (UFS), Agriculture Building, Campus UFS, 9301 Bloemfontein, South Africa

Agromet Vision, Groenestraat 13, 5314AJ Bruchem, The Netherlands
e-mail: cjstigter@usa.net

Y.T. Winarto • M. Wicaksono
Cluster Response Farming to Climate Change, Faculty of Social and Political Sciences, Department of Anthropology, Center for Anthropological Studies, University of Indonesia (UI), Fl.6/Building H—SeloSoemardjan Room, Kampus UI, 16424 Depok, Indonesia

© Springer International Publishing Switzerland 2016
W. Leal Filho et al. (eds.), *Implementing Climate Change Adaptation in Cities and Communities*, Climate Change Management, DOI 10.1007/978-3-319-28591-7_21

climate change and how we can jointly fight them in such and otherwise changing environments. Anthropology and climatology are combined.

Keywords Extension • Adaptation • Climate change • Science Field Shops • Indonesia

Introduction

Basic Knowledge and Literature

Farmers from all over the world have reported that both the timing of rainy seasons and the patterns of rains within these seasons are changing. These perceptions of change are striking in that they are geographically widespread and because the changes are described in remarkably consistent terms (e.g. Jennings and Magrath 2009). They are also in line with what science found: recent trends can be correlated with a change in the timing of seasons and increasingly unpredictable rainfall patterns in Indonesia (World Bank 2011), although data scarcity makes it sometimes difficult to verify (Marjuki et al. 2014). Yamauchi et al. (2012) reported "how change in rainfall patterns induced autonomous adaptation of farmers and affected their rice production. Based on recently collected household data from seven provinces in Indonesia, this analysis clearly demonstrated delays in the onset of rainy seasons and increased uncertainty in rainfall patterns in the region. Farmers made sequential decisions: adjusting planting time in response to delays in the onset of rainy season while changing crop variety responding to delays in the end of the previous year's rainy season. In the case of rice production: (i) delay in the onset significantly decreased land productivity growth in rice production; 1 month delay offsets the average growth observed in 1999–2007, and (ii) though irrigation share significantly explained the growth of land productivity, delayed onset increasingly constrains the role of irrigation" (Yamauchi et al. 2012).

Already in 2007, Case et al. wrote an extensive summary report that indicated for Indonesia among others: (i) mean annual temperature has increased by about 0.3 °C; (ii) warming from 0.2 to 0.3 °C per decade is projected; (iii) the seasonality of precipitation (wet and dry seasons) has changed; the wet season rainfall in the southern region of Indonesia has increased; (iv) in southern Indonesia rainfall is projected to decline by up to 15 %. Aldrian and Djamil (2008) added to this "that the increased ratio of the rainfall in the wet season has led to an increased threat of drought in the dry season and extreme weather in the wet season in recent decades". So it may be assumed based on these early data that our farmers in NW Java and Lombok do suffer and will suffer from increasing temperatures as well as from decreasing rainfall.

In Boer and Suharnoto (2012) it is referred to that for coastal areas of East Java, the number of extreme dry months increased to 4 months in the last 10 years and in 2002 it reached 8 months, which was considered as the longest dry season for the last five decades. They also referred to the strong relationship between the El-Niño Southern Oscillation (ENSO) and rainfall variability in most of Indonesia. They confirmed as well that climate change will have impact on the occurrence and strength of both droughts and floods and that the onset of the monsoon and the lengths of the dry and wet season will be influenced (Boer and Suharnoto 2012). This was also already confirmed explicitly for Indonesia by the World Bank (2011). A presently common problem for farmers in Southeast Asia are false starts of the rainy season (e.g. Marjuki et al. 2014), which are isolated rainfall events followed by long dry spells, preceding the onset date that brings enough rainfall to keep the seedlings growing. For many farmers in Indonesia this was a relatively new experience (Winarto and Stigter 2011). This all together is a sufficient knowledge basis for our work of making farmers more resilient to a changing climate.

The seasonality has of course been observed by our farmers much longer and it is clear that in coastal West Java the rains start these days more often in November, and often late in November, in El-Niño neutral years, instead of in September/October 50 years ago. This trend may continue. As to changing amounts in parts of rainy and dry seasons, the above earlier data have been strengthened with recent observations of less wetter days and more drier days but more intensive rainfall on the wet days (e.g. Shah (2015), in which Indonesia is explicitly mentioned, Oskin (2014), where the totals change less than the distributions).

Farmers have always responded to climatic variability, particularly to changes in rainfall distributions and patterns, by adapting their practices throughout the season. This involves adapting their choices of crops, crop varieties, planting and other cultural measures, while at the same time managing and manipulating the soil, water and microclimate where possible. Climate change complicates this so-called "response farming" (e.g. Stigter and Winarto 2012a), but it does not change the principles of the approach (Stigter 2008, 2010; Winarto et al. 2008; Stigter and Winarto 2011). However, it has been found that some local knowledge systems to forecast weather and predict climate patterns were becoming unreliable, and that some traditional emergency preparedness systems needed to be adapted to accommodate the unprecedented contemporary climate change (Winarto and Stigter 2015).

But in Indonesia (Simamora 2010a, b) and elsewhere (Stigter 2010) there are also social and technical constraints to crop adaptation in response to climate change. A lack of flexibility of farming systems may just prevent temporary or permanent changes of crops or cropping patterns. Adequate government policies for the agricultural sectors can help to improve this situation (Stigter et al. 2007; Stigter 2010).

Rainfall Measurements for a Start

One of the recommendations of a review of the successful agrometeorological pilot projects on operational meteorological assistance to rural areas in Mali, West

Africa, over the past decades, was to continue the promotion of farmer rain gauges to get them at the disposal of each and every farmer (Diarra and Stigter 2008). In 2008 such trials were started with farmers growing dryland rice and/or rainfed rice/maize intercropping systems in Gunungkidul, Yogyakarta, Indonesia, using American farmer rain gauges (Winarto and Stigter 2011), baptized by these Indonesian farmers as "Obama rain gauges".

Since 2010, local farmers in West Java, Indonesia, were stimulated to measure rainfall in their own plots, on a daily routine basis, using homemade cylindrical rain gauges, following routines that were proposed earlier (Stigter et al. 2009). This has never been a goal in itself. It should now serve other purposes in a rural response to climate change (Stigter and Winarto 2011).

A reason for advocating rainfall measurements by farmers in their fields is that official data are very often not of much use, due to high differences in rainfall that exist over relatively small distances. Official data are often deficient and what exists is also not made available free of charge, even for comparisons (Stigter et al. 2009). Climate change makes it even more necessary to do such measurements properly and with high spatial measurement densities.

What farmers gain by measuring rainfall is very explicitly spelled out in our paper and is of value annually and for comparisons between the increasing numbers of years that the data are measured within their fields. That is of much more value than to give them general trends, but the direction of temperature is given and that of rainfall is particularly of value when measured in the ongoing season and compared with recent years.

Methodology

Rainfall

Doing the rainfall measurements with an organized group of well instructed farmers in a region as part of an extension approach, has the methodological advantages that

- each participating farmer can create a record over the years in a "climate logbook";
- derivatives as monthly, seasonal and annual totals, maxima and minima can be easily obtained, graphically compared and understood as consequences of climate realities;
- higher than usual measurement densities can be obtained;
- measurements can be compared and discussed in (preferably) monthly meetings;
- measurements can be part of a larger extension routine in which other data are collected as well;
- measurements can serve as an input to understanding yield differences between areas, farmers, seasons and years;

– measurements can form a basis for attempts of adaptation to climate change, particularly in relation to increasing climate (including rainfall) variability and the occurrence of more (and sometimes more severe) meteorological and climatological extreme events (including droughts, heavy rains and floods that are changing in behavior with the change of climate).

This is the way a group of farmers, organized in the Indramayu Rainfall Observers Club (IROC) (Winarto et al. 2010), developed a new attitude towards climate realities in Indramayu region, coastal West Java, Indonesia, for the past 6 years. This is part of a new extension approach made necessary because the Indonesian extension systems have not or inadequately been prepared for the consequences of a changing climate (Winarto and Stigter 2011). For the same reasons we are now extending this to the island of Lombok, West Nusa Tenggara, Indonesia.

Other Parameters

In addition to daily rainfall measurements, these rice farmers do make and write down agro-ecosystem observations regarding sowing methods, sowing/planting dates, crop varieties, crop stages and development, soil properties and soil moisture, including irrigation situations where applicable, pests and diseases and their developments (including measures they can take in initial stages), the results of fertilizer use and pesticide use. The observations made are noted down on fact sheets that with the "climate logbook" form the historical farm plot records.

Climate Crises in the Livelihood of Indonesian Farmers

Stigter (2014) distinguishes three climate crises in the livelihood of Indonesian farmers:

The first, infrastructure related crisis is not new. The scientific as well as the journalistic literature is full of the following issues (e.g. Stigter 2007; Stigter et al. 2007):

– Indonesia should get away from self-sufficiency in rice through a crop diversification programme. There are nutritional arguments, there are clear economic arguments, there are environmental arguments and there are now also clear climatological arguments.

It is climate that endangers the rice production (Stigter and Winarto 2013a) and crop diversification with less endangered and/or more flexible crops as to water requirements will release some of the climate crisis of rice. More efforts to create new short maturing rice varieties (Boer and Suharnoto 2012) and more heat resistant varieties (IRRI 2015) should also be in place to anticipate the shorter wet season.

- the irrigation infrastructure has not been timely maintained and improved. In NW Indramayu the management of irrigation schedules and insufficient irrigation water cause a delay in planting rice in both dry and rainy seasons. That means part of cropping seasons missed.

If the rice-rice cropping pattern is maintained, development or improvement of water storage and irrigation facilities will be required for balancing increased rainfall in AMJ with decreased rainfall in JAS so that irrigation water is still available during the dry season (Boer and Suharnoto 2012).

- other infrastructure, like roads, bridges, coastal defense plantings/ structures, is far from up to date, causing problems of coastal erosion, sea water intrusion, market underdevelopment, food insecurity etc.

This is a climate crisis also related to neglected infrastructure (Stigter 2014).

- extension services do hardly function, partly related to the above (for comparable problems in Africa, see Stigter and Ofori 2014).

We are these days convinced that progress will come from knowledge intensive production basics and interventions. This will not work without a science backed well trained extension service (e.g. IAASTD 2009).

Climate change will worsen all these issues, so the longer government waits, the more crucial do these infrastructural points and the related climate crisis become.

The second climate crisis is a rice crisis "pure sang". Rice is an endangered crop (e.g. Stigter and Winarto 2013a) and all Indonesian researchers agree here. Higher temperatures at night, in daytime and on average, already decrease rice yields of the present varieties in many places. Furthermore

- new heat resistant rice varieties will take at least another 10 years to develop and to be introduced;
- famers should be encouraged to experiment in the high temperature dry season to find out which of the available rice varieties do best;
- any attempts to provide mild shade should be encouraged: leguminous trees, fruit trees, other economically valuable trees;
- management of rice pests, fertilizers, water and soils, leaves much to be desired and is sometimes particularly wrong (Stigter 2012), because extension is virtually absent or wrongly informed.

The third climate crisis is a crisis due to a badly working extension (policies and practices) on rice growth under conditions of a changing climate. Here are again several aspects (from information collected by Yunita T. Winarto). It appears that:

- presently Indonesian extension intermediaries are only relatively shortly on a post before being nominated elsewhere. No time to get to know the region and build up a relationship with the farmers;

- such extension people are badly paid. There are also transport problems;
- pest and disease observers are selling pesticides themselves;
- they are all badly trained and wrongly trained.

Moreover, the statistics used to create the extension planting calendar "package" (with varieties, fertilizers, pesticides included) are basically not suitable to predict the highly variable rainfall in time and space (Stigter and Winarto 2014).

Although Indonesia recently got a new government, these crises may be partly insufficiently recognized and may partly need unavailable high sums of money and high numbers of suitably trained trainers to find solutions in the immediate future (Stigter 2014). A new bottom-up solution must therefore be designed locally that can be scaled up from within farmer communities without involvement of extension services as long as their situations as outlined above have not changed and their large scale training for conditions of a changing climate has not taken place.

Science Field Shops

The above forms the context in which the work with farmers in Indramayu and on Lombok must take place. Although Climate Field Schools have been and are tried out in Indonesia (e.g. Winarto et al. 2008; WMO 2014), they have been very critically evaluated (Siregar and Crane 2011; Winarto and Stigter 2011; Stigter and Winarto 2013b; Leippert 2014). The Field School concept is not what we have applied. In the recent range of attempts to reach out to farmers (May et al. 2013) and bring climate services to farmers (Stigter 2011; Tall et al. 2014), the very much needed scaling up has not got the serious attention that it needs.

At the request of our farmers, our Science Field Shops (SFSs) are Shops, not Schools. We have no curricula for knowledge exchange with farmers and we have no curricula for the training of trainers. For implementation of climate change adaptation, in the Indramayu communities the IROC farmers are part of, therefore a new extension approach was designed and carefully further developed while establishing it. The design was based on the existing climate crises as explained above and the observations discussed in the monthly evaluation meetings, that we have called SFSs (Winarto and Stigter 2011; Winarto et al. 2011; Stigter and Winarto 2012b; Winarto et al. 2013; Stigter and Winarto 2013b; Winarto and Stigter 2013; Nurkilah et al. 2014; Winarto and Stigter 2015). The name was chosen with an eye on the "Law Shops" established in the sixties and the seventies, in among others England and the Netherlands, to assist poor people in finding justice free of charge (e.g. Smith 1975). The idea for example lingers on strongly in the paralegals, the community-based advice centers, the law clinics etc. in South Africa (Ameermia 2015).

In these "Shops", farmers, scientists/scholars and (in the course of time where possible, so ideally) local extension officers meet to discuss consequences of vulnerabilities and to contribute to solving actual local problems expressed by farmers. First and for all, in a new "farmer first" paradigm, listening to farmers precedes dialogues between participants in the SFSs (e.g. Stigter 2010; Stigter and Winarto 2012a). In such context, agricultural extension is defined best as: "bringing new knowledge to farmers" and this is coordinated at these SFSs (Stigter and Winarto 2012b). This bringing of new knowledge is used in first instance to understand the yields obtained, including yield differences between varieties, farmers, fields, seasons and years.

Now the critical scaling-up was approached by having the Indramayu farmers of the IROC select Farmer Facilitators (FFs) from within their own group that we subsequently gave and still give additional training ("training of trainers") that made them good enough to train their fellow farmers better in the many details of response farming as an answer to the (still changing) conditions related to a changing climate. We presently try to involve these FFs in the set-up of satellite IROCs in "satellite villages" in Indramayu but also to have a beginning of SFSs in Lombok. The "satellite villages" are (circling) around villages where farmers already participated.

It follows from this approach that we want to use SFSs to temporarily bridge the gap in availability and training of extension intermediaries by using FFs. Farmers are confused by consequences of climate change and want answers on questions that are related to local climate problems but also many other issues in growing their crops, among which rice is most important (Stigter and Winarto 2012c). Giving and discussing answers (Appendix) and predictions (Stigter et al. 2015) demand real dialogues in an agrometeorological learning approach to response farming. The only comparison between CFSs and SFSs that we are aware of (Leippert 2014) is convinced of the intrinsic values that the SFSs have.

Issues discussed with farmers in our Science Field Shops as to be carried out in more farmer field trials of growing rice, for the near future, are:

- comparing existing varieties and new ones becoming available;
- use of less nitrogen;
- use of less inorganic pesticides and more organic pesticides;
- trying alternatives for using pesticides on BPH;
- trying SRI modes of growing rice (e.g. ILEIA 2013);
- using mid-season drying and/or alternate wetting and drying to reduce methane emissions without yield depressions and with economic advantages related to saving water;
- using composted instead of fresh plant material for ploughing under, to reduce methane emissions;
- using direct planting instead of making nurseries, to reduce methane emissions and shortening the growing season;
- using rural radio to disseminate knowledge obtained with the "SFSs".

Finally, increasing the reception, understanding and use of monthly seasonal rainfall scenarios was a further issue discussed.

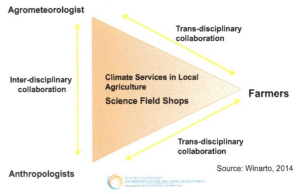

Fig. 21.1 Innovation employed in the project: providing services to farmers

Conclusions

Farmers start to believe in their attempts to understand and reduce yield differences with the past and between them, by actively learning about consequences of climate change and how we can jointly fight them in such and otherwise changing environments. With the present varieties we cannot expect yield increases (Stigter and Winarto 2012c), but keeping yields stable would already be a great success. We have succeeded in combining anthropological and climatological knowledge in the "SFSs" as an inter- and trans-disciplinary educational commitment for adaptation to climate change by providing climate services to farmers/agriculture (Winarto and Stigter 2013, 2015; Nurkilah et al. 2014). The inter-disciplinary collaboration links the agrometeorologist and anthropologists, whereas the trans-disciplinary collaboration links the scientists (agrometeorologist and anthropologists) with the farmers. The relationship of the three parties is pictured in Fig. 21.1.

They are all engaged in jointly establishing services while also jointly learning from implementing SFSs together.

Acknowledgements We are grateful for the financial supports we received from the Academy Professorship Indonesia in Social Sciences & Humanities (under the auspices of the Royal Netherlands Academy of Arts and Sciences, KNAW and Indonesian Academy of Sciences, AIPI) in the period of 2008–2011; the Directorate for Research and Community Development, Universitas Indonesia (2011); SAMDHANA Institute (2012–2014); and the Australian Research Council Grant provided to the University of Western Australia (Lyn Parker and Greg Acciaioli) for the research on: "Fostering Pro-Environment Consciousness and Practice: Environmentalism, Environmentality and Environmental Education in Indonesia"(2013–2015). Our sincere thanks are also due to the administration and leadership of the Center for Anthropological Studies, Department of Anthropology, Faculty of Social and Political Sciences, Universitas Indonesia, the undergraduate students that participated, and the members of the Indramayu Rainfall Observers Club, Indramayu Regency, for their much appreciated collaboration. The work is now also continued under the Universities (and Councils) Network on Innovation for Inclusive Development in Southeast Asia (UNIID-SEA Program, 2015).

Appendix

Some examples of climate change related questions of Indramayu farmers (2011–2014) and our answers [Examples of farmer questions and our replies from before 2011 may be found in Winarto and Stigter (2011).]

2011: *By measuring rainfall only, can we determine the weather without measuring air temperature?*

Answer 1: In the tropics, air temperature is much less variable in space and much more logically (and predictably) varying in time than rainfall. Over one season it is hardly influencing yields in the lowland tropics while rainfall and rainfall distribution in space and time are all determining. So measuring temperature is not necessary on-farm. The most determining for rice is the night minimum temperature but also that one hardly is very different over a season as far as influence on yields are concerned.

However, what is important is to keep track of the consequences of global warming over time. The night minimum temperatures have risen over the last decades and this is ongoing. The time will not be far that this will influence rice yields negatively if it not already has started to do so. So in the long run temperature in tropical lowlands, such as the Indonesian coastal areas, will negatively influence rice yields and it would be wise to diversify agriculture before this is doing real damage.

2011: *In this situation of global warming, what are its effects on food crops? If there are some effects/implications on food crops, what are the rules for planting: should we move ahead or delay planting in relation to pranatamangsa (the Javanese cosmology)?*

Answer 2: In Java and on Bali, one of the issues is a later start of the rainy seasons. Ideally Farmer or Climate Field Schools should be established with collaboration of Universities/Research Institutes/Weather Services in which well-trained facilitators (farmers or extension intermediaries) can discuss with farmers what is the best approach just before, during and just after the rainy season or seasons, making use of available information from the past or the present, if any. But measuring the rainfall on-farm and observing the consequences is part of that School approach, as well as water management by various means: dykes, ridges, ponds, drainage, staggered planting, changing varieties, changing crops etc.

As to *pranatamangsa* (the Javanese cosmology), it is likely that new rules would have to be derived but that these rules would not be valid for a long time. Creating a higher flexibility into the farming/cropping systems, more diversification, may be expected to be more successful.

2012: *Can we solve or prevent the uncertainty of weather?*

Answer 3: Only by improving response farming, which means to be better prepared for uncertainties and calamities, by being ready for changing planned decisions, can we fight uncertainties, but they cannot be prevented.

Of course, if we would be able to reduce greenhouse gasses in the atmosphere, we would at least prevent ever increasing uncertainties!

2012: *When we measure rainfall, can we then predict what is going to happen following the rainfall measurement?*

Answer 4: No. Measuring rainfall daily over long numbers of years has two advantages:

- if done together with observations of the consequences in the agro-ecosystem(s) of the fields where the rains were measured, farmers can more easily understand the connections with the rains and therefore can make decisions more easily;
- by noting down the data and making graphs, farmers can see the differences between years but also the similarities between years! If you make averages of two, three, four etc. years, on a day by day, week by week and month by month basis, you can see whether the actual rainfall of the ongoing season is clearly larger or smaller than or rather similar to these averages. That increases the power of decision making, knowing the history of that decision making over the years. Additionally, simple climate predictions can also help in such decision making.

[We can discuss the recent case of the dry season of 2012 to illustrate this.]

2013. *Is it true that uncertain changes in weather (rain/hot) could induce diseases on people/human? What are the causal factors?*

Answer 5: These changes are not uncertain as far as temperature is concerned but they are uncertain as to rains and their distributions. Such changes as due to global warming (causing climate change) could induce more virulent diseases on people/animals/plants that have themselves survived these changes. This means that climate change is not only influencing people/animals/plants/trees but also organisms with which they interact. Examples for Indonesia could be malaria and dengue fever becoming more serious because the mosquitos like the warmer water to breed.

2013: *Usually in the rainy season, in December (month 12) and January (month 1), also in February, there are lots of rains and wind. Farmers used to name it as "wet season", but why in this rainy season of 2012/2013 there are only winds and lack of rains, even at the generative stage (pregnancy)? Now, there are lots of "bacterial leaf blight" infestations, though we apply enough fertilizers? (especially in Kalensari village, Widasari district) Why?*

Answer 6: The rainy season of 2012/2013 is a so called ENSO-neutral period, which means that there is no El Niño nor a La Niña period that gives its climate signal. In such periods it is, among others, the convection above the western Pacific that determines rainfall. So, the rains should be somewhere from the lower end to the higher end of normal and this is difficult to forecast for Indonesia as a whole and for a certain place in Indonesia in particular. This convection changed character over the months. The skill of the forecast suffers. I really have to see the rainfall figures from all places first. What is of particular importance is the rainfall distribution that determines the periods of leaf wetness (necessary for rice blast to spread).

For the **bacterial blight** (= **rice blast**) see the answer to the previous question. Fertilizers can even be too much and not too little in this case. The reason of the **rice**

blast we have to discuss with experience of the past. Temperatures becoming higher could be involved. We also have to get advice on previous **rice blast** attacks.

2014: *Why will the dry season start early?*

Answer 7: Be careful, the seasonal scenario is a prediction expressed as probabilities. By the first week of March 2014, the general situation in the atmosphere for Indonesia showed reduced convection in the higher layers of the atmosphere. There were also developments of variations in the temperature of the tropical eastern ocean surface in the Pacific that have taught us that an El Niño might come into existence during the northern hemisphere summer. Also by experience we know that this means drought conditions for Indonesia and that made us conclude that the dry season might well be starting early. By early April the predictions were even stronger and the chances of an El Niño starting in June were even rated higher.

2014: *Why are the rains at present different from the old days?*

Answer 8: The global warming also applies to the atmosphere as a whole that therefore can hold more water vapour, which leads to heavier rains in general. However, these atmospheric processes have also changed in the sense that the atmosphere takes longer to produce such rains. So we end up with more dry days and heavier rains in the on average lower number of rainy days. Rainfall measurements at various places have confirmed this. Because of the additional increasing variability of rainfall patterns, we do not recognize the seasons that we knew.

References

Aldrian E, Djamil YS (2008) Spatio-temporal climatic change of rainfall in East Java Indonesia. Int J Climatol 28:435–448

Ameermia M (2015) Being able to access justice is a human right. Mail & Guardian: 24 (February 20–26)

Boer R, Suharnoto Y (2012) Climate change and its impact on Indonesia's food crop sector. Paper presented at the sixth Executive Forum on Natural Resource Management: water and food in a changing environment on 11–13 Apr at SEARCA headquarters, Los Baños

Case M, Ardiansyah F, Spector E (2007) Climate change in Indonesia: implications for humans and nature. World Wildlife Fund. http://assets.wwf.org.uk/downloads/indonesian_climate_ch.pdf. Accessed 6 Jul 2015

Diarra DZ, Stigter K (2008) Operational meteorological assistance to rural areas in Mali. Development-results-perspectives: conclusions and recommendations. Translated from the French. http://www.agrometeorology.org/topics/accounts-of-operational-agrometeorology/operational-meteorological-assitance-to-rural-areas-in-mali/. Accessed 10 Feb 2015

IAASTD (2009) Agriculture at a crossroads. Synthesis report. International Assessment of Agricultural Knowledge, Science and Technology for Development. http://apps.unep.org/publications/pmtdocuments/-Agriculture%20at%20a%20crossroads%20-%20Synthesis%20report-2009Agriculture_at_Crossroads_Synthesis_Report.pdf. Accessed 9 Jul 2015

ILEIA (2013) SRI: much more than more rice. Farming Matters 29 (1), Special issue. http://www.agriculturesnetwork.org/magazines/global/sri. Accessed 19 Jun 2015

IRRI (2015) Climate change-ready rice. http://irri.org/our-work/research/better-rice-varieties/climate-change-ready-rice. Accessed 7 Jul 2014

Jennings S, Magrath J (2009) What happened to the seasons? OXFAM Research Report, October. http://www.oxfam.org.uk/resources/policy/climate_change/downloads/research_what_happened_to_seasons.pdf. Accessed 8 Mar 2014

Leippert F (2014) "Climate field schools": a suitable approach for climate change adaptation? The Indonesian Case. http://www.nadel.ethz.ch/Essays/MAS_2012_Leippert_Fabio.pdf

Marjuki, van der Schrier G, Klein Tank AMG, van Den Besselaar EJM, Nurhayati, Swarinoto YS (2014) Observed climatic trends and variability relevant for crop yields in southeast Asia. In: Past, present and future; a display of climate science and services in South East Asia. Proceedings of the International ASEAN SACA&D Conference and Workshop 2014, pp 22–23. http://sacad.database.bmkg.go.id/documents/IASCW_Proceedings.pdf. Accessed 7 Jul 2015

May S, Hansen J, Tall A (2013) Workshop report: developing a methodology to communicate climate services for farmers at scale. CGIAR Research Program on Climate Change, Agriculture and Food Security (CCAFS). Copenhagen, Denmark. https://cgspace.cgiar.org/bitstream/handle/10568/33443/WorkshopReport-ClimateServicesAtScale.pdf Accessed 6 Jul 2015

Nurkilah, Wicaksono MT, Paramitha Sri BU, Stigter K. (C.J.), Winarto YT, Afiff S (2014) Science Field Shops in Indonesia: Agrometeorological learning and the provision of climate services to rice farmers in Indramayu. Conference on Innovation for Inclusive Development: Making Innovation for All, by All, Happen in the ASEAN Region. Universities (and Councils) Network for Innovative and Inclusive Development (UNIID), South East Asea (SEA). http://inclusiveinnovationhub.org/collections/266. Accessed 25 Feb 2015

Oskin B (2014) How dry will it get? New climate change predictions. Livescience. http://www.livescience.com/44065-climate-change-means-less-rainfall.html. Accessed 6 Jul 2015

Shah V (2015) Unchecked emissions spell hotter, wetter days for Singapore. http://www.eco-business.com/news/unchecked-emissions-spell-hotter-wetter-days-for-singapore/. Accessed 6 Jul 2015

Simamora AP (2010a) Farmers feel the pinch of climate change. Jakarta Post, Wednesday 3 November: 4

Simamora AP (2010b) Farmers guessing weather trends to manage crops: study. Jakarta Post, Wednesday 3 November: 4

Siregar R, Crane TA (2011) Climate information and agricultural practice in adaptation to climate variability: the case of climate field schools in Indramayu, Indonesia. http://onlinelibrary.wiley.com/doi/10.1111/j.2153-9561.2011.01050.x/abstract. Accessed 7 Jul 2015

Smith G (1975) Law shops. London post of 22 April. Reprinted in: Free life. J Libertarian Alliance 1(1), Winter 1979

Stigter K (2007) New cropping systems to help farmers. Jakarta Post, 22 Jan: 7

Stigter K (2008) Agrometeorological services under a changing climate: old wine in new bags. WMO Bull 57(2):114–117

Stigter K (ed) (2010) Applied agrometeorology. Springer, Berlin

Stigter CJ (2011) Agrometeorological services: reaching all farmers with operational information products in new educational commitments. CAgM Report 104. WMO, Geneva, Switzerland. http://www.wamis.org/agm/pubs/CAGMRep/CAGM104.pdf. Accessed 17 Mar 2015

Stigter C(K)J (2012) Climate-smart agriculture can diminish plant hopper outbreaks but a number of bad habits are counterproductive. http://ricehoppers.net/2012/02/cimate-smart-agriculture-can-diminish-planthopper-outbreaks-but-a-number-of-bad-habits-are-counterproductive/. Accessed 19 Feb 2015

Stigter C(K)J (2014) Climate crises in the livelihood of Indonesian rice farmers. In: Memulihkan Ketangguhan Ekosistemdan Produksi Beras: Mengatasi "Ancaman Krisis Pangan" danMenanggulangi "Keliru Pikir"? [Restoring ecosystem sustainability and rice production: managing "threats of food crisis" and changing "wrong mindsets"?] AIPI (Indonesian Academy of Sciences), Jakarta, Indonesia

Stigter C(K)J, Ofori E (2014) What climate change means for farmers in Africa. A triptych review. Middle panel: introductional matters and consequences of global warming for African farmers. Afr J Food Agric Nutr Dev 14(1):8420–8444

Stigter K, Winarto Y (2011) Science field shops may precede climate field schools but simple adaptations to climate should be validated as part of both. http://www.agrometeorology.org/topics/educational-aspects-of-agrometeorology/science-field-shops-may-precede-climate-field-schools-but-simple-adaptation-to-climate-should-be-validated-as-part-of-both. Accessed 3 Mar 2015

Stigter C(K)J, Winarto YT (2012a) Coping with climate change: an active agrometeorological learning approach to response farming. APEC Climate Symposium 2012 "Harnessing and using climate information for decision making: an in-depth look at the agriculture sector", St. Petersburg, Russia. Extended Abstract in Proceedings. http://www.apcc21.org/eng/acts/pastsym/japcc0202_viw.jsp. Accessed 15 Jan 2015

Stigter C(K)J, Winarto YT (2012b) Considerations of climate and society in Asia (II): our work with farmers in Indonesia. Earthzine 4(6). http://www.earthzine.org/2012/04/17/considerations-of-climate-and-society-in-asia-farmers-in-indonesia/. Accessed 22 Nov 2014

Stigter C(K)J, Winarto YT (2012c) Considerations of climate and society in Asia (I). What climate change means for farmers in Asia. Earthzine 4(5). http://www.earthzine.org/2012/04/04/what-climate-change-means-for-farmers-in-asia/. Accessed 22 Nov 2014

Stigter K, Winarto YT (2013a) Rice and climate change: adaptation or mitigation? Facts for policy designs. A choice from what recent summaries say and some critical additions for use with Indonesian farmers. In: Soeparno H, Pasandaran E, Syarwani M, Dariah A, Pasaribu SM, Saad NS (eds) Politik Pembangunan Pertanian Menghadapi Perubahan Iklim [Politics of agricultural development facing climate change], pp 474–485. http://www.agrometeorology.org/topics/climate-change/rice-and-climate-change-adaption-or-mitigation-facts-for-policy-designs/. Accessed 18 Sept 2013

Stigter C(K)J, Winarto YT (2013b) Science Field Shops in Indonesia. A start of improved agricultural extension that fits a rural response to climate change. J Agric Sci Appl 2(2):112–123

Stigter C(K)J, Winarto YT (2014) Climate prediction for farmers: should we laugh or should we cry? Memulihkan Ketangguhan Ekosistemdan Produksi Beras: Mengatasi "Ancaman Krisis Pangan" danMenanggulangi "Keliru Pikir"? [Restoring ecosystem sustainability and rice production: managing "threats of food crisis" and changing "wrong mindsets"?] AIPI (Indonesian Academy of Sciences), Jakarta, Indonesia

Stigter K, Das HP, Van Viet N (2007) On farm testing of designs of new cropping systems will serve Indonesian farmers. http://www.agrometeorology.org/topics/needs-for-agrometeorological-solutions-to-farming-problems/on-farm-testing-of-designs-of-new-cropping-systems-will-serve-indonesian-farmers. Accessed 8 Feb 2014

Stigter K, Winarto YT, Stathers T (2009) Rainfall measurements by farmers in their fields. http://www.agrometeorology.org/topics/accounts-of-operational-agrometeorology/rainfall-measurements-by-farmers-in-their-fields. Accessed 15 Apr 2013

Stigter C(K)J, Winarto YT, Wicaksono M (2015) Monthly updated seasonal rainfall "scenarios" as climate predictions for farmers in Indonesia. World Symposium on Climate Change Adaptation, Manchester, UK, 2–4 Sept 2015, Session 4: Climate Change Adaptation, Resilience and Hazards (in print)

Tall A, Hansen J, Jay A, Campbell B, Kinyangi J, Aggarwal, PK, Zougmoré R (2014) Scaling up climate services for farmers: mission possible. Learning from good practice in Africa and South Asia. CCAFS Report No. 13. Copenhagen: CGIAR Research Program on Climate Change, Agriculture and Food Security (CCAFS). https://cgspace.cgiar.org/bitstream/handle/10568/42445/CCAFS%20Report%2013%20web.pdf. Accessed 8 Jul 2015

Winarto YT, Stigter K (eds) (2011) Agrometeorological learning: coping better with climate change. LAP LAMBERT Academic Publishing, Saarbrucken

Winarto Y, Stigter K (2013) Science Field Shops to reduce climate vulnerabilies: an inter- and trans-disciplinary educational commitment. Australian Anthropological Society's Panel on Collaborative Processes across Disciplinary Boundaries, University of Queensland, Brisbane,

26–29 Sept 2012. In: Collaborative Anthropology (University of Nebraska, USA) 6, pp 419–441

Winarto YT, Stigter K (2016) Incremental learning and gradual changes: "Science Field Shops" as an educational approach to coping better with climate change in agriculture. In: Wilson L, Stevenson C (eds) Promoting climate change awareness through environmental education. IGI Global, Hershey, PA 60–95

Winarto YT, Stigter K, Anantasari E, Hidayah SN (2008) Climate field schools in Indonesia: coping with climate change and beyond. LEISA Mag 24(4):16–18

Winarto YT, Stigter K, Anantasari E, Prahara H, Kristyanto (2010) "We'll continue with our observations": agro-meteorological learning in Indonesia. Farming Matters (formerly LEISA Magazine) 26(4): 12–15

Winarto YT, Stigter K, Anantasari E, Prahara H, Kristyanto (2011) Collaborating on establishing an agrometeorological learning situation among farmers in Java. Anthropol Forum 21(2): 175–197

Winarto YT, Stigter K, Dwisatrio B, Nurhaga M, Bowolaksono A (2013) Agrometeorological learning increasing farmers' knowledge in coping with climate change and unusual risks. Southeast Asian Stud 2(2):323–349 (Kyoto University, Japan)

WMO (2014) News. Indonesian Climate Field School supports food security. https://www.wmo.int/pages/mediacentre/news/IndonesianClimateFieldSchool_en.html. Accessed 18 Jun 2015

World Bank (2011) Climate risk and adaptation country profile. http://sdwebx.worldbank.org/climateportalb/doc/GFDRRCountryProfiles/wb_gfdrr_climate_change_country_profile_for_IDN.pdf. Accessed 6 Jul 2015

Yamauchi F, Takeshima H, Sumaryanto S, Dewina R, Haruna H (2012) Climate change, perceptions and the heterogeneity of adaptation and rice productivity: evidence from Indonesian villages. http://ageconsearch.umn.edu/bitstream/126473/2/ID_16331_IAAE.pdf. Accessed 7 Jul 2015

Lightning Source UK Ltd.
Milton Keynes UK
UKHW05f0951190718
325953UK00002B/24/P